Lecture Notes in Business Information Processing 444

More information about this series at https://link.springer.com/bookseries/7911

Witold Abramowicz · Sören Auer ·
Milena Stróżyna (Eds.)

Business Information Systems Workshops

BIS 2021 International Workshops
Virtual Event, June 14–17, 2021
Revised Selected Papers

Editors
Witold Abramowicz
Poznań University of Economics and Business
Poznan, Poland

Sören Auer
TIB Leibniz Information Center Science and Technology and University of Hannover
Hannover, Germany

Milena Stróżyna
Poznań University of Economics and Business
Poznan, Poland

ISSN 1865-1348 ISSN 1865-1356 (electronic)
Lecture Notes in Business Information Processing
ISBN 978-3-031-04215-7 ISBN 978-3-031-04216-4 (eBook)
https://doi.org/10.1007/978-3-031-04216-4

This Springer imprint is published by the registered company Springer Nature Switzerland AG
The registered company address is: Gewerbestrasse 11, 6330 Cham, Switzerland

Preface

In 2021 we had a great opportunity to organize the 24th edition of the International Conference on Business Information Systems (BIS 2021). The conference has grown to be a well-renowned event for scientific and business communities, and this year the main topic was "Enterprise Knowledge and Data Spaces". The conference was jointly organized by the BIS Steering Committee, the TIB Leibniz Information Center Science and Technology, and the University of Hannover, Germany. Due to the COVID-19 pandemic, the conference was organized fully online.

During each edition of the BIS conference series we make the effort to provide an opportunity for discussion about up-to-date topics from the area of information systems research. However, there are many topics that deserve particular attention. Thus, a number of workshops and accompanying events are co-located with the BIS conference series. The workshops give researchers the possibility to share preliminary ideas and initial experimental results, and to discuss research hypotheses from a specific area of interest.

Eighth workshops took place during BIS 2021. We were pleased to host well-known workshops such as AKTB (12th edition), BITA (11th edition), iCRM (6th edition), BSCT (4th edition), and QOD (4th edition), as well as relatively new initiatives such as DigEx, BisEd, and DigBD. Each workshop focused on a different topic: knowledge-based business information systems (AKTB), the challenges and current state of business and IT alignment (BITA), integrated social CRM (iCRM), blockchain (BSCT), digital customer experience (DigEx), data quality (QOD), digitization in the area of big data (DigBD), and new trends and challenges in education (BisEd).

The workshop authors had the chance to present their results and ideas in front of a well-focused audience; thus, the discussion gave the authors new perspectives and directions for further research. Based on the feedback received, authors had the opportunity to update the workshop articles for the current publication. This volume contains 31 articles that are extended versions of papers accepted for BIS workshops. In total, there were 67 submissions for all mentioned events. Based on the reviews, the respective workshop chairs accepted 31 in total, yielding an acceptance rate of 46%.

We would like to express our thanks to everyone who made the BIS 2021 workshops successful: our workshops chairs, members of the workshop Program Committees (PCs), authors of submitted papers, and all workshops participants. We cordially invite you to visit the BIS website at http://bisconf.info and to join us at future BIS conferences.

June 2021

Witold Abramowicz
Sören Auer
BIS 2021 PC Co-chairs

Milena Stróżyna
BIS 2021 Workshops OC Chair

Contents

AKTB Workshop

BIS Education Workshop

QOD Workshop

BITA Workshop

BSCT Workshop

AKTB Workshop

AKTB 2021 Workshop Chairs' Message

The 12th Workshop on Applications of Knowledge-Based Technologies in Business (AKTB 2021) was organized online in conjunction with the 24th BIS 2021 conference. The workshop invited researchers, practitioners, and policy makers to gather for the discussion on the digitalization topics affecting business and society and share the newest research knowledge of the efficient computational intelligence methods for implementing business information systems in finance, healthcare, e-business, and other application domains. The workshop called for papers which suggest solutions for providing advanced services for the information systems users or propose innovative approaches for smart business and process modeling, especially targeting digital transformation issues.

A total of 11 articles were submitted to the AKTB 2021 workshop. Each paper was evaluated by two or three independent reviewers of the Program Committee. The five highest ranked papers, prepared by authors from six different countries representing nine research institutions, were accepted for presentation during the conference and the second stage of reviewing, pending inclusion in the post-conference proceedings.

In total, 12 outstanding researchers representing prestigious scientific institutions from five countries formed the Program Committee and served as paper reviewers. They evaluated the research level of the articles by taking into account the criteria of relevance to the workshop topics, originality, novelty, and quality of presentation.

The unique characteristic of the AKTB workshop is highlighting the application aspect of investigations in a wide variety of domain areas, exploring artificial intelligence and knowledge-based technologies for modeling and experimental validation of the research. Researchers submitted their works based on theoretical and experimental research in the application domains of energy, insurance, healthcare, and knowledge-based processes in car rental businesses.

We appreciate the expertise of the Program Committee members, whose reviews provided deep analysis of the submitted research works and highlighted valuable insights for the authors. Their expertise ensured the high quality of the workshop event, excellent presentations, intensive scientific discussions, and added value to the post-conference workshop proceedings.

We would like to express our gratitude for the joint input to the success of AKTB 2021 to all authors of the submitted papers, members of the Program Committee, research departments of Vilnius University, and the Department of Information Systems of the Poznan University of Economics and Business, and we acknowledge the outstanding efforts of Organizing Committee of the 24th International conference BIS 2021.

Virgilijus Sakalauskas
Dalia Kriksciuniene

Organization

Chairs

Virgilijus Sakalauskas	Vilnius University, Lithuania
Dalia Kriksciuniene	Vilnius University, Lithuania

Program Committee

Dumitru Dan Burdescu	University of Craiova, Romania
Elpiniki I. Papageorgiou	Technological Educational Institute of Central Greece, Greece
Ferenc Kiss	Budapest University of Technology and Economics, Hungary
Dalia Kriksciuniene	Vilnius University, Lithuania
Dariusz Krol	Wrocław University of Science and Technology, Poland
Roman Lewandowski	University of Social Sciences, Poland
Saulius Masteika	Vilnius University, Lithuania
Justyna Patalas-Maliszewska	University of Zielona Góra, Poland
Tomas Pitner	Masaryk University, Czech Republic
Giedrius Romeika	Vilnius University, Lithuania
José Raúl Romero	University of Cordoba, Spain
Vytautas Rudzionis	Vilnius University, Lithuania
Virgilijus Sakalauskas	Vilnius University, Lithuania
Darijus Strasunskas	HEMIT, Norway
Sebastián Ventura	University of Cordoba, Spain
Leonard Walletzky	Masaryk University, Czech Republic
Danuta Zakrzewska	Technical University of Lodz, Poland

Analysis of the Structure of Germany's Energy Sector with Self-organizing Kohonen Maps

Irina Potapenko[1], Vladislav Kukartsev[1,2], Vadim Tynchenko[1,2,3(✉)], Anton Mikhalev[2], and Evgeniia Ershova[2]

[1] Reshetnev Siberian State University of Science and Technology, Krasnoyarsk, Russia
vadimond@mail.ru

[2] Siberian Federal University, Krasnoyarsk, Russia
EErshova@sfu-kras.ru

[3] Marine Hydrophysical Institute, Russian Academy of Sciences, Sevastopol, Russia

Abstract. The purpose of the research in this article is to analyze the structure of energy in Germany and compare the obtained data with events occurring in the country and the world. The article reviews the world energy sector and considers the rating of regions by gross energy production. The analysis helps to identify the leading regions in terms of energy production: Asia and Oceania, North America and Europe. The German economy and energy sector were considered, as well as the development of nuclear power in particular and the gradual abandonment from nuclear power plants because of the occurred radiation accidents in the world. It also describes the relevance of data analysis in the energy sector, especially in working with renewable energy sources due to their instability and unpredictability. Using self-organizing Kohonen maps, the data on German energy indicators was analyzed. Basing on the analysis it was concluded that these maps correspond to the changes in the energy policy of Germany.

Keywords: Energy · Germany · Kohonen maps · Neural networks · Intellectual analysis · Clustering

1 Introduction

The amount of resources consumed by a person increases every year. It is primarily caused by the constant development of technologies, as well as the growth of the world's population [1]. One of the main resources is electricity. Annual electricity consumption is steadily increasing, and most of the energy is generated from exhaustible sources. This fact directly has influenced the world economy and the environment.

The most important problem in this situation is the problem of efficiency [2]. This problem is faced by absolutely all countries that generate electricity for domestic use or for export [3, 4]. Besides there is an urgent problem how to distribute electricity generation considering the available resources of the country, its ecology and

W. Abramowicz et al. (Eds.): BIS 2021 Workshops, LNBIP 444, pp. 5–13, 2022.
https://doi.org/10.1007/978-3-031-04216-4_1

geographical location. At the same time, it is necessary to cover all electricity needs [5]. Data analysis is used to solve these types of problems.

2 Review of World Energy

According to the information analytical site EES EAEC for 2017, the rating of regions for electricity generation is presented in Table 1 [6].

Based on the table data, it is possible to identify the leaders in energy production; they are Asia and Oceania, North America and Europe. This is due to the size of the GDP of the countries included in the selected regions. At the moment the leader in terms of GDP is Asia and Oceania, due to the fact that this region includes the largest volume of production worldwide.

Table 1. Rating of regions by energy production-gross for 2017.

№	Name of countries	The volume of energy produced-gross in 2017, billion kWh
1	Asia and Oceania (Australia, Hong Kong, India, China, Malaysia, Singapore, Taiwan, Thailand, Trinidad and Tobago, South Korea, Japan)	11510
2	North America	5267
3	Europe	3867
4	Eurasia (the totality of post-Soviet countries)	1577
5	Central and South America	1273
6	Middle East	1201
7	Africa (Egypt, Nigeria, South Africa, etc.)	812

3 Germany and Its Energy Sector

At the end of 2018 the population of Germany is slightly more than 83 million people. The area of the country is 357 408.74 km^2. Germany is one of the leading countries in a number of technological and industrial sectors. The economy is based on services (70%–78%) and manufacturing (23%–28%). There are reserves of coal and brown coal, amounting to 40.5 billion tons in the cities of Westphalia, Brandenburg and Saxony.

Germany's main energy sources of 2018 are:

- 40.4%–renewable energy sources (20.4% – wind power, 3.1% – hydro power, 8.3%–bioenergy, 8.4% – solar power).
- 24.1%–brown coal.
- 13.5%–coal.
- 13.3%–nuclear energy.
- 7.7%–gas.

According to the analytical and informational site of energy economic statistics EES EAEC for 2017, 66.7% of the world's energy is produced at thermal power plants, including TPPs per organic fuel type. About 16.0% of the world's energy comes from hydroelectric power plants, and 10.3% is produced by nuclear power plants. The rest of the energy is shared between wind, solar, hydro-accumulating, geothermal and other power plants [6].

Comparing the data for Germany with the world's data, one can see that the global trend of electricity generation is not reflected in the country's energy sector. It's connected with German policy, which is aimed at environmental protection. Also, because of the world's disasters, Germany decided to abandon the use of nuclear power plants gradually. That's why it is necessary to examine the trends in German politics and the environmental situation in more details.

The ongoing open-pit mining caused the water pollution in some German rivers because of the water leaks around the mines. Also, after the work completion, due to rising water levels, the Spree river was filled with brown sludge, which led to the destruction of the ecosystem and wildlife of the nearby nature reserve. It is worth to mention the problem of air pollution, which is relevant for many industrialized countries. This problem became particularly acute for the public after the Fukushima nuclear disaster in 2011, when the German government decided to phase out the nuclear power plant. For this reason, it was allowed to burn more coal, which led to extremely high levels of air pollution.

Germany's environmental policy is primarily related to the fight against global warming and carbon emissions [7]. Also, Germany has aimed to use raw materials more efficiently maintaining the existing level of welfare at the same time. An equally important decision of the government was to focus on the production and sale of renewable energy. This is already noticeable at the moment, because 40.4% of all German energy is produced from renewable sources.

4 Development of Nuclear Power in Germany

After Germany joined NATO in 1955, a number of restrictions connected with the development of certain industries were lifted for the country. Thanks to this, the first nuclear power plant was launched in Germany in the early 60s. Nuclear power was a priority in the country's energy policy in the 60s and 70s. The chosen development vector allowed to minimize losses during the sharp rise in world oil prices, as well as during the global energy crisis.

Starting in the late 1990s, the Green party came to power in Germany and actively promoted the abandonment of nuclear energy. In 2000, the "nuclear consensus" was achieved, where the German government planned to abandon the use of nuclear power plants by 2021. Faced with problems of commissioning of replacement capacity, the government was forced to push back the deadline for complete abandonment of nuclear power by 2035, which caused strong public unrest. The political situation in Germany was aggravated by the accident at Fukushima, after which the German authorities stopped the operation of all nuclear power plants built before 1980 for three months and conducted urgent inspections of all nuclear power plants. In spite of no critical violations were detected, but the German authorities ignored the opinions of experts and preferred to listen to the opinion of the Ethics Commission, which promoted the political ideology of Green party. As a result, it was decided to terminate the operation of eight power units out of the seventeen available, and the other nine had to be closed by 2022.

Such a sharp abandons nuclear energy entailed a number of certain consequences which affected the country. Already in 2012, the growth rate of greenhouse gas emissions increased significantly by the low growth in the volume of the German economy. It's related to an abrupt increase in the amount of burned coal for producing replacement electricity. This severe environmental degradation in the country led to an additional 1,100 deaths per year from cardiovascular and respiratory diseases. Also, residents of Germany suffered from a sharp rise in electricity prices because of the abandonment of nuclear power plants. The use of wind farms was not an enough efficient solution for generating electricity to replace nuclear power. Moreover, wind farms in Germany kill up to 220 thousand birds per year.

5 Data Analysis

Data analysis is a set of methods for processing data to obtain useful information and make decisions based on it. The most popular methods are multidimensional types of data analysis, since they allow studying the dependencies of several input parameters simultaneously. One of these methods is cluster analysis.

Cluster analysis allows analyzing multidimensional data without imposing restrictions on its representation. The general scheme of the clustering algorithm is shown in Fig. 1 [1].

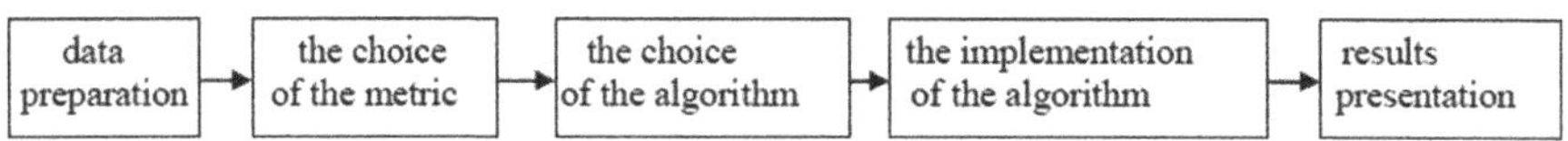

Fig. 1. General scheme of the clustering algorithm.

The essence of the method is to combine data into groups with a common closely correlated feature of grouped data. The purpose of cluster analysis is to identify a small number of data groups that are the most similar to each other and also differ from other groups.

Cluster analysis has found its application in all areas of activity to solve a variety of problems [8–12]. The world energy sector is no exception [13].

6 Application of Data Analysis in the Energy Sector

Data analysis in the energy sector plays an important role, starting with the optimization of energy production, ending with the efficient distribution of received energy. Data analysis is typically used in work with non-traditional types of energy, such as solar, wind, geothermal, and marine wave and tidal energy. A special feature of work with these types of energy is unpredictability, due to the dependence on a variety of natural factors. Research data in this area allows avoiding many risks.

Every year the international energy Agency IEA issues a report about the energy efficiency market that shows progress in energy efficiency worldwide. The IEA is the world's leading Agency for data analysis and energy efficiency consulting [14].

7 Transition to Available Data and Analysis of Kohonen Maps

In Germany there is a system of official statistics that provides information collected in a functionally independent and methodological manner for making decisions. Kohonen maps were constructed basing on data from the German Federal statistical office [15]. Data was taken:

- About electric power generation, transmission and distribution.
- About production and distribution of gaseous fuels.
- About the supply of steam and air condensate.
- About collecting, processing and supplying water.

Each data type was taken in the sections described in Table 2.

Table 2. Description of German energy data sections.

Name	Description
Local units	Local departments of energy and water companies with 20 or more employees and local departments from other industries with 20 or more employees. The values of the local departments are average They are considered as institutions: – in power supply: thermal power plants, nuclear power plants, hydroelectric power plants, wind, solar, geothermal and fuel cell power plants. Smaller power plants in a regionally restricted area (such as power plant chains) can be combined into a single company; – in gas supply: gas production, extraction, conversion and storage plants; – in heating and cooling: heating installations, thermal power plants; –in water supply: installations for extraction, treatment and storage of water
Engaged people	The number of engaged people is taken as average annual. They are: – active owners and active co-owners / employees (only from partnerships); – unpaid family members, if they work in the company not less than 1/3 of the normal working time; – people who have an employment relationship with the company (for example, Directors, employees in a business trip, volunteers, interns, and students) The following key points should be fully taken into consideration: – sick people, vacationers, people on maternity and paternity leave (less than 1 year) and all others who are temporarily absent; – strikers and those who are affected by the lockout until the employment relationship is terminated;

(*continued*)

Table 2. (*continued*)

Name	Description
	– seasonal workers and temporary workers, part-time workers, employees and short-term employees, persons with age regulation of working hours; – staff from construction and installation sites, vehicles, etc.; – people working abroad temporarily (less than 1 year) The following points aren't taken into consideration: – people working abroad permanently (at least 1 year); – workforce of companies performing installation or repair work; – people who get advance payments; – labor force from employment agencies, etc
Hours which are worked out	Hours which are worked out are working hours which are paid (not paid) for all employed people. For multi-level businesses, it is customary to specify the sum of all hours in all levels. Overtime, nighttime, Sunday, and holiday hours are also taken into consideration All defective working hours, even if they are paid for, as well as working hours for installation and repair work by representatives of other companies are not counted as hours which are worked out
Remuneration	Remuneration is amount of all gross payments (cash and benefits in any kind) without any deductions. These amounts include the share of payments for employees without deductions to the health insurance fund, for pensions, unemployment and long-term care insurance Remunerations also include amounts paid to actual employees in their social institutions (for example, factory doctors) Remunerations include the following points: – any bonuses (for example, for piecework, installation work, shift and Sunday work, productivity, dirty work and unpleasant bonuses); – payment for state holidays, vacation, lost working days etc.; – continued hospital payments, including sickness benefit supplements; – special bonuses, additional monthly remunerations, vacation pay and other one-time rewards; – vacation pay is not provided; – housing allowances, child allowances, some other family allowances and educational allowances; – payments for nutrition, travel time compensation, employee's travel allowances to and from work, if the relevant payroll tax has been paid; – daily allowance in case of paid payroll taxes; – employer-provided benefits, which are defined in Sect. 3 of the fifth law about the assistance to capital formation by employees; – commissions and royalties paid to employees; – severance pay for employees, after deduction of subsidy paid by the Federal employment Agency (for example, short-term work benefit) Points that should not be included: the imputed entrepreneurial income and expenses to employees hired for work; remuneration for temporary employment in enterprises and similar agencies according to Employment law

Kohonen maps of electric power industry indicators in Germany are presented in Fig. 2.

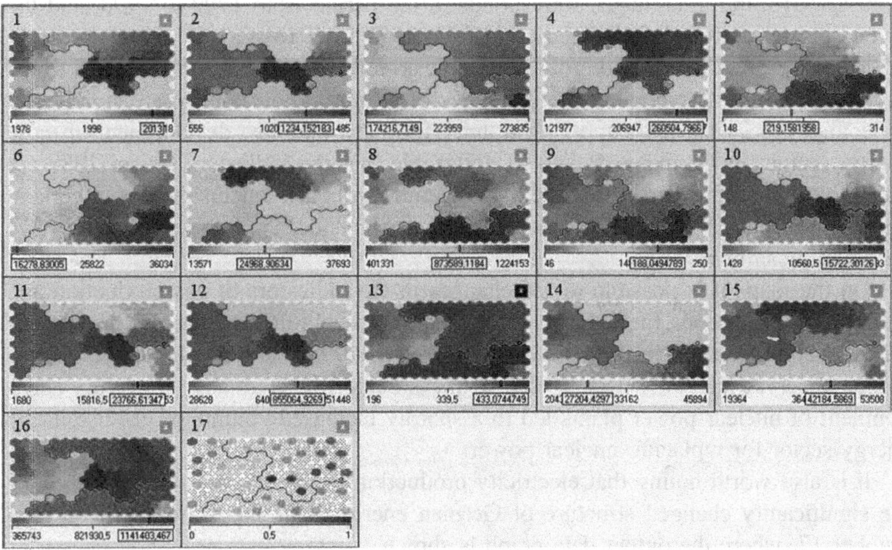

Fig. 2. Kohonen maps of electric power industry indicators in Germany.

Description of the maps presented in Fig. 2:

- 1–year.
- 2–electric power generation, transmission and distribution (local units).
- 3–electric power generation, transmission and distribution (engaged people).
- 4–electric power generation, transmission and distribution (hours which are worked out (thousand)).
- 5–production and distribution of gaseous fuels (local units).
- 6–production and distribution of gaseous fuels (engaged people).
- 7–production and distribution of gaseous fuels (hours which are worked out (thousand)).
- 8–production and distribution of gaseous fuels (remuneration (thousand euros)).
- 9–supply of steam and air condensate (local units).
- 10–supply of steam and air condensate (engaged people).
- 11–supply of steam and air condensate (hours which are workd out (thousand)).
- 12–supply of steam and air condensate (remuneration (thousand euros)).
- 13–collecting, processing and supplying water (local units).
- 14–collecting, processing and supplying water (engaged people).
- 15–collecting, processing and supplying water (hours which are worked out (thousand)).
- 16–collecting, processing and supplying water (remuneration (thousand euros)).

- 17–electricity generation, transmission and distribution (remuneration (thousand euros)).

Based on the constructed maps, the zero cluster corresponds to the period from 2000 to 2018, the first cluster corresponds to the period from 1990 to 1998, and the second cluster corresponds to the period from 1978 to 1988.

Moreover, it is worth noting the low indicators from the beginning of 1978 to the end of 1988 in maps showing the dynamics of steam and air condensate supply and collecting, processing and supplying water. This shows the low development of these energy sectors in Germany. It is also noticeable that the indicators before 1988 are much lower than the same after 1988. Thus, one can make conclusions about certain, rather sharp changes in the use of steam and water, but since 2000 these indicators have been significantly reduced.

On the maps it is possible to see changes in the indicators of gas production and distribution of gaseous fuels. In this case, it is noticeable that the top productivity of this energy industry is in the period from 1990 to 1998, but these indicators are no longer observed in 2000. This is due to the situation in Germany, when the abandonment of nuclear power plants led to a sharply increased volume of gas use in the energy sector for replacing nuclear power.

It is also worth noting that electricity production in the country was increased by the significantly changed structure of German energy. This fact is reflected on map number 17, where the output data graph is shown.

8 Conclusion

Based on the constructed maps, it can be concluded that these maps correspond to the changes in the energy policy of Germany. The features of changes in the structure of the country's electricity production were identified in the period from 1988 to 2018.

The obtained data revealed low rates of steam and air condensate supplies, as well as collecting, processing and supplying water from 1978 to 1988. In the period from 2000 to 2018, the parameters were slightly higher, namely from 1988 to 2000 the indicators reached the highest values. The dynamics of production and distribution of gaseous fuels, which fully describes the features of the period of Germany's rejection of nuclear power and the search for replacement energy resources, is also revealed in this article.

During the period under review, the German energy sector has undergone major structural changes. The country abruptly abandoned nuclear power, replacing it with gas and coal fuel, and switched to renewable energy sources due to the deteriorating environmental situation. As the result of its Germany has become one of the leading countries in the use and investment in renewable energy sources.

References

1. Hara, T.: Introduction: the sustainability of the world population. In: An Essay on the Principle of Sustainable Population. SPS, pp. 1–10. Springer, Singapore (2020). https://doi.org/10.1007/978-981-13-3654-6_1
2. Sreewirote, B., Noppakant, A., Pothisarn, C.: Increasing efficiency of an electricity production system from solar energy with a method of reducing solar panel temperature. In: 2017 International Conference on Applied System Innovation (ICASI). IEEE, pp. 1308–1311 (2017)
3. de Lima, L.P., de Deus Ribeiro, G.B., Perez, R.: The energy mix and energy efficiency analysis for Brazilian dairy industry. J. Clean. Prod. **181**, 209–216 (2018)
4. Fei, F., Wen, Z., Huang, S., De Clercq, D.: Mechanical biological treatment of municipal solid waste: energy efficiency, environmental impact and economic feasibility analysis. J. Clean. Prod. **178**, 731–739 (2018)
5. Cornelis, M.: Energy efficiency, the overlooked climate emergency solution. Econ. Policy **2**, 48–67 (2020)
6. EES EAEC. http://www.eeseaec.org. Accessed 28 Jan 2021
7. Salygin, V.I., Meden, N.K.: On the issue of methodology of energy policy research (German example) MGIMO Univ. Bullet. **6**(45), 181–192 (2015)
8. Kukartsev, V.V., Beletskaya, O.D., Fabrichkina, M.O., Tynchenko, V.S., Mikhalev, A.S.: Kohonen maps to organize staff recruitment and study of workers' absenteeism. J. Phys. Conf. Series **1399**(3), 033108 (2019)
9. Caruso, G., Gattone, S.A., Fortuna, F., Di Battista, T.: Cluster analysis for mixed data: an application to credit risk evaluation. Socio-Econ. Plann. Sci. **73**, 100850 (2021)
10. Jasiński, M., et al.: A case study on data mining application in a virtual power plant: cluster analysis of power quality measurements. Energies **14**(4), 974 (2021)
11. Vedernikov, M., Zelena, M., Volianska-Savchuk, L., Litinska, V., Boiko, J.: Management of the social package structure at industrial enterprises on the basis of cluster analysis. TEM J. **9** (1), 249–260 (2020)
12. Tynchenko, V.S., Tynchenko, V.V., Bukhtoyarov, V.V., Kukartsev, V.A., Eremeev, D.V.: Application of Kohonen self-organizing maps to the analysis of enterprises' employees certification results. IOP Conf. Series Mater. Sci. Eng. **537**(4), 042010 (2019)
13. Ermolaev, D.V.: Clustering as a factor in industrial development. Bulletin Tula State Univ. Econ. Legal Sci. **3**(1), 82–95 (2016)
14. IEA. https://www.iea.org. Accessed 28 Jan 2021
15. German Federal statistical office. https://www.destatis.de/EN/Home/_node.html. Accessed 28 Jan 2021

Time-to-Event Modelling for Survival and Hazard Analysis of Stroke Clinical Case

Dalia Kriksciuniene[1], Virgilijus Sakalauskas[1(✉)], Ivana Ognjanovic[2], and Ramo Sendelj[2]

[1] Institute of Applied Informatics, Vilnius University, Universiteto Street 3, Vilnius, Lithuania
{dalia.kriksciuniene, virgilijus.sakalauskas}@knf.vu.lt

[2] University of Donja Gorica, Oktoih 1, 81000 Podgorica, Montenegro
{ivana.ognjanovic, ramo.sendelj}@udg.edu.me

Abstract. The problems of "time-to-event' data analysis explore data of processes with the defined end point of time when an explored event occurs. The methods of survival modelling, hazard analysis, risk evaluation are combined for understanding and exploring these data. In this article the time-to-event modelling addresses in-hospital mortality of stroke patients. The clinical data of stroke cases is explored based on the historical records of neurology department of Clinical Centre in Montenegro. The main aim of the research is to explore survival techniques for getting insights from the stroke clinical data with the goal of identifying impact of the variables for the predictive research. The time-to-event data does not follow the normal distribution, thus limiting application of major methods of analysis. In the article the survival analysis techniques such as Life table, Kaplan–Meier survival plot and the Cox proportional hazards regression model are applied for exploring stroke data.

Keywords: Survival analysis · Stroke clinical trial · Life table · Kaplan-Meier method · Cox's Proportional Hazard Model

1 Introduction

In this article, we apply method of survival modelling for exploring risks and their factors for the patients impacted by stroke. In general, survival analysis is defined as a set of analysis methods for time-to-event trial data. For instance, an event can be death, heart attack, relapse, divorce or violation of parole in technical systems. These examples illustrate a wide application of survival analysis in many different fields. Specifically, survival analysis is utilized in biology, medicine, engineering, marketing, social sciences or behavioural sciences [3–7].

The application of survival analysis requires some specific research methods due to the data origin: survival data does not correspond to normal distribution, as the time-to-event implies that at the explored moment of time part of the individuals have not yet had the event of interest making their true time to event unknown, therefore this part of individuals is censored [8].

W. Abramowicz et al. (Eds.): BIS 2021 Workshops, LNBIP 444, pp. 14–26, 2022.
https://doi.org/10.1007/978-3-031-04216-4_2

This article aims not only to introduce the basic concepts of survival analysis, but also seeks to apply them for exploring stroke clinical data collected at Clinical Centre in Montenegro from the time-to-event perspective. Stroke is considered to be one of the leading causes of disability and mortality all over the world. According to the report of World Stroke Organization 13, 7 million people suffer stroke worldwide each year. Of these, 5 million die and another 5 million are permanently disabled, resulting to a total of 80 million stroke survivals, who live with the impact and burden of stroke. Almost 60% of all stroke cases affect people under their 70 years of age [1, 2]. These are just few facts demonstrating the impact of stroke to our society.

The presented research aims to apply computational methods to explore of in-hospital mortality factors by analysis stroke patient data and provide insights for enhancing value of data.

The techniques of survival analysis include generation of life table, Kaplan–Meier (KM) plots, log-rank tests, and Cox (proportional hazards) regression. Life table enables to calculate survival function, probability density and hazard rate. Log-rank test quantifies and tests survival differences between two or more groups of patients. Kaplan and Meier have introduced a method for evaluation of survival probabilities [9], Cox Proportional Hazard Model, designed as a regression model for predicting hazard rate from covariates was proposed by Cox [10].

All these methods are applied for Stroke clinical data using STATISTICA software and its Survival Analysis module functions [https://www.statistica.com/]. For the statistical visualisation of initial stroke data, we utilize the MS EXCEL charts.

The paper is organized as follows. In Sect. 2 we provide characteristics of stroke clinical data. Section 3 presents statistical analysis and visualisation of Stroke data to highlight the links among the variables. Section 4 researches survival functions by applying life table, Kaplan and Meier evaluation method and log-rank test. In Sect. 5 the hazard function is explored by the Cox proportional hazard and regression analysis. The results are summarized in conclusion section.

2 Stroke Clinical Data for Time-to-Event Analysis

The database applied for the experimental research consist of stroke patient clinical data records registered by the neurology department of Clinical Centre of Montenegro, operating in Podgorica, Montenegro. The original database consists of the structured 944 records of stroke patients, 58 variables, where 50 of them are coded by scale values of {1, 2, 3} corresponding to "Yes, No, Unspecified", and 8 variables consisting of the demographic data, admission date and discharge date from hospital. The data was collected between 02/25/2017 and 12/18/2019. The demographic data of stroke patients varies by age (from 13 to 96 years), and gender (485-male, 427-female). For survival modelling we have cleansed the initial stroke database, recoded some variables and finally got the 10 variables database for research. The example of database structure and data records is presented in Table 1.

Table 1. Sample of data records and variables of the stroke cases database.

Days at Hospital	Vital Status	Stroke Type	Treatment methods	Health Status	Age	Gender	Past Stroke	Stroke Symptoms	Health complications
8	0	3	0	1	17	2	0	0	0
23	0	3	4	2	16	1	0	12	0
12	1	2	4	0	13	2	1	12	0
13	1	1	24	0	96	1	0	23	4
3	1	1	24	9	94	2	0	23	4
3	1	4	4	0	93	2	0	123	2
3	1	1	24	0	94	1	0	12	4
14	1	1	24	1	93	2	1	123	23

The variables of the stroke database are coded for applying the survival modelling methodologies, their values of the are explained in Table 2.

Table 2. The definition of stroke clinical database variables

Variable name	Meaning and coding of data
Days at Hospital	– the number of days spent in hospital after stroke
Vital Status	– 1:Event (death), 0: Alive/censored
Stroke Type	– 1: Ischemic, 2: Hemorag, 3: SAH, 4: Unspecified
Treatment Methods	– 0:No treatment,1:Anticoagulation, 2:Dual Antiplatelet Therapy, 3: Thrombolysis, 4:Others, Two digit codes: mean combined treatment methods, e.g. 24:means 2 and 4 are applied
Health Status	– Health score before stroke from 0:best to 9:worst: 0: Without symptoms; 1: Without significant disability despite symptoms; 2: Minor disability; 3: Moderate disability, but able to walk independency; 4: Moderate disability, not able to walk independency; 5: Major disability; 9: Unknown
Age	– Patient age, years
Gender	– 1:Male, 2:Female, 9:Unspecified
Past Stroke	– Stroke in past. 1:Yes, registered in patient health record, 0:No
Stroke Symptoms	0:No symptoms, 1:Impaired consciousness, 2:Weakness/paresis, 3: Speech disorder (aphasia), Several digit codes: 123-means all 3 symptoms
Health Complications	0:unspecified, 1: other CV (cardiovascular) complications 4:other complications, Several digit codes: 23:means 2 and 3

A stroke is a medical condition where there is an interruption in blood flow to the brain. In general, the stroke is classified into two primary types of strokes: a hemorrhagic (Hemorag) stroke and an ischemic stroke (Ishemic) [13]. However, the definition of symptoms enables to define more stroke types and subtypes. The origin of stroke implies causes linking to different morbidities of patients, such as diabetes and heart diseases, therefore numerous potentially significant variables are registered in the stroke data sets.

Ischemic strokes happen when the blood vessels carrying blood to the brain become clogged. This type of stroke makes about 87% of all stroke cases [13]. There are two different types of ischemic strokes: thrombotic strokes and embolic strokes, which are considered to have different underlying reasons. A thrombotic stroke is caused by a clot forming in a blood vessel of the brain, which is often related to atherosclerosis. An ischemic stroke can be embolic, where blood clot travels from another part of body to the brain. An estimated 15% of embolic strokes are due to a condition called atrial fibrillation. A hemorrhagic stroke is caused by bleeding which can happen within the brain or in the area between the brain and skull. This is the cause of about 20% of all strokes [1]. Hemorrhagic strokes make two categories based on the place where the bleeding occurs and its cause: Intracerebral hemorrhage (Hemorag) and Subarachnoid hemorrhages (SAH). Intracerebral hemorrhages are caused by a broken blood vessel located in the brain. A very high blood pressure can cause weakening of the small blood vessels in the brain. It may be related to anticoagulant therapy. The other category -Subarachnoid hemorrhages (SAH) occur when a blood vessel gets damaged, leading to bleeding in the area between the brain and the thin tissues that cover it. SAH can be caused by a ruptured aneurysm, AVM, or head injury. SAH is a less common type of hemorrhagic stroke, approximately 5–6% of all strokes. Due to different causes it is allocated to the separate category of stroke SAH. The other types of stroke include various cases such as Cryptogenic Stroke, Brain Stem Stroke, and others.

The survival and process of recovery after stroke are affected with numerous conditions. The health status by applying Modified ranking score is evaluated during the admission and post-treatment. The length of stay in the hospital and mortality rate is affected not only by severity of the stroke case, treatment interventions, but also complications, such as pneumonia, neural, cardiovascular and other cases [1, 13].

The stroke tends to occur repeatedly, which makes nearly 25% of all stroke cases. The effect of age and the modifiable life-style factors, such as smoking, hypertension, obesity is among the common predictors of stroke and its outcomes [1, 13].

The variables for the research data set were prepared according to the provided characteristic of stroke types.

3 Statistical Analysis and Visualisation of Stroke Data

The dataset was explored in order to characterize data distribution and suitability for survival modelling as a Time-to-Event stroke research, where the Time-To-Event is an in-hospital mortality of the stroke patients. Firstly, we try to see the mortality percentage for different Types of stroke. The results we got is presented as a frequency table (Table 3). It corresponds to the general statistics of distribution of stroke cases, slightly less for the Ischemic and Hemorag cases, and higher rate of SAH (Sect. 2).

Table 3. Percent of death from different stroke type

	No. of Patients	Percent of cases	Percent of Death
Ischemic	643	68	74.5
Hemorag	115	12	59.1
SAH	153	16	51.6
Unspecified	33	3	48.5
Grand Total	944	100	68.0

The highest number of patients is affected by Ischemic stroke type. The biggest percent of in-hospital death cases is also in this group of patients.

The Fig. 1 visualizes the number of deaths related to the patients age.

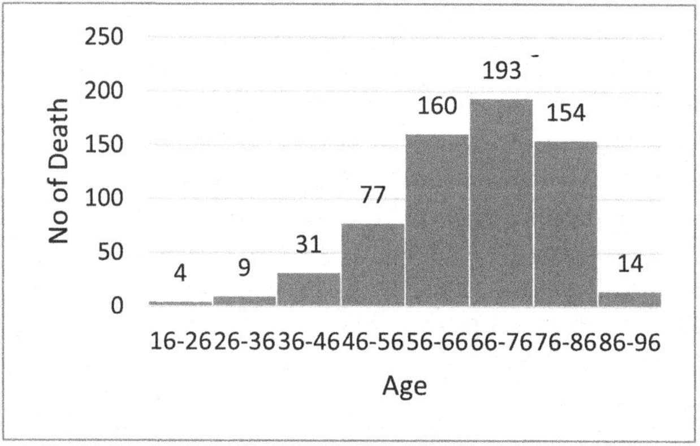

Fig. 1. Histogram of death by patients age

The result is fully in line with the general statistics: the most dangerous age for stroke is in between 66 and 76 [1].

The influence of the Health status before stroke is differently distributed among the Stroke types (Fig. 2). In Fig. 2 the patients with the worst health symptoms (the values approaching to 9) are most likely to experience the SAH stroke. The Ischemic stroke type is most dangerous for people who do not have any serious health problem before stroke.

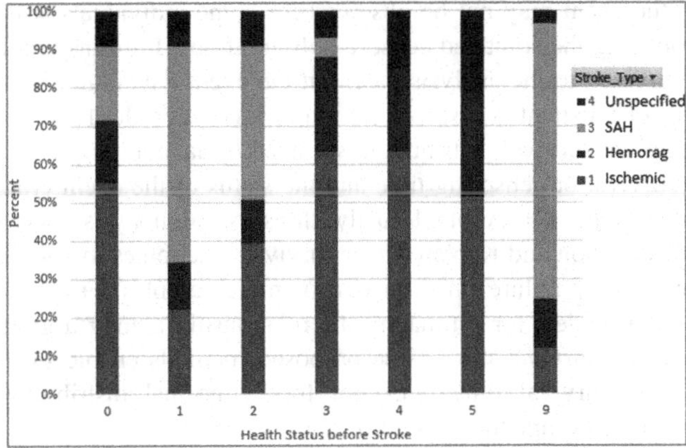

Fig. 2. Relationship among Health status and Stroke type

The stroke symptoms are revealed in different ways for specific stroke types (Fig. 3). The symptoms 2, 3 and jointly 2 and 3 are more peculiar to the ischemic stroke, whereas SAH implies 1, 12 and 13 symptoms. All symptoms are similarly distributed for the Hemorag stroke type. The selected data follows general distribution of the stroke patient variable values, which makes it possible for comparative evaluation of the survival modelling research results to the existing studies and explore possibilities of new insights.

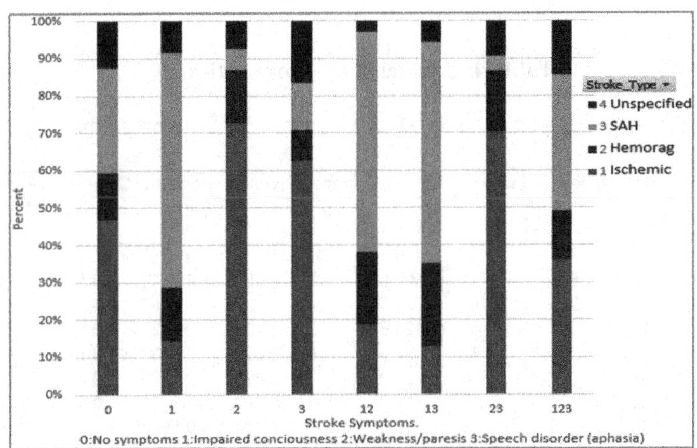

Fig. 3. Relationship among Stroke symptoms and Stroke type

4 Survival Modelling from Stroke Clinical Trial

Survival analysis involves the modelling of the expected duration of time until occurrence of the events, such as death, disease or other incidence of interest.

Sometimes, the event may not be observed for some individuals within the study time period, producing the so-called censored observations. It means that the observed individual has survived to the analysis date, but the event data is not yet known, or it will never occur, or a patient is lost to follow-up observation during the study period.

For estimating the survival distribution, we need to have a sample of observations either with the survival or censoring time and the status of the event (variable values: 0 = no event (censored), 1 = event). Usually, the survival analysis starts with the life table, survival distribution and Kaplan-Meier survival function estimation. Also we can compare the survival or failure times in two or more samples. Finally, the Survival Analysis regression models for estimating the relationship among a group of continuous variables to the survival times. The proposed methods enable to overcome the conditions that the survival time does not follow normal distribution and some observations will be incomplete (censored).

Survival Distribution Modelling by the Life Table

The method of the Life table enables to summarize the events and the proportion of individuals surviving at each event time point. The origins of this method can be traced back to the 1950s [11, 12]. For such calculations we use STATISTICA for Windows (https://www.statistica.com/) Survivor Analysis module.

We select for analysis only the data records with Days at Hospital <=40, and exclude 17 other cases as outliers. In order to construct a life table, we firstly distribute the Days at Hospital variable into equally spaced intervals. As it is shown in the Table 4 the 5-day intervals are chosen, and the Life table characteristics are calculated for the start of each interval: the number of alive patients, the number of patient who have died, and the number of censored patients.

Table 4. Life table for stroke patients

	Life Table Interval Width=5 Days										
Interval	Interval Start	Interval End	Number Entering N_t	Number Withdrwn C_t	Number Exposed N_t^*	Number Dying D_t	Proportn Dead Q_t	Proportn Survivng P_t	Cum.Prop Survivng S_t	Problty Density F_t	Hazard Rate H_t
1	0	5	927	104	875.0	149	0.1703	0.8297	1.0000	0.0341	0.0372
2	5	10	674	232	558.0	87	0.1559	0.8441	0.8297	0.0259	0.0338
3	10	15	355	161	274.5	34	0.1239	0.8761	0.7004	0.0173	0.0264
4	15	20	160	63	128.5	13	0.1012	0.8988	0.6136	0.0124	0.0213
5	20	25	84	40	64.0	6	0.0938	0.9063	0.5515	0.0103	0.0197
6	25	30	38	14	31.0	3	0.0968	0.9032	0.4998	0.0097	0.0203
7	30	35	21	8	17.0	8	0.4706	0.5294	0.4515	0.0425	0.1231
8	35	40	5	4	3.0	1	0.3333	0.6667	0.2390		

In Table 4, only for the first 5-day interval the number of deaths exceeds the number of patients discharged (withdrawn) from hospital. As the events of interest (deaths) are assumed to occur at the end of the interval and censored events are assumed to distribute uniformly throughout the interval, usually for survival calculations we use the variable N_t^*-*Number Exposed*. It is defined as the adjustment of *Number Entering* variable N_t and is calculated by formula $N_t^* = N_t - \frac{C_t}{2}$ to reflect the average number of participants at risk during the interval. Here C_t, denotes number of censored events (*Number Withdrawn*).

Assume that D_t means the number of participants who have died during interval t. Then the ratio of dying during interval t, is calculated by formula: $Q_t = \frac{D_t}{N_t^*}$, and the ratio of surviving by $P_t = 1 - Q_t$.

The survival function S_t – sometimes is called a cumulative survival probability, which means probability that a subject survives longer than time t. The $S_0 = 1$ and the mean value of surviving past time 0 (all participants are alive at time zero or study start). The Surviving function value for each subsequent interval is computed according to formula: $S_{t+1} = S_t \cdot P_{t+1}$. From the Table 3 we can notice that the probability to survive from stroke after 40 days (interval 8) is equal 0.239.

The estimated probability of failure in the respective interval is computed per unit of time. It is called Probability density and can be calculated by formula: $F_t = (P_t - P_{t+1})/W_t$. Here W_t is the width of the respective interval, in our case it is 5 days. From Table 4 it is seen that probability of the failure (Probability density F_t) takes a highest value in the interval 1 (the first 5 days) and interval 7 (from 30 to 35 days).

The hazard function in survivor analysis is understood as the probability per time unit that a patient that has survived to the beginning of the respective interval will fail during that interval. Specifically, it is computed as the number of failures per time units in the respective interval, divided by the average number of surviving cases at the mid-point of the interval. Or we can evaluate it by the relationship $H_t = F_t/S_t$. The Table 4 shows the highest hazard rate for the last interval.

The life table evaluates the sample distribution of failures over time. However, for practical purposes, we need to know the underlying shape of survival distribution in the population. To model the survival time usually are utilized the exponential, Weibull or Gompertz distributions. The fitting survival function parameters are estimated using a least squares method. The Fig. 4 presents the observed and fitted exponential distribution with parameter lambda equal 0.0403.

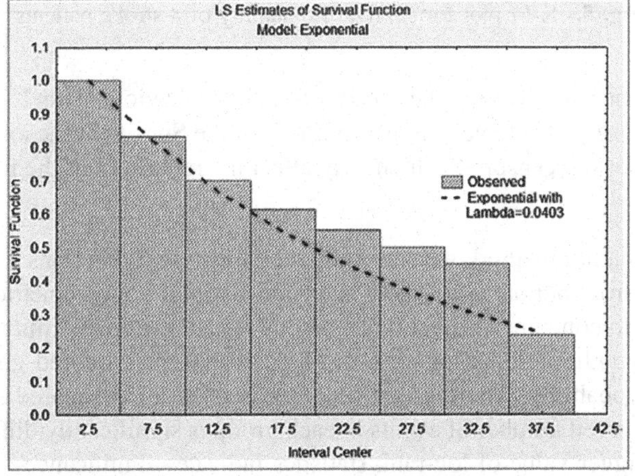

Fig. 4. Observed and fitted Survival function

The main assumptions of selected distribution and the length of selected interval may have influence to the results, which can be investigated by combining to other survival modeling approaches.

Kaplan-Meier Method
The Kaplan-Meier (K-M) method is a non-parametric technique to estimate the survival function from observed survival times (Kaplan and Meier, 1958). The calculated probability S_t is a step function that changes value each time the observed event changes. We can imagine creating a life table with time intervals containing exactly one individual case. The survival probability at time t, S_t can be found by using the same formula as in case of life table $S_{t+1} = S_t \cdot P_{t+1}$, assuming that the interval length is equal to 1. The advantage of the K-M over the life table method is that the resulting survival function does not depend on the grouping of the data. The survival function for stroke data estimated using K-M method is visualized in Fig. 5.

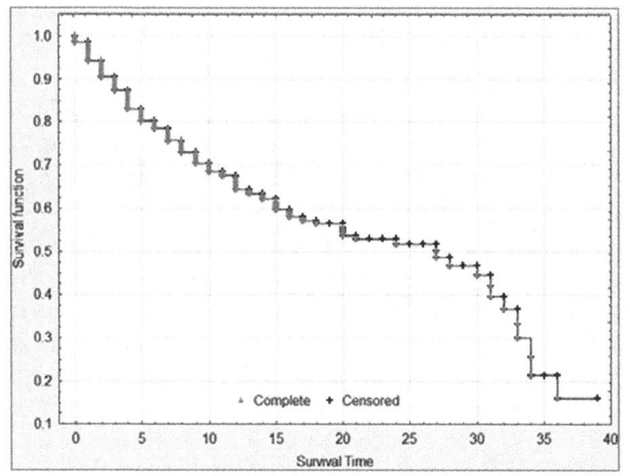

Fig. 5. K-M plot for survival probability of a stroke patients

As we can notice, the shape of the survival function behaviour using K-M plot is more smooth than in case of life table. On the right part of the Survival time axis a significant part of the samples are censored, which may affect the reliability of the results.

Log-rank Test
Using K-M estimation method, we can compare the survival functions for two or more different groups from our data set. There is a wide range of nonparametric tests that can be used in order to compare survival times, in this research we will limit to the log-rank test. The null hypothesis for a log-rank test asserts that the explored groups have the same survival probability. The log-rank test is based on a chi-squared statistic which checks if the observed number of events in each group is significantly different from the expected. The bigger value of log-rank statistics indicates significance of difference in the survival times among the groups.

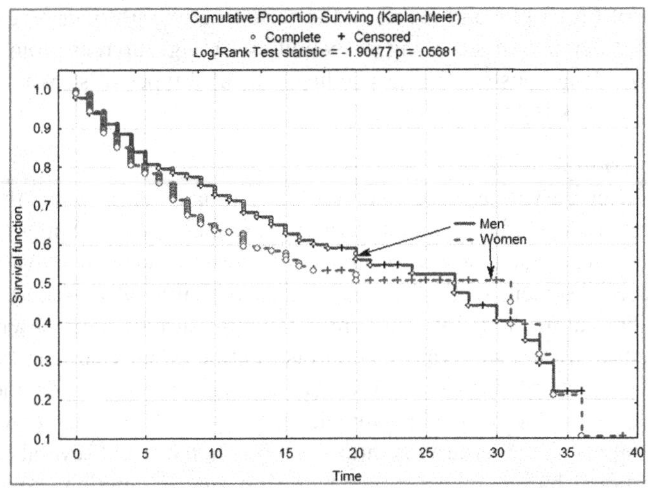

Fig. 6. K-M plot for survival probability of men and women groups of stroke patients

In order to compare the survival probabilities of men and women who have experienced stroke, we calculated the log-rank statistics and p value for our data set. According to Fig. 6, there is no significant influence of gender to survival probability, assuming an alpha level of 0.05. The p value is close to the threshold equal to 0.05681, and visually we can indicate a little higher survival chance from stroke for men's, especially in the interval of [5, 20] days.

STATISTICA tool enables us to compare K-M estimations for multiple groups. The plot in Fig. 7 illustrates the survival probability values for Ischemic, Hemorag, SAH and Unspecified stroke groups.

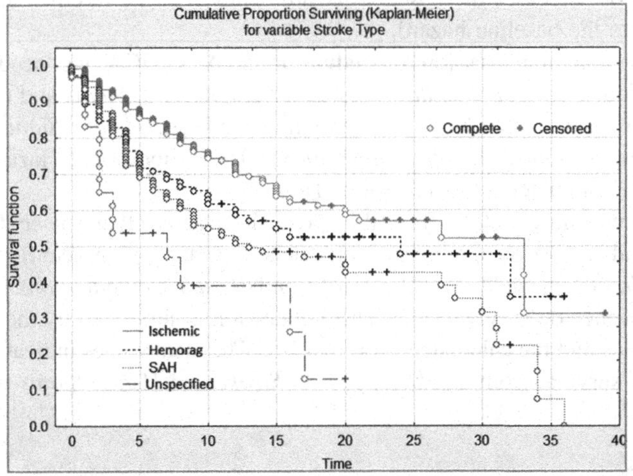

Fig. 7. K-M plot for stroke survival probability for Stroke Type groups

The results of Fig. 7 plot and values of the corresponding p values reveal that there is a significant difference in surviving probabilities among different groups of Stroke Types. The most dangerous are the unspecified and SAH types of stroke, and the best survival probability is in the Ishemic stroke group.

Survival Regression Models
The Kaplan–Meier curves and log-rank tests are usually used when the predictor variable is categorical. If the predictor variable is quantitative we apply the survival regression, which estimated regression for covariates of independent variables (e.g. age, illness type, treatment methods, etc.) against the survival or failure times. We cannot use traditional methods like linear regression because of some observations are incomplete (censored) and survival function distribution is not Gaussian (normal).

For this type of analysis, we explored several survival regression models: Proportional Hazard (Cox) model, exponential, normal and lognormal regression. All models aim to represent the hazard rate h(t|x) as a function of t and several covariates x.

The proportional hazard model is not based on any assumptions concerning the nature or shape of the underlying survival distribution. The Cox model may be considered as a nonparametric method and requires the underlying hazard rate to be a function of the independent variables (covariates). The idea of Cox's regression is that the log-hazard of an individual is a linear function of the selected covariates and a population-level baseline hazard that changes over time. The model may be formalized as:

$$h(t|x) = b_0(t) \cdot \exp(\sum_{i=1}^{n} \beta_i \cdot (x_i - \overline{x}_i))$$

The term $b_0(t)$ is called the baseline hazard. It is the hazard for the respective individual when all independent variable values are equal to zero. Next part of this equation is called partial hazard. The partial hazard is a time-invariant scalar factor that only increases or decreases the baseline hazard. Thus changes in covariates will only inflate or deflate the baseline hazard.

To use the Cox's regression procedure in STATISTICA we need to set the dependent variable *Days at Hospital*, censoring variable *Vital Status* and to choose the independent covariates: *Stroke Type*, *Treatment methods*, *Health Status*, *Age*, *Past Stroke*, *Stroke Symptoms*, *Health complications*. The Proportional Hazard (Cox) regression model parameter estimates are in Table 5.

Therefore, we can conclude from the spreadsheet above that *Stroke Type*, *Health Status*, *Age* and *Stroke Symptoms* are the most important (significant) predictors of hazard. The *Treatment methods*, *Past Stroke* and *Health Complications* don't have a significant influence on survival time. This conclusion is supported by the significance of Wald statistics and the calculated p-probability. The Beta values indicate the highest dependence of survival from *Stroke type*, *Past Stroke* and *Health Status* before stroke variables.

Table 5. Cox regression model parameter estimates.

	Dependent Variable: Days at Hospital Censoring var.: Vital Status Chi^2 = 148.259 df = 7 p = 0.0000					
N=927	Beta	Standard Error	t-value	exponent beta	Wald Statist.	p
Stroke Type	0.317580	0.071538	4.43929	1.373799	19.70732	0.000009
Treatment methods	-0.007548	0.006290	-1.19986	0.992481	1.43966	0.230203
Health Status	0.109808	0.019864	5.52803	1.116064	30.55916	0.000000
Age	0.028722	0.005421	5.29817	1.029138	28.07061	0.000000
Past Stroke	-0.224203	0.156608	-1.43162	0.799153	2.04952	0.152264
Stroke Symptoms	0.005987	0.001109	5.39646	1.006005	29.12179	0.000000
Health complications	0.009313	0.005328	1.74778	1.009356	3.05475	0.080511

Taking a look at Beta coefficient for *Stroke Type* = 0.3176 we can conclude that a one unit increase in *Stroke Type* means the baseline hazard will increase by a factor of exp(0.3176) = 1.3738-about a 14% increase. Recall, in the Cox proportional hazard model, a higher hazard means more at risk of the event occurring.

Combining several methods for the research enable to explore a set of variables for their influence to in –hospital survival. The research results may be compared to the outcomes of other research. The main variables age, Health status, complications (pneumonia) are most important for Iran study of 1990 stroke patient cases [14]. For the Saudi Arabian 1249 stroke cases [15] most influential factors were gender (men), morbidities, stroke types and severity of symptoms. Severity of stroke, stroke type (Hemorag), pneumonia and other complications were most significant for Tanzania 224 stroke cases [16]. The research of 1732 patient cases (specified for SAH stroke type) explored in Norway [17] defined age and complications (aneurism) as the major factors. The interchanging significance of the variables imply necessity to search for data and methods leading to higher quality of research outcomes.

5 Conclusion

The study was conducted according to time-to-event stroke clinical trial records of the neurology department of Clinical Centre in Montenegro. The original database consists of the structured 944 records to ensure adequate significance of the obtained results.

A simple analysis of baseline data showed a meaningful relationship between the type of stroke symptoms experienced and the type of stroke. We found the SAH impact on the symptoms of Impaired consciousness and a significant influence of Ischemic stroke to patient's weakness/paresis. A significant difference in the number of deaths among different stroke types is also identified. Most dangerous are Ischemic and Hemorag strokes.

The compiled life table evaluated survival probability after stroke affection and allowed the survival function to be approximated by an exponential distribution.

Using the Log-rank test, we examined survival probabilities for different stroke types. We found that the most risky are SAH and unidentified stroke types.

The Proportional Hazard (Cox) regression model highlighted the most important factors influencing the magnitude of the hazard rate. Stroke Type, Health Status, Age and Stroke Symptoms was seen as the most important (significant) predictors of hazard.

References

1. World Stroke Organization (WSO) Annual report, (2019). https://www.world-stroke.org/assets/downloads/WSO_2019_Annual_Report_online.pdf. Accessed 2 Apr 2021
2. Feigin, V., et al.: Global, regional, and country-specific lifetime risks of stroke, 1990 and 2016. New Eng. J. Med. **379**(25), 2429–2437 (2018)
3. Emmert-Streib, F., Dehmer, M.: Introduction to survival analysis in practice, machine learning and knowledge extraction. Open Access J. **1**(3), 1013–1038 (2019)
4. Gross, S.R., O'Brien, B., Hu, C., Kennedy, E.H.: Rate of false conviction of criminal defendants who are sentenced to death. Proce. Natl. Acad. Sci. USA **111**, 7230–7235 (2014)
5. Ancarani, A., Di Mauro, C., Fratocchi, L., Orzes, G., Sartor, M.: Prior to reshoring: a duration analysis of foreign manufacturing ventures. Int. J. Prod. Econ. **169**, 141–155 (2015)
6. Jung, E.Y., Baek, C., Lee, J.D.: Product survival analysis for the App Store. Mark. Lett. **23**, 929–941 (2012)
7. Rapport, F.: Summative analysis: a qualitative method for social science and health research. Int. J. Qual. Method. **9**, 270–290 (2010)
8. Clark, T.G., Bradburn, M.J., Love, S.B., Altman, D.G.: Survival analysis part i: basic concepts and first analyses. Br. J. Cancer **89**, 232–238 (2003). https://doi.org/10.1038/sj.bjc.6601118
9. Kaplan, E.L., Meier, P.: Nonparametric estimation from incomplete observations. J. Am. Stat. Assoc. **53**, 457–481 (1958)
10. Cox, D.R.: Regression models and life-tables. J. R. Stat. Soc. Ser. B (Meth.) **34**, 187–202 (1972)
11. Berkson, J., Gage, R.P.: Calculation of survival rates for cancer. Mayo Clin. Proc. **25**, 270–286 (1950)
12. Cutler, S.J., Ederer, F.: Maximum utilization of the life table method in analyzing survival. J. Chron. Dis. **8**(53), 457–481 (1958)
13. Brennan, D.: What Is the Difference Between Ischemic Stroke and Hemorrhagic Stroke? 2021. https://www.medicinenet.com/difference_ischemic_stroke_and_hemorrhagic_stroke/article.htm. Accessed on 3 Jan 2021
14. Deljavan, R., Farhoudi, M., Sadeghi-Bazargani, H.: Stroke in-hospital survival and its predictors: the first results from Tabriz Stroke Registry of Iran. Int. J. Gen. Med. **11**, 233–240 (2018)
15. Alhazzani, A.A., et al.: In hospital stroke mortality: rates and determinants in southwestern saudi arabia. Int. J. Environ. Res. Public Health **15**(5), 927 (2018). https://doi.org/10.3390/ijerph15050927
16. Okeng'o, K., Chillo, P., Gray, W.K., Walker, R.W., Matuja, W.: Early mortality and associated factors among patients with stroke admitted to a large teaching hospital in tanzania. J. Stroke Cerebrovasc. Dis. **26**(4), 871–878 (2017). https://doi.org/10.1016/j.jstrokecerebrovasdis.2016.10.037
17. Øie, L.R., et al.: Incidence and case fatality of aneurysmal subarachnoid hemorrhage admitted to hospital between 2008 and 2014 in Norway. Acta Neurochir. **162**(9), 2251–2259 (2020). https://doi.org/10.1007/s00701-020-04463-x

Automatically Extracting Insurance Contract Knowledge Using NLP

Alexandre Goossens(✉), Laure Berth, Emilia Decoene, Ziboud Van Veldhoven, and Jan Vanthienen

Katholieke Universiteit Leuven, Naamsestraat 69, Leuven, Belgium
alexandre.goossens@kuleuven.be

Abstract. Vanbreda Risk & Benefits, a large Belgian insurance broker and risk consultant, allocates a substantial amount of time and resources to answer contract related questions from customers. This requires employees to manually search the relevant parameters in the contracts. In this paper, a solution is proposed and evaluated that automatically extracts insurance parameters from contracts using regular expressions and Natural Language Processing. While Natural Language Processing has been used in insurance for optimising premiums, detecting fraudulent claims, or underwriting, limited work has been done regarding parameter extraction. The proposed solution has been developed on 127 different contracts and two different contract types in terms of accuracy and time performance. Moreover, the automatic parameter extraction has been compared to manual parameter extraction. We conclude that automatic parameter extraction using regular expressions achieves better accuracy than manual extraction on top of being significantly faster, allowing Vanbreda Risk & Benefits to invest more time into providing better customer service.

Keywords: Service automation · Insurance industry · NLP · Regular expressions

1 Introduction

The development and improvement of text mining algorithms have given rise to a number of new applications. More specifically in the insurance industry, natural language processing (NLP) is used nowadays for example to fine-tune premiums [1], detect fraudulent claims [2], or perform sentiment analysis on tweets [3]. These applications deal with optimising profit for insurance companies. Insurance companies also need to provide information to customers related to their products. Customers need to know under which conditions they are insured or the rate at which they are covered in case something happens. These questions occur without necessarily having an insurance claim filed by the customers.

This paper will deal with the automatic extraction of insurance parameters from contracts using regular expressions and has been implemented at Vanbreda Risk &

W. Abramowicz et al. (Eds.): BIS 2021 Workshops, LNBIP 444, pp. 27–38, 2022.
https://doi.org/10.1007/978-3-031-04216-4_3

Benefits[1] a large Belgian insurance broker and risk consultant. A solution based on regular expressions is proposed and formally evaluated on a test set of 42 real contracts in terms of accuracy and time-performance compared to manual work.

We proceed as followed: Sect. 2 provides background information concerning social security in Belgium and the company where the solution has been implemented. Section 3 states the problem statement. In Sect. 4, related work is discussed and in Sect. 5 the methodology is explained. The results are presented in Sect. 6 and the evaluation and future work are presented in Sect. 7. We conclude the paper in Sect. 8.

2 Context

2.1 Social Security in Belgium

In Belgium, the social security provisions consist of three parts or pillars. Each one of these pillars is written down in forms, regulations, or contracts describing under which conditions a person is eligible for provisions. Due to their complex nature, these forms or contracts are long and not straightforward to extract all relevant information by a non-expert. The three pillars of Belgian social security are: statutory provisions such as child benefits, illness benefits, retirement, unemployment; supplementary provisions provided by the employer or also known as employee benefits [4] and mainly exists because the first pillar alone is not able to guarantee living standards; individual provisions allowing individuals to build up their provisions on their own.

2.2 Current Situation at a European Insurance Broker Vanbreda Risk and Benefits

In the subsequent paragraphs, the current parameter extraction process at Vanbreda Risk & Benefits will be explained. Vanbreda Risk & Benefits is an independent insurance broker and risk consultant mainly operating within Belgium. Vanbreda Risk & Benefits has a department called Employee Benefits providing assistance in analysing and managing employee insurances of other companies. Due to the complex nature of these contracts, expertise knowledge is required to interpret these contracts. Whenever a client employee or customer company requires information (simple or complex), they can contact Employee Benefits to ask their questions. These questions require the employees to analyse the contracts of the customer. Vanbreda Risk & Benefits currently has approximately 2000 contracts of 25 pages long on average without counting the attachments.

Recently, a technical summary of a contract has been introduced in the company. This technical summary summarises the most relevant parameters of a contract such as a formula, a definition, or a date. A technical summary consists of a parameter column and a value column.

The employees of Employee Benefits can use the data of these technical summaries to answer most of the queries. Thanks to this technical summary, the question-answering

[1] https://www.vanbreda.be/en/.

process is sped up. Whenever a new contract is received, a technical summary is created. The creation of a technical summary requires between 15 and 20 min for a single contract. Since these technical summaries have been introduced recently, not all contracts have a technical summary yet. The extraction of parameters from contracts has always been manual work at Vanbreda Risk & Benefits whether the parameters were extracted directly from the contracts or more recently with the introduction of technical summaries.

3 Problem Statement: From Contract to Automatic Technical Summary

As stated previously, the creation of technical summaries is currently time-consuming. The employee queries the contract in the correct database and manually extracts the required parameters. In case the parameter has an assigned list of default values, one of them is chosen. If not, the parameter values do not have a standard form. This lack of standardisation means that a technical summary also depends of the employee creating it and therefore some variation exists between technical summaries. Moreover, contracts are subject to changes due to changing interest rates, changing market conditions, preferences and such.

It would be highly beneficial for insurance companies to have a digital solution to (partially) automate contracts into technical summaries. Hence, this research is interested in a scalable and applicable solution for different types of contracts (in this case: guaranteed income and waiver of premiums) a company is dealing with. Moreover, the proposed solution needs to be independent of the insurance provider. In short, the solution has to reduce the creation time of a technical summary and capable of handling different contract types and different insurance providers.

4 Related Work

4.1 Applications of NLP and AI in the Insurance Industry

The impact and applications of AI in data-intensive domains such as finance, insurance, and public services is a widely studied domain, e.g. [5–8] For the insurance industry, the authors of [7] say that robots and people can fully create benefits when people focus on building customer relationships and robots perform the repetitive tasks. This principle can indeed be seen as one of the main drivers for research in this direction by insurance companies.

There are numerous potential NLP applications in the insurance industry [9]. One application is the cost predictions of insurance claims using text mining on the injury and incident descriptions in conjunction with more structured data such as demographics [10]. NLP can also be used to fine-tune premiums [1] or to detect fraudulent claims [2]. Furthermore, these techniques can be used to gain customer insights for example by analysing customer calls [11].

With the rise of social media platforms, insurance companies have much more data at hand. In [3], tweets are analysed revealing the most common topics and their feelings towards an insurance company. This provides the insurance company the ability to provide better customer service and to reach potential new customers. The use of chatbots in the insurance industry has also been investigated in [12]. The so-called Intellibot is able to answer specific questions dealing with insurance provided it is given the correct insurance knowledge.

In short, current NLP applications mainly deal with fine-tuning premiums, detecting fraudulent claims, sentiment analysis or customer interaction, and so forth. However, limited academical attention has been paid to automatically extracting insurance parameters from contracts using NLP.

4.2 Document Segmentation

Within the widely studied field of NLP, information extraction is a possible application [13]. The goal of information extraction is to extract structured information from (semi)-unstructured documents.

In most organisations, legal documents in a digital format are often available. Unfortunately, this is often in a semi-structured form. Even though humans can interpret such documents easily, it remains a challenge for machines. This limitation inhibits the performance of information extraction [14].

Table 1. Overview of segmentation techniques.

Contract Segmentation Techniques			
Rules-based approach	Paragraph boundary detection	Conditional random fields	Generic named entity recognition
Search start of new paragraph by set of rules	Search paragraph boundaries by sentence detection	Classifying words by looking at the features the previous and next word (sequential classifier)	Identifying and recognizing specific entities in a text (names of persons, percentages, indications of time, etc.)

Some research has already been done to segment documents. The authors of [14] propose to segment documents into smaller parts to process legal documents easier hence allowing the program to have knowledge of the contract structure. These smaller parts could be sentences, paragraphs, or pages. The authors of [15] provide an overview of segmentation techniques. These are shortly summarised in Table 1. Since contracts are legal documents, knowing the structure of a contract can help in extracting contract elements [16].

5 Methodology

To start the automatic summarisation process, the contract must be read in. Every contract is received in a format called smart PDF. To read these contracts, the Python package PyMuPDF [17] or PDFMiner [18] allows for the text extraction from the PDF contract, next the text is stored as a string variable allowing us to perform the three necessary processing operations. Firstly, the contract is segmented, next the relevant parameters are searched and extracted, to finally be filled into the technical summary. The segmentation and extraction methods can both be used for other contract types. The parameter extraction needs to be fine-tuned to fit the correct purpose.

5.1 Segmentation

In general, a contract is segmented into two parts. The general terms and the special terms. The special terms deal with the insurance conditions and premiums and such. Every parameter that needs to be extracted is located in the special terms of the contract. Prior to the extraction, these special terms are divided into different parts allowing for the extraction to be more efficient. Dividing a contract into smaller parts allows for a more targeted search of a parameter instead of going through the full contract. In short, this means that the contract will be segmented into paragraphs. An overview of the segmentation steps is given in Fig. 1.

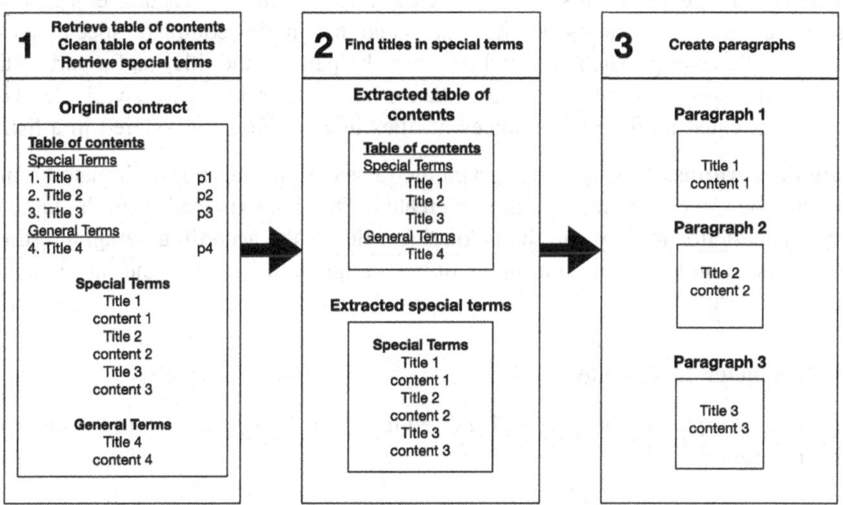

Fig. 1. Overview of the segmentation process.

Retrieve the Table of Contents: A contract structure remains more or less the same for all contract types and insurance providers for Vanbreda Risk & Benefits. In our case, the table of contents itself will be located between the words 'table of contents' and the very first title that reoccurs after the table of contents.

These exact words are looked up in the contract string by using regular expressions. The authors of [19] define a regular expression or regex as "a sequence of characters (letters, numbers and special characters) that form a pattern that can be used to search text to see if that text contains sequences of characters that match the pattern (p.257)". For each matching pattern, the start indices are saved resulting in a substring that only consists of the table of contents.

Clean up the Table of Contents: Only the titles are needed in our solution but a table of contents also contains the page number, title number and so forth. Therefore, the table of contents needs to be cleaned. Each line of the table of contents is placed in a list and considered as a different element. For each list element, the title number, page number, and other punctuation marks are removed, leaving only the title in the list. At the end of this step, only a list of contract titles remains. These steps are also visualised in Fig. 1.

Retrieve the Special Terms: The same procedure as for the table of contents is used to retrieve the special terms. The special terms are located between "special terms" and "general terms" and these words are used as patterns to further search the contract. A separate substring is created that contains these special terms.

Find Titles in Special Terms: The extracted titles are put into regular expressions. These exact titles or patterns are then searched for in the complete special terms substring. Whenever a match is found between the pattern (the title) and a part of the substring (the special terms), the start index of the title is saved in a variable. This process is repeated until all start indices of titles in the contract are stored in a list.

Create Paragraphs: Finally, a paragraph dictionary is created. For each match found in the previous step a dictionary entry is created. The title extracted from the table of contents is a dictionary key and the belonging value is the according paragraph. Each paragraph is delimited by the start index of the paragraph title and by the start index of the next paragraph title.

5.2 Parameter Extraction

The parameter extraction process is visualised in Fig. 2. This part needs to be finetuned for each parameter.

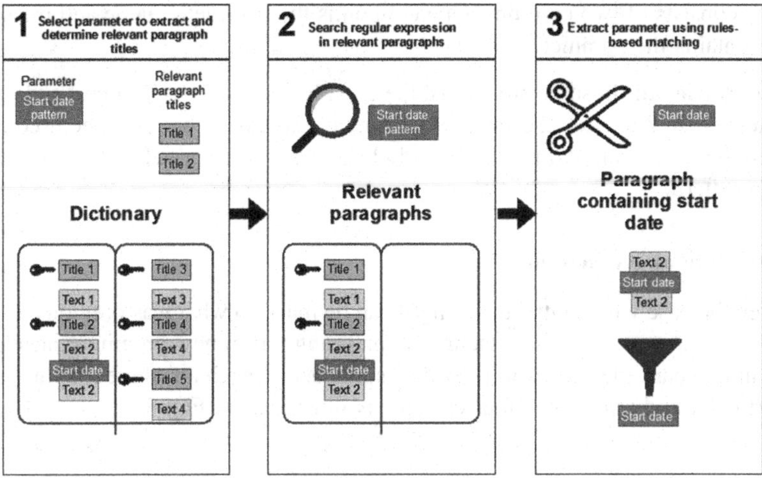

Fig. 2. Parameter extraction process.

Select Parameter to Extract: Determine which parameters need to be extracted for the creation of a technical summary.

Determine Relevant Paragraphs: In this step, the program needs to know in advance which paragraphs might contain the parameter. The most common relevant paragraph titles are given to the parameter extraction function. To increase the robustness, the title of the previous paragraph (according to the table of contents) is given in case there has been a spelling mistake. This is not a problem as for each insurance type and insurance provider the contract follows the same scheme. Once the pattern is determined it is scalable to all other contracts of the same type and insurance provider.

Extract Relevant Paragraphs: The required paragraphs have already been extracted following the steps explained in Sect. 5.1 and have been stored in a string variable.

Transform Text: In this step, the natural language toolkit (NLTK) is required [20]. The NLTK library allows us to preprocess a string to perform further analysis. The extracted paragraphs are parsed through a so-called sentence tokenizer. This will transform a given text into a list of sentences. In this case study, the Punkt sentence tokenizer has been chosen and as the contracts are in Dutch, the Dutch sentence tokenizer is used to transform the text.

Rule-Based Matching: To retrieve the parameters, rule-based matching is used. The idea of rule-based matching is to use regular expressions and to return a value matching the pattern in a given text. Suppose the desired parameter value is the start date of a contract. A date can either be written as DD/MM/YYYY or as DD-MM-YYYY. The following regular expression is able to detect both of these formats: \d{1,2}[\/\-]\d{1,2}[\/\-]\d{4}. This process of rule-based matching is repeated until all desired parameters have been extracted. To finalise, a pandas dataframe [21, 22] is constructed. This dataframe consists of two columns. One column is the parameter name and the second column contains the parameter value.

Sentence Retrieval: In some situations, the complete sentence is required and not just one parameter. In that case, the same procedure as explained above can be used except that when the patterns find a match, the whole sentence is returned instead of only the matching pattern.

5.3 Attachment Extraction

As is often the case with contracts, changes can be made to what was previously agreed upon. These changes are put in attachments and range from new premium rates to new clauses. In this part, the goal is to provide a summary of each attachment by giving the title, start date, and relevant sentences. This is illustrated in Fig. 3.

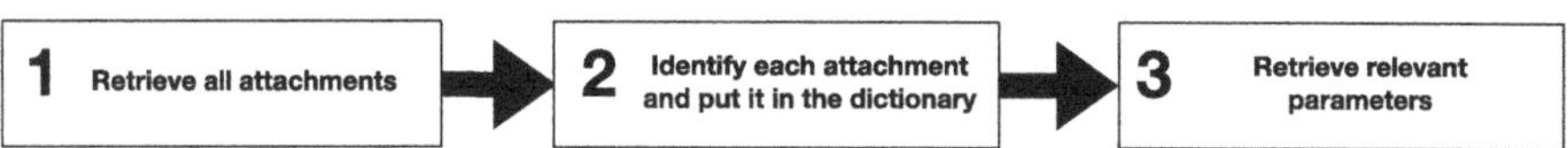

Fig. 3. Overview extraction attachments.

Retrieve All Attachments: For the contracts of Vanbreda Risk & Benefits, the attachments can always be found just before the contract. But it is safe to assume for other problems that the attachments will either be before or after the contract. Once the attachments have been found, a subset is created consisting of only the attachments.

Identify Each Attachment and Put It in the Dictionary: Once again regular expressions are used to find the title of each attachment. This regular expression will consist of the words "attachment", "nr" and a digit. Once a match is found, the corresponding index is stored in a list. These indices allow for the identification of each attachment and the creation of a corresponding dictionary. Once again the keys equal the title of attachment and the corresponding value the content of the attachment.

Retrieve Relevant Parameters: Each title of an attachment is considered a parameter and the second parameter is the start date on which the attachment is valid. Regular expressions are used to find the start dates. In case more information is required later, the body of the attachment is also retrieved and stored as a third parameter. This subset starts at the end of the introduction and ends at the start of the conclusion. To not always retrieve the full content of the attachments, it is possible to identify and store key phrases containing words such as 'the following has been changed'.

Creation of the Technical Summary: During the extraction process all parameters are stored in a dataframe. At the end of the process, the contents of this dataframe are automatically put into a spreadsheet file such as an Excel file. This Excel file consists of two columns containing parameter name and parameter value. For each contract, one technical summary is created containing the extracted parameters.

6 Experiments

6.1 Dataset

The proposed approach has been trained on 127 real-life contracts and has been evaluated on the two contract types: guaranteed income and waiver of premiums. The amount of time necessary for the creation of a technical summary by an employee was provided by Vanbreda Risk & Benefits. To determine whether a parameter has been extracted correctly, an expert has checked each one of them individually by going through each contract of the test set. The technical summaries created by humans have also been subject to control by the same expert.

It is only possible to perform parameter extraction that was correctly segmented by the program. The test set of type guaranteed income consisted of 21 real contracts out of which 15 were correctly segmented. The test set of type waiver of premiums also consisted of 21 real contracts and 17 contracts were correctly segmented.

6.2 Results

In Table 2, we report the number of parameters for each type of contract, the number and percentage of correctly extracted parameters, and finally the average processing time per contract. These results are reported for both manual and automatic parameter extraction.

Note that the number of parameters differs between automatic and manual extraction because automatic extraction also extracts attachments and some other parameters offering more information for the employees.

Table 2. Performance comparison: manual vs automatic.

	Guaranteed income		Waiver of premiums	
	Manual	Automatic	Manual	Automatic
Number of parameters per contract	14	17	7	10
Average number of parameters extracted correctly	10.8	13.9	4.5	8.1
% parameters correct	77.14	81.57	63.87	81.18
Average processing time per contract	15–20 min	4.91 s	15–20 min	5.23 s

7 Discussion, Limitations, and Future Work

7.1 Discussion

Results Analysis. From Table 2 we can conclude that automatic parameter extraction has achieved promising results. The time required to create technical summaries got reduced by approximately 99%. In addition, a high level of accuracy is maintained as automatic parameter extraction achieves an accuracy of above 80%. This method can be used by insurance companies to create technical summaries automatically with a high level of accuracy, thereby reducing the workload on the employees with a significant amount.

Overall, we conclude that even though some manual fine-tuning is necessary to get the automatic extraction started, this is offset by the improved time necessary to create the technical summaries. The proposed solution has proven itself to be scalable, provided that most of the contracts follow the same template.

Automatic Extraction Errors. The errors produced by the automatic parameter extraction process are mainly due to the smart scanner Vanbreda Risk & Benefits is using and due to the packages used to read in the PDF file. We identified the cause of each misread parameter individually and found that approximately 10% of the errors were due to wrong scanning or manual handwriting on the original contract. If the PDF files were read in more accurately and if the contracts were not marked with handwriting, it is safe to assume those 10% could be extracted correctly. Other mistakes can be attributed to the fact that specific parameters were not present in the special terms part of the contract but located somewhere else in the contract.

Manual Extraction Errors. The manual extraction errors can mainly be attributed to three reasons. These reasons are inherently human mistakes.

1. Certain parameters were not filled in by the employees in the technical summaries.
2. The extracted parameter did not contain all necessary information to be classified as correctly extracted.
3. The extracted parameter was wrong.

7.2 Limitations

One limitation of our results is that only Dutch documents that already had pdf versions were used. To analyse different languages, the pipeline must be adapted by using different tokenizers for the respective language. Fortunately, this feature is already supported in the used NLTK package for prevalent languages such as English, Spanish, French, and German. It is possible that the results would improve on, for example, English documents as this language is more researched. Hence, this methodology is possible in different languages on the condition that the respective tokenizer exists and that the regular expressions are expressed in the corresponding language.

Both python packages, PyMyPDF and PDFMiner, can be used for reading PDF files and both have advantages and disadvantages. The PyMyPDF package is better at reading regular sentences whilst the PDFMiner packages is better at reading tables for example. Hence, careful consideration must be made when choosing which python package to utilise.

Extracting parameters automatically from attachments is inherently difficult as here the problem shifts from being structured to unstructured. It is challenging to know in advance in which attachments the parameters will be located. Moreover, attachments are not as standardised as contracts. Due to the unstructured nature of attachments, employees are better at capturing the relevant parameters. Therefore, it has been decided to provide a list of attachments with certain common parameters in the technical summary such as start date or title of the attachment.

7.3 Future Work

As future work, the next steps include fine-tuning the approach to more contract types and insurance providers. We are also planning on improving the digitisation of contracts so that more contracts can be summarised automatically. This will lead to improving customer service at Vanbreda Risk & Benefits.

Moreover, the pattern-based approach works best for structured documents. When dealing with more variable documents, our approach becomes unstable. More research is needed into the usage of complex NLP algorithms to deal with parameter extraction in unstructured documents.

8 Conclusion

In this paper, insurance contract summarisation is performed using NLP. The proposed solution has been implemented at Vanbreda Risk & Benefits a large Belgian insurance broker on 127 contracts. From the results, we conclude that regular expressions are faster and perform as well as the employees at Vanbreda Risk & Benefits. The authors would like to express their gratitude to Vanbreda Risk & Benefits for the collaboration.

References

1. Zappa, D., Borrelli, M., Clemente, G.P., Savelli, N.: Text Mining in Insurance: From Unstructured Data to Meaning. Variance, Press. https://www.variancejournal.org/articlespress/. Accessed 23 March 2021 (2019)
2. Wang, Y., Xu, W.: Leveraging deep learning with LDA-based text analytics to detect automobile insurance fraud. Decis. Support Syst. **105**, 87–95 (2018)
3. Mosley Jr., R.C.: Social media analytics: data mining applied to insurance Twitter posts. In: Casualty Actuarial Society E-Forum. p. 1 (2012)
4. Benavides, T.: Practical Human Resources for Public Managers: A Case Study Approach. CRC Press (2011)
5. Donepudi, P.K.: AI and machine learning in banking: a systematic literature review. Asian J. Appl. Sci. Eng. **6**, 157–162 (2017)
6. Kankanhalli, A., Charalabidis, Y., Mellouli, S.: IoT and AI for smart government: a research agenda. Gov. Inf. Q. **36**, 304–309 (2019). https://doi.org/10.1016/j.giq.2019.02.003
7. Lamberton, C., Brigo, D., Hoy, D.: Impact of robotics, RPA and AI on the insurance industry: challenges and opportunities. J. Financ. Perspect. **4** (2017)
8. Balasubramanian, R., Libarikian, A., McElhaney, D.: Insurance 2030—The Impact of AI on the Future of Insurance. McKinsey Co. (2018)
9. Ly, A., Uthayasooriyar, B., Wang, T.: A survey on natural language processing (nlp) and applications in insurance. arXiv Prepr. arXiv2010.00462 (2020)
10. Kolyshkina, I., Rooyen, M.: Text mining for insurance claim cost prediction. In: Williams, G.J., Simoff, S.J. (eds.) Data Mining. LNCS (LNAI), vol. 3755, pp. 192–202. Springer, Heidelberg (2006). https://doi.org/10.1007/11677437_15

11. Liao, X., Chen, G., Ku, B., Narula, R., Duncan, J.: Text mining methods applied to insurance company customer calls: a case study. North Am. Actuar. J. **24**, 153–163 (2020)
12. Nuruzzaman, M., Hussain, O.K.: IntelliBot: a dialogue-based chatbot for the insurance industry. Knowl.-Based Syst. **196**, 105810 (2020)
13. Yogish, D., Manjunath, T.N., Hegadi, R.S.: Review on natural language processing trends and techniques using NLTK. In: Santosh, K.C., Hegadi, R.S. (eds.) Recent Trends in Image Processing and Pattern Recognition. pp. 589–606. Springer Singapore, Singapore (2019)
14. Loza Mencía, E.: Segmentation of legal documents. In: Proceedings of the 12th International Conference on Artificial Intelligence and Law. pp. 88–97. Association for Computing Machinery, New York, NY, USA (2009). https://doi.org/10.1145/1568234.1568245
15. Shah, P., Joshi, S., Pandey, A.K.: Legal clause extraction from contract using machine learning with heuristics improvement. In: 2018 4th International Conference on Computing Communication and Automation (ICCCA), pp. 1–3 (2018)
16. Chalkidis, I., Androutsopoulos, I., Michos, A.: Extracting contract elements. In: Proceedings of the 16th Edition of the International Conference on Articial Intelligence and Law, pp. 19–28 (2017)
17. McKie, J.X., Liu, R.: PyMuPDF. https://pypi.org/project/PyMuPDF/
18. Shinyama, Y., Guglielmetti, P., Marsman, P.: PDFMiner. https://pdfminersix.readthedocs.io/en/latest/
19. Hunt, J.: Regular expressions in python. In: Advanced Guide to Python 3 Programming. UTCS, pp. 257–271. Springer, Cham (2019). https://doi.org/10.1007/978-3-030-25943-3_22
20. Bird, S., Klein, E., Loper, E.: Natural language processing with Python: analyzing text with the natural language toolkit. O'Reilly Media, Inc. (2009)
21. McKinney, W.: Data structures for statistical computing in python. In: van der Walt, S. and Millman, J. (eds.) Proceedings of the 9th Python in Science Conference, pp. 56–61 (2010). https://doi.org/10.25080/Majora-92bf1922-00a
22. pandas development team, T.: pandas-dev/pandas: Pandas (2020). https://doi.org/10.5281/zenodo.3509134

Analyzing Medical Data with Process Mining: A COVID-19 Case Study

Marco Pegoraro[1(✉)], Madhavi Bangalore Shankara Narayana[1], Elisabetta Benevento[1,3], Wil M.P. van der Aalst[1], Lukas Martin[2], and Gernot Marx[2]

[1] Chair of Process and Data Science (PADS), Department of Computer Science, RWTH Aachen University, Aachen, Germany
{pegoraro,madhavi.shankar,benevento,vwdaalst}@pads.rwth-aachen.de

[2] Department of Intensive Care and Intermediate Care, RWTH Aachen University Hospital, Aachen, Germany
{lmartin,gmarx}@ukaachen.de

[3] Department of Energy, Systems, Territory and Construction Engineering, University of Pisa, Pisa, Italy

Abstract. The recent increase in the availability of medical data, possible through automation and digitization of medical equipment, has enabled more accurate and complete analysis on patients' medical data through many branches of data science. In particular, medical records that include timestamps showing the history of a patient have enabled the representation of medical information as sequences of events, effectively allowing to perform process mining analyses. In this paper, we will present some preliminary findings obtained with established process mining techniques in regard of the medical data of patients of the Uniklinik Aachen hospital affected by the recent epidemic of COVID-19. We show that process mining techniques are able to reconstruct a model of the ICU treatments for COVID patients.

Keywords: Process mining · Healthcare · COVID-19

1 Introduction

The widespread adoption of Hospital Information Systems (HISs) and Electronic Health Records (EHRs), together with the recent Information Technology (IT) advancements, including e.g. cloud platforms, smart technologies, and wearable sensors, are allowing hospitals to measure and record an ever-growing volume and variety of patient- and process-related data [7]. This trend is making the most innovative and advanced data-driven techniques more applicable to process analysis and improvement of healthcare organizations [5]. Particularly, *process*

We acknowledge the ICU4COVID project (funded by European Union's Horizon 2020 under grant agreement n. 101016000) and the COVAS project for our research interactions.

W. Abramowicz et al. (Eds.): BIS 2021 Workshops, LNBIP 444, pp. 39–44, 2022.
https://doi.org/10.1007/978-3-031-04216-4_4

mining has emerged as a suitable approach to analyze, discover, improve and manage real-life and complex processes, by extracting knowledge from event logs [2]. Indeed, healthcare processes are recognized to be complex, flexible, multidisciplinary and ad-hoc, and, thus, they are difficult to manage and analyze with traditional model-driven techniques [9]. Process mining is widely used to devise insightful models describing the flow from different perspectives—e.g., control-flow, data, performance, and organizational.

On the grounds of being both highly contagious and deadly, COVID-19 has been the subject of intense research efforts of a large part of the international research community. Data scientists have partaken in this scientific work, and a great number of articles have now been published on the analysis of medical and logistic information related to COVID-19. In terms of raw data, numerous openly accessible datasets exist. Efforts are ongoing to catalog and unify such datasets [6]. A wealth of approaches based on data analytics are now available for descriptive, predictive, and prescriptive analytics, in regard to objectives such as measuring effectiveness of early response [8], inferring the speed and extent of infections [3,10], and predicting diagnosis and prognosis [11]. However, the process perspective of datasets related to the COVID-19 pandemic has, thus far, received little attention from the scientific community.

The aim of this work-in-progress paper is to exploit process mining techniques to model and analyze the care process for COVID-19 patients, treated at the Intensive Care Unit (ICU) ward of the Uniklinik Aachen hospital in Germany. In doing so, we use a real-life dataset, extracted from the ICU information system. More in detail, we discover the patient-flows for COVID-19 patients, we extract useful insights into resource consumption, we compare the process models based on data from the two COVID waves, and we analyze their performance. The analysis was carried out with the collaboration of the ICU medical staff.

The remainder of the paper is structured as follows. Section 2 describes the COVID-19 event log subject of our analysis. Section 3 reports insights from preliminary process mining analysis results. Lastly, Sect. 4 concludes the paper and describes our roadmap for future work.

2 Dataset Description

The dataset subject of our study records information about COVID-19 patients monitored in the context of the COVID-19 Aachen Study (COVAS). The log contains event information regarding COVID-19 patients admitted to the Uniklinik Aachen hospital between February 2020 and December 2020. The dataset includes 216 cases, of which 196 are complete cases (for which the patient has been discharged either dead or alive) and 20 ongoing cases (partial process traces) under treatment in the COVID unit at the time of exporting the data. The dataset records 1645 events in total, resulting in an average of 7.6 events recorded per each admission. The cases recorded in the log belong to 65 different variants, with distinct event flows. The events are labeled with the executed activity; the log includes 14 distinct activities. Figure 1 shows a dotted chart of the event log.

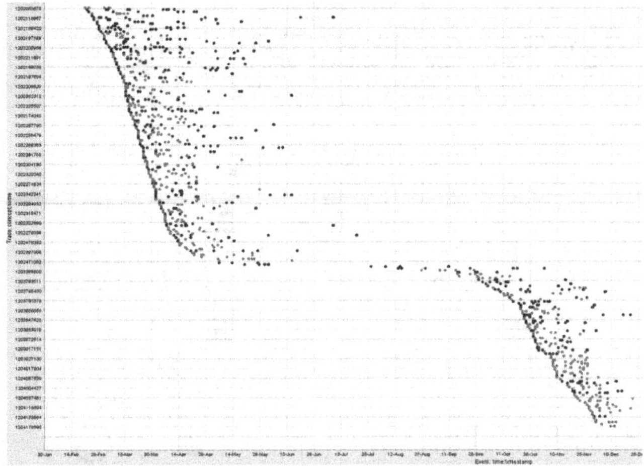

Fig. 1. Dotted chart of the COVAS event log. Every dot corresponds to an event recorded in the log; the cases with Acute Respiratory Distress Syndrome (ARDS) are colored in pink, while cases with no ARDS are colored in green. The two "waves" of the virus are clearly distinguishable. (Color figure online)

3 Analysis

In this section, we illustrate the preliminary results obtained through a detailed process mining-based analysis of the COVAS dataset. More specifically, we elaborate on results based on control-flow and performance perspectives.

Firstly, we present a process model extracted from the event data of the COVAS event log. Among several process discovery algorithms in literature [2], we applied the Interactive Process Discovery (IPD) technique [4] to extract the patient-flows for COVAS patients, obtaining a model in the form of a Petri net (Fig. 2). IPD allows to incorporate domain knowledge into the discovery of process models, leading to improved and more trustworthy process models. This approach is particularly useful in healthcare contexts, where physicians have a tacit domain knowledge, which is difficult to elicit but highly valuable for the comprehensibility of the process models.

The discovered process map allows to obtain operational knowledge about the structure of the process and the main patient-flows. Specifically, the analysis reveals that COVID-19 patients are characterized by a quite homogeneous high-level behavior, but several variants exist due to the possibility of a ICU admission or to the different outcomes of the process. More in detail, after the hospitalization and the onset of first symptoms, if present, each patient may be subject to both oxygen therapy and eventually ICU pathway, with subsequent ventilation and ECMO activities, until the end of the symptoms. Once conditions improve, patients may be discharged or transferred to another ward.

We evaluated the quality of the obtained process model through conformance checking [2]. Specifically, we measured the token-based replay fitness between the Petri net and the event log, obtaining a value of 98%. This is a strong indication

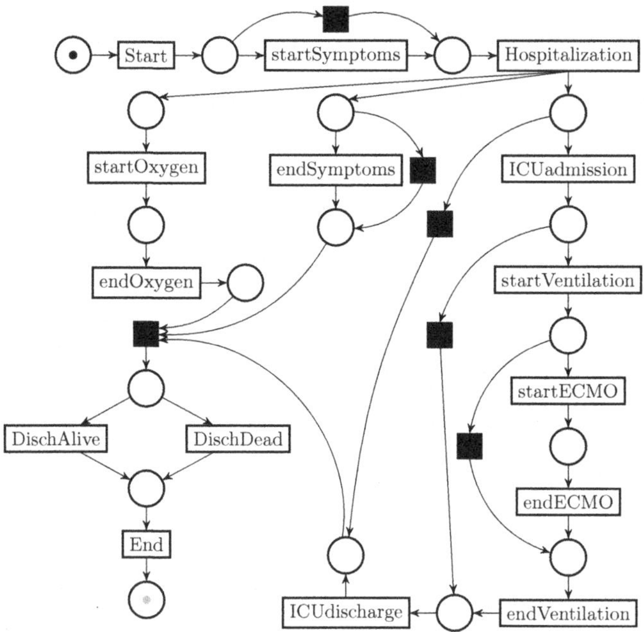

Fig. 2. A normative Petri net that models the process related to the COVAS data.

of both a high level of compliance in the process (the flow of events does not deviate from the intended behavior) and a high reliability of the methodologies employed in data recording and extraction (very few deviations in the event log also imply very few missing events and a low amount of noise in the dataset).

From the information stored in the event log, it is also possible to gain insights regarding the time performance of each activity and the resource consumption. For example, Fig. 3 shows the rate of utilization of ventilation machines.

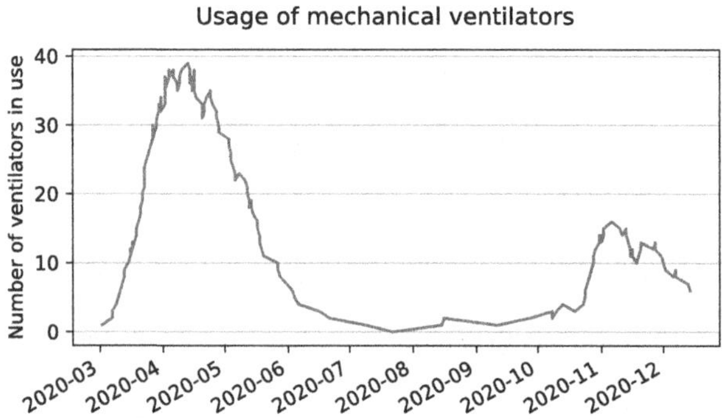

Fig. 3. Plot showing the usage of assisted ventilation machines for COVID-19 patients in the ICU ward of the Uniklinik Aachen. Maximum occupancy was reached on the 13th of April 2020, with 39 patients simultaneously ventilated.

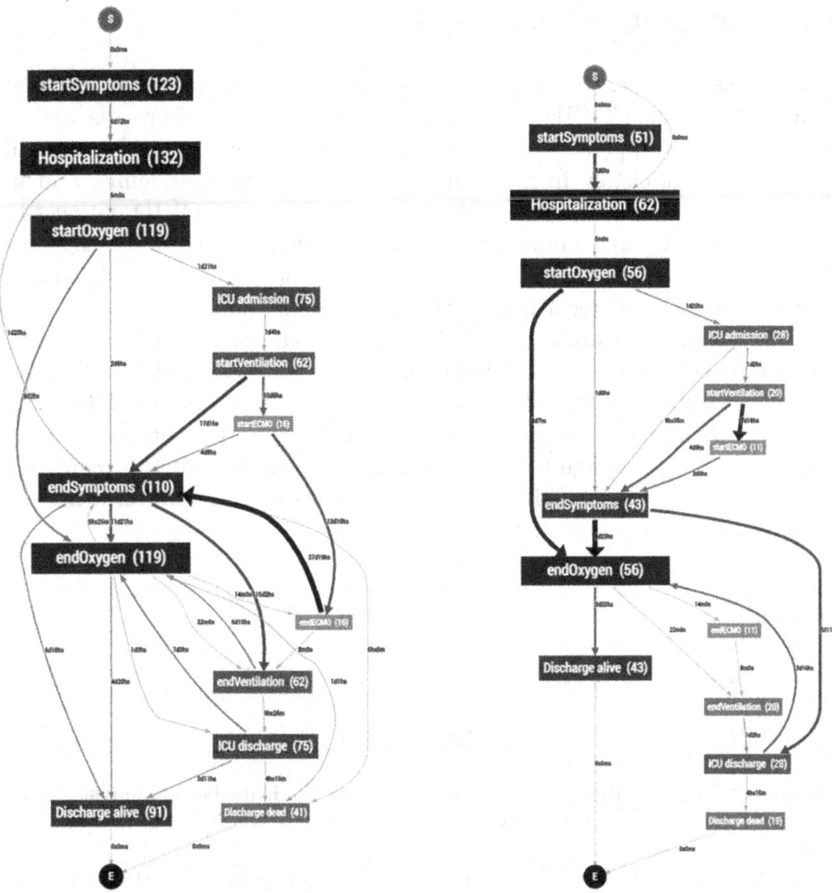

Fig. 4. Filtered directly-follows graph related to the first wave of the COVID pandemic.

Fig. 5. Filtered directly-follows graph related to the second wave of the COVID pandemic.

This information may help hospital managers to manage and allocate resources, especially the critical or shared ones, more efficiently.

Finally, with the aid of the process mining tool Everflow [1], we investigated different patient-flows, with respect to the first wave (until the end of June 2020) and second wave (from July 2020 onward) of the COVID-19 pandemic, and evaluated their performance perspective, which is shown in Figs. 4 and 5 respectively. The first wave involves 133 cases with an average case duration of 33 days and 6 hour(s); the second wave includes 63 patients, with an average case duration of 23 days and 1 hour(s). The difference in average case duration is significant, and could have been due to the medics being more skilled and prepared in treating COVID cases, as well as a lower amount of simultaneous admission on average in the second wave.

4 Conclusion and Future Work

In this preliminary paper, we show some techniques to inspect hospitalization event data related to the COVID-19 pandemic. The application of process mining to COVID event data appears to lead to insights related to the development of the disease, to the efficiency in managing the effects of the pandemic, and in the optimal usage of medical equipment in the treatment of COVID patients in critical conditions. We show a normative model obtained with the aid of IPD for the operations at the COVID unit of the Uniklinik Aachen hospital, showing a high reliability of the data recording methods in the ICU facilities.

Among the ongoing research on COVID event data, a prominent future development certainly consists in performing comparative analyses between datasets and event logs geographically and temporally diverse. By inspecting differences only detectable with process science techniques (e.g., deviations on the control-flow perspective), novel insights can be obtained on aspects of the pandemic such as spread, effectiveness of different crisis responses, and long-term impact on the population.

References

1. Everflow Process Mining. https://everflow.ai/process-mining/. Accessed 17 May 2021
2. Van der Aalst, W.M.P.: Process Mining: Data Science in Action. Springer, Cham (2016). https://doi.org/10.1007/978-3-662-49851-4
3. Anastassopoulou, C., Russo, L., Tsakris, A., Siettos, C.: Data-based analysis, modelling and forecasting of the COVID-19 outbreak. PLoS ONE **15**(3), e0230405 (2020)
4. Dixit, P.M., Verbeek, H.M.W., Buijs, J.C.A.M., Van der Aalst, W.M.P.: Interactive data-driven process model construction. In: Trujillo, J.C., et al. (eds.) ER 2018. LNCS, vol. 11157, pp. 251–265. Springer, Cham (2018). https://doi.org/10.1007/978-3-030-00847-5_19
5. Galetsi, P., Katsaliaki, K.: A review of the literature on big data analytics in healthcare. J. Oper. Res. Soc. **71**(10), 1511–1529 (2020)
6. Guidotti, E., Ardia, D.: COVID-19 data hub. J. Open Source Softw. **5**(51), 2376 (2020)
7. Koufi, V., Malamateniou, F., Vassilacopoulos, G.: A big data-driven model for the optimization of healthcare processes. In: MIE, pp. 697–701 (2015)
8. Lavezzo, E., et al.: Suppression of a SARS-CoV-2 outbreak in the Italian municipality of Vo'. Nature **584**(7821), 425–429 (2020)
9. Mans, R.S., Van der Aalst, W.M.P., Vanwersch, R.J.B.: Process Mining in Healthcare: Evaluating and Exploiting Operational Healthcare Processes. Springer, Cham (2015). https://doi.org/10.1007/978-3-319-16071-9
10. Sarkar, K., Khajanchi, S., Nieto, J.J.: Modeling and forecasting the COVID-19 pandemic in India. Chaos Solitons Fractals **139**, 110049 (2020)
11. Wynants, L., et al.: Prediction models for diagnosis and prognosis of COVID-19: systematic review and critical appraisal. Br. Med. J. **369**, m1328 (2020)

Problem Domain Example of Knowledge-Based Enterprise Model Usage for Different UML Behavioral Models Generation

Ilona Veitaite[1(✉)] and Audrius Lopata[2]

[1] Kaunas Faculty, Institute Social Sciences and Applied Informatics, Vilnius University, Muitines Street 8, 44280 Kaunas, Lithuania
ilona.veitaite@knf.vu.lt

[2] Faculty of Informatics, Kaunas University of Technology, Student Street 50, 51368 Kaunas, Lithuania
audrius.lopata@ktu.lt

Abstract. The main purpose of this paper is to represent how knowledge-based Enterprise Model (EM) as problem domain data storage may be used in Information Systems (IS) engineering process. Enterprise Meta-Model (EMM) presented more than two decades ago justifies EM structure. EM stores problem domain data gathered by analyst and this EM structure can be used for project models creation in IS design phase. Unified Modeling Language (UML) models are one of the possible models, which can be generated from EM. These models can be generated through transformation algorithms. To present possibility of UML models generation from EM particular problem domain example is defined in this paper. Presented example demonstrates that data stored in EM is enough for different UML models generation and this paper presents that different UML behavioral models can be generated from EM and can illustrate same problem domain from different perspectives.

Keywords: IS engineering · Transformation algorithm · Enterprise modeling · Knowledge-based · UML

1 Introduction

Nowadays IS engineering process is still challenging for all IT professionals: analysts, designers, developers. They have same goal but all are responsible for different phase of IS engineering process [1, 5, 7]. Beginning of this process is most important, because problem domain analysis is performed and final result will depend from quality of gathered data. Duration of IS engineering process and number of errors during it may also impact the final result. So it is very important how gathered data of problem domain will be used, where it will be stored and what project models standards will be chosen [1, 4].

There are a wide number of models to store problem domain data, also there are a lot of modeling standards and notations. UML models are widely applied in IS engineering process and are used as by IT professionals in IS design phase as by not this field professionals for better understanding of final IS result [4, 5, 7]. UML latest version 2.5 [6, 8]. UML models may be designed one by one according to collected

W. Abramowicz et al. (Eds.): BIS 2021 Workshops, LNBIP 444, pp. 45–55, 2022.
https://doi.org/10.1007/978-3-031-04216-4_5

data by analyst, they can be created by using certain data storages assigned for modeling and design phase and also they can be generated from particular EM [5, 7]. EM used in this analysis is created more than two decades ago and the main goal of this article is to present its efficiency for UML models generation process. For this purpose, there are created transformation algorithms defined in previous researches. By using these algorithms UML models generation from EM is possible [2, 3, 10, 12]. Car Rental Company example and generated UML models of this example, which defines problem domain information from different perspectives presented in this research depict how knowledge stored in EM can be used for UML models generation.

2 Knowledge-Based EM Definition

EMM is formally defined EM structure, which consists of a formalized EM in line with the general principles of control theory. EM is the main source of the necessary knowledge of the particular business domain for IS engineering and IS re-engineering processes [2, 3, 10, 12].

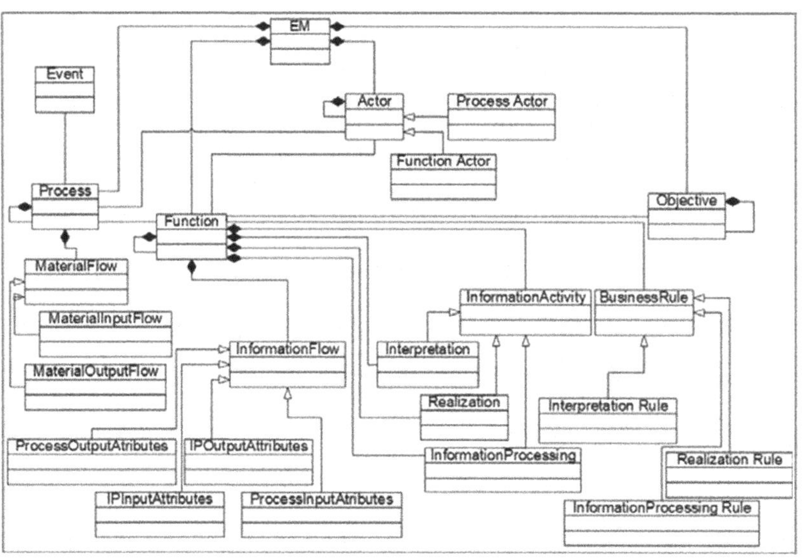

Fig. 1. Class diagram of Enterprise Meta-Model [2, 3, 9]

EM consists of twenty-three classes (Fig. 1). Essential classes are Process, Function and Actor. Class Process, Function, Actor and Objective can have an internal hierarchical structure. These relationships is presented as aggregation relationship. Class Process is linked with the class MaterialFlow as aggregation relationship. Class MaterialFlow is linked with the classes MaterialInputFlow and MaterialOutputFlow as generalization relationship. Class Process is linked with Classes Function, Actor and

Event as association relationship. Class Function is linked with classes InformationFlow, InformationActivity, Interpretation, InformationProcessing and Realization as aggregation relationship. These relationships define the internal composition of the Class Function. Class InformationFlow is linked with ProcessOutputAtributes, ProcessInputAtributes, IPInputAttributes and IPOutputAttributs as generalization relationship. Class InformationActivity is linked with Interpretation, InformationProcessing and Realization as generalization relationship. Class Function linked with classes Actor, Objective and BusinessRule as association relationship. Class BusinessRule is linked with Interpretation Rule, Realization Rule, InformationProcessing Rule as generalization relationship. Class Actor is linked with Function Actor and Process Actor as generalization relationship [2, 3, 10, 12].

During IS engineering analysis phase analyst gather all necessary information and stores it in EM for further IS engineering design phase, when system project models are created. There are a lot of standards and different notations of project model and UML models are most commonly used by IT professionals. With the EM usage, IS design phase project models can be generated.

3 UML Models Transformation Algorithm

Each of structural or behavioral UML models can be generated through transformation algorithm and each of models has separate transformation algorithm [9–11]. These transformation algorithms are presented in previous researches. Main focus of researches is dedicated for generation behavioral or dynamic UML models, because they are more complex and variable [12–14]. To have better understanding of transformation algorithm itself, top level transformation algorithm of UML models generation from EM process is described step by step [9–14]:

- Step 1: Particular UML model for generation from EM process is identified and selected.
- Step 2: If the particular UML model for generation from EM process is selected then algorithm process is continued, else the particular UML model for generation from EM process must be selected.
- Step 3: First element from EM is selected for UML model, identified previously, generation process.
- Step 4: If the selected EM element is initial UML model element, then initial element is generated, else the other EM element must be selected (the selected element must be initial element).
- Step 5: The element related to the initial element is selected from EM.
- Step 6: The element related to the initial element is generated as UML model element.
- Step 7: The element related to the previous element is selected from EM.
- Step 8: The element related to the previous element is generated as UML model element.

- Step 9: If there are more related elements, then they are selected from EM and generated as UML model elements one by one, else the link element is selected from EM.
- Step 10: The link element is generated as UML model element.
- Step 11: If there are more links, then they are selected from EM and generated as UML model elements one by one, else the Business Rule element is selected from EM.
- Step 12: The Business Rule element is generated as UML model element.
- Step 13: If there are more Business Rules, then they are selected from EM and generated as UML model elements one by one, else the generated UML model is updated with all elements, links and constraints.
- Step 14: Generation process is finished.

Usually main problem domain knowledge necessary for IS engineering process is stored in particular EM elements (commonly mandatory for most used UML models) [12–14]:

- Actor: in actor element can be stored information related with process or function executor. Actor element is responsible of information related with the process or function participant, it can be person, group of persons, subject such as an IS, subsystem, module and etc.
- Process or Function: in process or function elements can be stored all information related with any user, entity, object, subject and its behavior. Process or function element is responsible of information related with any operation, activity, status change, movement which is implemented by any actor, entity, participant and etc.
- InformationFlow: in Information Flow element can be stored diverse information flow types, such as Information input and output attributes or/and process input and output attributes. Information Flow element is responsible of information related with each element input and output attributes, details which make impact on other elements, their state or status.
- MaterialFlow: in Material Flow element can be stored two types of material information Material input and Material output. Material Flow element is responsible of information related with any material flows of the described process or function.
- BusinessRule: in Business Rule element can be stored different rules such as interpretation, realization or/and information processing. Business rule element is responsible of information about how different elements in IS design phase are related; what restrictions and restraints are applied to these elements.

4 Generated UML Models of Car Rental Company

To present sufficiency of knowledge-based EM example of particular problem domain – Car Rental Company – is analyzed. Car Rental process may be defined as car rental process management system that manages rental process with participant of two users: client and manager. Client may enquire for a rental car, if after client verification, documents check and car availability check all requirements are satisfied, client may rent a car; after the usage client returns car and pays for services, and manager controls this process: calculates fees, receives payment and maintains returned car.

4.1 UML Use Case Model

UML Use Case Model can be generated from EM, because all necessary elements for UML Use Case Model are stored in it. Generation process follows the steps of transformation algorithm. Connections between EM and UML Use Case Model elements are described in Table 1.

Table 1. EM and UML use case model elements of car rental company [6, 8, 9, 11, 12]

EM element -> UML element	Car Rental Company example
Actor -> Actor	There are two participants: Client, who wants to rent a car, enquires for it and pays for the service and Manager, who controls car renting process, verifies the client, calculates payment and etc
Process, Function -> Use Case	There are eleven Use Cases, part of them are performed by Client, part – by Manager. There are also common for both participant Use Cases, only functionality is different
BusinessRule -> Association, Include, Extend	Use Cases are related with particular actors. Also there is defined which Use Cases are necessary to perform and which aren't

Table 1 defines EM and UML Use Case Model elements and defines their meaning in Car Rental Company example. Generated UML Use Case Model is presented in Fig. 2.

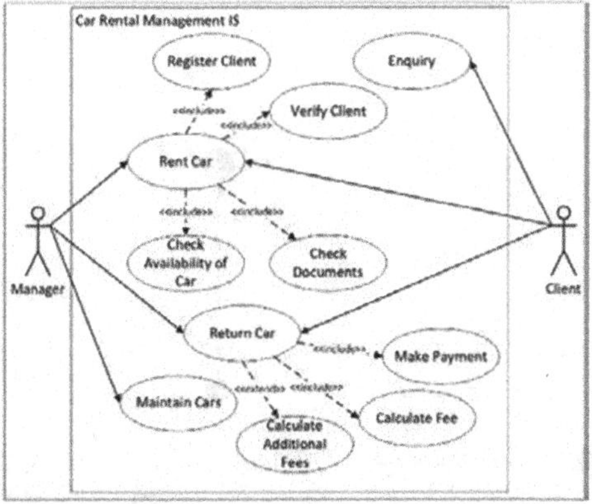

Fig. 2. Generated UML use case model of car rental company

Figure 2 presents UML Use Case Model generated from EM through transformation algorithm. It defines how two participants – Actors perform their activities – Use Cases linked by Association relationship, which of the must be performed – Include relationship and which not – Extend relationship.

4.2 UML Activity Models

UML Activity Models (only for two processes (rent and return car) in this research. Note: number of UML Activity Models of this example can be higher) can be generated from EM, because all necessary elements for UML Activity Models are stored in it. Generation process follows the steps of transformation algorithm. Connections between EM and UML Activity Models elements are described in Table 2.

Table 2. EM and UML Activity model elements of Car Rental Company [6, 8, 9, 12]

EM element -> UML element	Car Rental Company example
Actor -> Actor, Swimlane	In Fig. 3 UML Activity Model of renting car, there are two participants: Manager and Client. In Fig. 4 UML Activity Model there is only one participant: Manager
Process, Function -> Activity	Figure 3 presents activities performed by two participants: Client and Manager and Fig. 4 only from the perspective of the Manager
MaterialFlow, InformationFlow -> Object Flows	In both UML Activity Models all activities are related through Object Flows
BusinessRule -> Control Nodes: Initial, Join, Decision, Final	Both UML Activity Models start from initial nodes and end with final nodes. Both models have Decision nodes and Join nodes

Table 2 defines EM and UML Activity Models elements of two diagrams and defines their meaning in Car Rental Company example. Generated UML Activity Models are presented in Fig. 3 and Fig. 4.

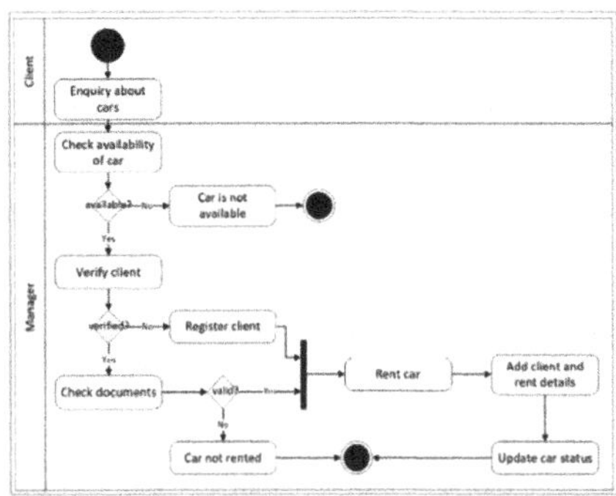

Fig. 3. Generated UML activity model of rent car process

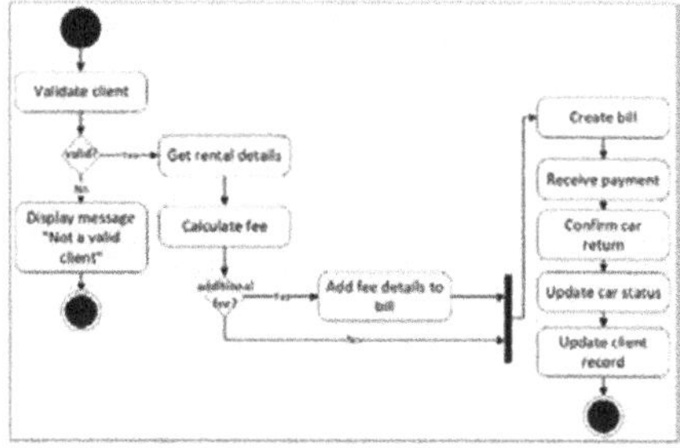

Fig. 4. Generated UML activity model of return car process

Figure 3 presents UML Activity Model of Rent Car Process generated from EM through transformation algorithm. It defines how two participants – Swimlanes: Manager and Client perform the activities related with Car renting process; there is presented beginning of the process, flow of the activities, decisions made during the process and possible endings of the process.

Figure 4 presents UML Activity Model of Return Car Process generated from EM through transformation algorithm. It defines how one participant – Manager performs the activities related with Car returning process; there is presented beginning of the process, flow of the activities, decisions made during the process and possible endings of the process.

4.3 UML State Models

UML State Models (only for two objects (car and manager) in this research. Note: number of UML State Models of this example can be higher) can be generated from EM, because all necessary elements for UML State Models are stored in it. Generation process follows the steps of transformation algorithm. Connections between EM and UML State Models elements are described in Table 3.

Table 3. EM and UML state model elements of car rental company [6, 8, 9, 12]

EM element -> UML element	Car rental company example
Process, Function -> Behavioral state machine	In Fig. 5 UML State Model there are defined states of the Car and in Fig. 6 UML State Model there are defined states of the Manager
InformationFlow -> Composite state	In both UML State Model information about each object/machine is provided as their state

Table 3 defines EM and UML State Models elements of two diagrams and defines their meaning in Car Rental Company example. Generated UML State Models are presented in Fig. 5 and Fig. 6.

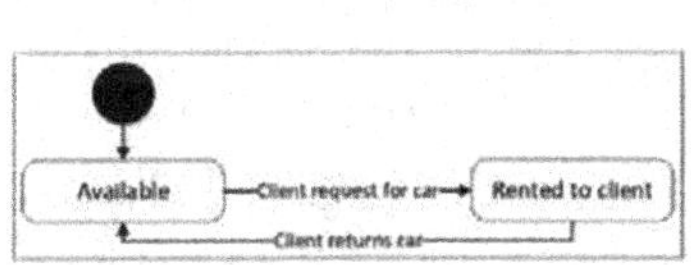

Fig. 5. Generated UML state model of car

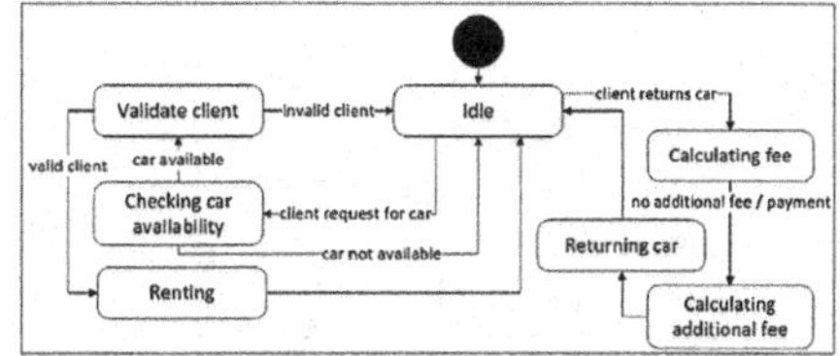

Fig. 6. Generated UML state model of manager

Figure 5 presents UML State Model of Rent Car Process generated from EM through transformation algorithm. It defines different states of the Car.

Figure 6 presents UML State Model of Return Car Process generated from EM through trans-formation algorithm. It defines different states of the Manager.

4.4 UML Communication Models

UML Communication Models (only for two processes (rent and return car) in this research. Note: number of UML Communication models of this example can be higher) can be generated from EM, because all necessary elements for UML Communication Models are stored in it. Generation process follows the steps of transformation algorithm. Connections between EM and UML Communication Models elements are described in Table 4.

Table 4. EM and UML communication model elements of car rental company [6, 8, 9, 12]

EM element -> UML element	Car rental company example
Actor -> Lifeline	In Fig. 7 UML Communication Model there are four Lifelines: ClientRecord, Manager, Transaction and Car and they are directly related with Car renting process. In Fig. 8 UML Communication Model there are five Lifelines, Invoice is new one and they are directly related with Car returning process
Process, Function -> Frame	Figure 7 presents Car renting process and Fig. 8 present Car returning process
InformationFlow -> Message	In both UML Communication Models messages between all Lifelines are presented
BusinessRule -> Sequence Expression	In both UML Communication Models Sequence Expressions define sequence of the messages

Table 4 defines EM and UML Communication Models elements of two diagrams and defines their meaning in Car Rental Company example. Generated UML Communication Models are presented in Fig. 7 and Fig. 8.

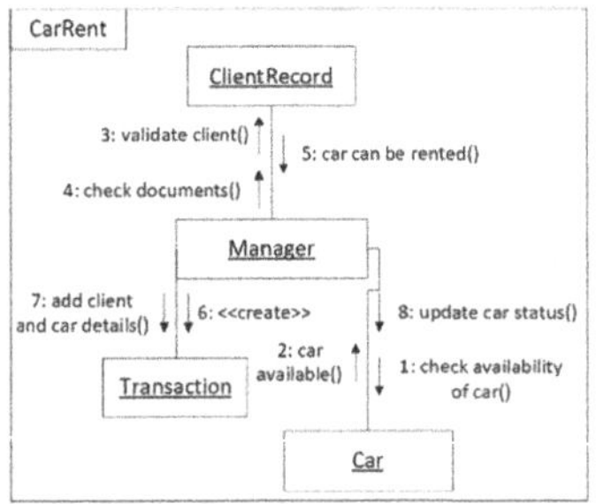

Fig. 7. Generated UML communication model of renting car

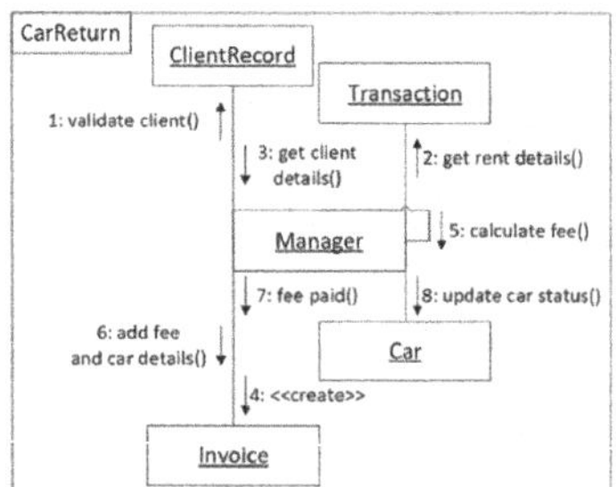

Fig. 8. Generated UML communication model of returning car

Figure 7 presents UML Communication Model of Rent Car Process generated from EM through transformation algorithm. It defines how four Lifelines: Manager, ClientRecord, Transaction and Car interacts during Car renting process.

Figure 8 presents UML Communication Model of Return Car Process generated from EM through transformation algorithm. It defines how five Lifelines: Manager, ClientRecord, Transaction, Car and Invoice interacts during Car returning process.

4.5 UML Sequence Models

UML Sequence Models (only for two processes (rent and return car) in this research. Note: number of UML Sequence Models of this example can be higher) can be generated from EM, because all necessary elements for UML Sequence Models are stored in it. Generation process follows the steps of transformation algorithm. Connections between EM and UML Sequence Models elements are described in Table 5.

Table 5. EM and UML sequence model elements of car rental company [6, 8, 9, 12]

EM element -> UML element	Car rental company example
Actor -> Lifeline	In Fig. 9 UML Sequence Model there are four Lifelines: ClientRecord, Manager, Transaction and Car and they are directly related with Car renting process. In Fig. 10 UML Sequence Model there are five Lifelines, Invoice is new one and they are directly related with Car returning process
Process, Function -> Message	In both UML Sequence Models messages between all Lifelines are presented
BusinessRule -> Execution Specification, Occurrence Specification	In both UML Sequence Models Execution Specifications define durations of the executions and occurrences of the messages

Table 5 defines EM and UML Communication Models elements of two diagrams and defines their meaning in Car Rental Company example. Generated UML Communication Models are presented in Fig. 9 and Fig. 10.

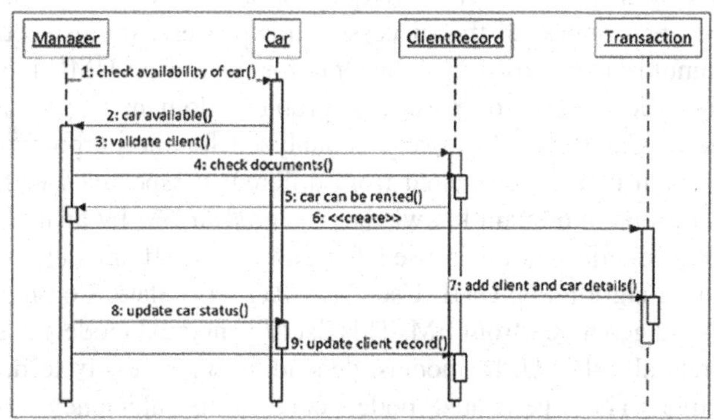

Fig. 9. Generated UML sequence model of renting car

Figure 9 presents UML Sequence Model of Rent Car Process generated from EM through transformation algorithm. It defines how four Lifelines: Manager, ClientRecord, Transaction and Car interacts during Car renting process, presents the sequence of Messages and duration of Executions between the Messages.

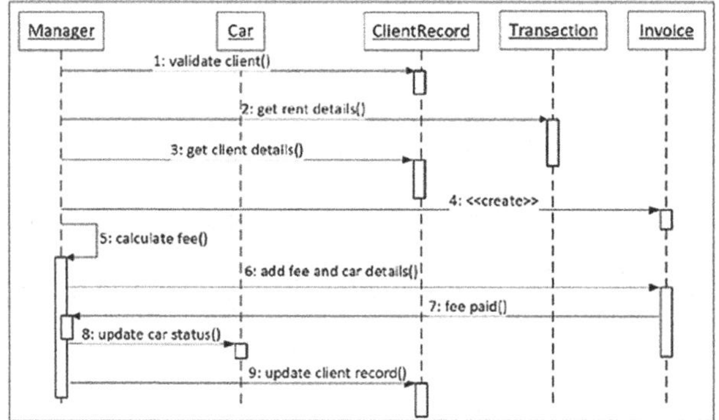

Fig. 10. Generated UML sequence model of returning car

Figure 10 presents UML Sequence Model of Return Car Process generated from EM through transformation algorithm. It defines how five Lifelines: Manager, ClientRecord, Transaction, Car and Invoice interacts during Car returning process, presents the sequence of Messages and duration of Executions between the Messages.

5 Conclusions

The first section of the article defines the Enterprise Model structure by presenting class diagram of EMM and listing all its elements also explaining their necessity in IS engineering process. Gathered problem domain data may be stored in EM and used to further IS development process.

The next section presents top level UML models transformation algorithm from EM depicted step by step. As UML is frequently used in IS design process, UML models may have great impact of the success of this process. In this section there are also defined commonly mandatory EM elements for most used UML models.

In final section description of particular problem domain is presented. In this research example of Car Rental Company is analyzed for the purpose to define that same problem domain may be presented from different perspectives with the help of UML models. All problem domain knowledge is stored in EM by analyst, and all this verified and validated information is used for different UML models generation by using transformation algorithms. UML Use Case, Activity, State, Communication and Sequence models are generated from EM. This list of generated models is not final, it is possible to generate all other UML models, depending on necessity to depict IS from different perspectives. These generated models confirm the sufficiency of EM in UML models generation process. Most important condition is that data stored in EM must be acknowledged by analyst, because this procedure ensures lower number of possible errors and allows to avoid increased IS engineering process duration.

References

1. Dunkel, J., Bruns, R.: Model-driven architecture for mobile applications. In: Abramowicz, W. (ed.) BIS 2007. LNCS, vol. 4439, pp. 464–477. Springer, Heidelberg (2007). https://doi.org/10.1007/978-3-540-72035-5_36
2. Gudas, S.: Architecture of knowledge-based enterprise management systems: a control view. In: Proceedings of the 13th world multiconference on systemics, cybernetics and informatics (WMSCI2009),) July 10–13, Orlando, Florida, USA, vol. III, pp.161–266 (2009). ISBN -10: 1-9934272-61-2 (Volume III). ISBN-13: 978-1-9934272-61-9
3. Gudas S.: Informacijos sistemų inžinerijos teorijos pagrindai/ Fundamentals of Information Systems Engineering Theory. (Lithuanian)Vilnius University (2012). ISBN 978–609–459–075–7
4. Jacobson, I., Rumbaugh, J., Booch, G.: Unified Modeling Language User Guide, p. 0321267974. Addison-Wesley Professional, The Second Edition (2005)
5. Jenney, J.: Modern Methods of Systems Engineering: With an Introduction to Pattern and Model Based Methods (2010). ISBN-13:978-1463777357
6. OMG UML: Unified Modeling Language version 2.5.1. Unified Modelling (2021). https://www.omg.org/spec/UML/About-UML/
7. Sajja, P.S., Akerkar, R.: Knowledge-based systems for development. Adv. Knowl. Based Syst. Model, Appl. Res. **1** (2010)
8. UML Diagrams: UML diagrams characteristic (2021). www.uml-diagrams.org
9. Veitaite, I., Lopata, A.: Transformation algorithms of knowledge based UML dynamic models generation. In: Abramowicz, W. (ed.) BIS 2017. LNBIP, vol. 303, pp. 59–68. Springer, Cham (2017). https://doi.org/10.1007/978-3-319-69023-0_6
10. Veitaite, I., Lopata, A.: Problem domain knowledge driven generation of UML models. In: Damaševičius, R., Vasiljevienė, G. (eds.) ICIST 2018. CCIS, vol. 920, pp. 178–186. Springer, Cham (2018). https://doi.org/10.1007/978-3-319-99972-2_14
11. Abramowicz, W., Corchuelo, R. (eds.): BIS 2019. LNBIP, vol. 373. Springer, Cham (2019). https://doi.org/10.1007/978-3-030-36691-9
12. Veitaite, I., Lopata, A.: Knowledge-based transformation algorithms of UML dynamic models generation from enterprise model. In: Dzemyda, G., Bernatavičienė, J., Kacprzyk, J. (eds.) Data Science: New Issues, Challenges and Applications. SCI, vol. 869, pp. 43–59. Springer, Cham (2020). https://doi.org/10.1007/978-3-030-39250-5_3
13. Veitaitė, I., Lopata, A.: Knowledge-based UML activity model transformation algorithm. Information society and university studies 2020. In: Proceedings of the Information Society and University Studies, 2020, Kaunas, Lithuania, April 23, 2020. Audrius, L., Vilma, S., Tomas, K., Ilona, V., Marcin, W.A: CEUR Workshop Proceedings. ISSN 1613-0073. pp. 114–120 (2020). (CEUR Workshop Proceedings, ISSN 1613–0073; vol. 2698)
14. Veitaite, I., Lopata, A.: Knowledge-based generation of the UML dynamic models from the enterprise model illustrated by the ticket buying process example. In: Lopata, A., Butkienė, R., Gudonienė, D., Sukackė, V. (eds.) ICIST 2020. CCIS, vol. 1283, pp. 26–38. Springer, Cham (2020). https://doi.org/10.1007/978-3-030-59506-7_3

BIS Education Workshop

BisEd 2021 Workshop Chairs' Message

The BisEd (BIS Education: Trends and Challenges) Workshop held at the 24th International Conference on Business Information Systems (BIS 2021) was the inaugural meeting of researchers and practitioners interested in the education of BIS. The workshop was organized with the intention to offer an opportunity to discuss ideas and share experiences related to the special challenges facing business information systems education under the recent surge in need for online learning and teaching.

Indeed, the recent global challenges reinforced by the COVID-19 pandemic have brought online education to the forefront of academic attention – both as a technological opportunity to maintain the continuity of teaching (at all levels of education) and as a challenge to innovate and apply new methodological approaches. To meet this challenge many institutions have put a lot of effort into both technology and teaching methodology improvements. Researchers and educators were thus invited to submit papers that could be of interest to the whole BIS community. There were seven proposals, of which four were accepted. These four covered case studies, theoretical discussion of a blended-learning compatible BIS curriculum based on shared values, and a review of BIS education programs worldwide.

The workshop was held on the afternoon of June 14, 2021, and had close to thirty participants from ten countries over the course of the program. In the focus was a recently launched Erasmus+ project titled 'BIPER' which had re-framed BIS education according to the TOGAF (The Open Enterprise Architecture Framework) approach using problem-based learning principles. This chapter contains the reviewed and updated paper versions of the four presentations that were delivered as part of the workshop. Based on the success of this first instalment of the workshop, there is a definite interest to continue next year at Bis 2022.

Csaba Csáki
Gaye Kiley
Zoltán Szabó

Organization

Chairs

Csaba Csáki	Corvinus University of Budapest, Hungary
Gaye Kiley	University College Cork, Ireland
Zoltán Szabó	Corvinus University of Budapest, Hungary

Program Committee

Md Shakil Ahmed	University of the West of Scotland, UK
Keshav Dahal	University of the West of Scotland, UK
András Gábor	Corvinus University of Budapest, Hungary
Blaz Gasperlin	University of Maribor, Slovenia
Marco Gilardi	University of the West of Scotland, UK
Gábor Kismihók	TIB Leibniz Information Centre for Science and Technology, Germany
Mirjana Kljajic Borstnar	University of Maribor, Slovenia
Marjeta Marolt	University of Maribor, Slovenia
Andreja Pucihar	University of Maribor, Slovenia

Reconsidering the Challenges of BIS Education in Light of the COVID Pandemic

Csaba Csáki(✉), Ildikó Borbásné Szabó, Zoltán Szabó, Olga Csillik, and András Gábor

Corvinus University of Budapest, Budapest, Hungary
csaki.csaba@uni-corvinus.hu

Abstract. So far, the biggest challenge for a comprehensive Business Information Systems (BIS) education curriculum was the fast-changing nature of its target market and the resulting demand for a combination of up-to-date technical knowledge, organization-centred mindset, and adaptive skills. However, advances in pedagogical methods, changes in the skills of high-school graduates, and widening online options in the wake of the COVID-19 pandemic brought on a new set of expectations. This situation may be considered an opportunity to address the threat of potentially increasing mismatch and misalignment between competences required by the IS industry labour market and current training contents offered and methods used by higher education institutions. This paper provides a systematic and comprehensive overview of the challenges BIS programs have to face and address. It considers everyday experiences of BIS educators and current best practices as starting point. Then provides an overview of employer and alumni opinion, as well as reviews up-to-date teaching methods related to teaching soft computer skills. It also considers the requirements and opportunities related to an increasingly online-centred situation. Based on these challenges the paper lays down the foundation for a potential curriculum design approach intended to address all of the above issues in an integrated framework.

Keywords: Business Information Systems · Higher education teaching methods · Online education · TOGAF · Curriculum design

1 Introduction

The academic field of Business Information Systems (BIS) is a complex area bridging business and organisational topics with questions of applied information technology. Teaching such a multidisciplinary domain which assumes not only knowledge of theoretical concepts and technical skills to use tools but also a problem centred mindset and related problem-solving abilities is a challenge in itself. However, with the heightened need for high-quality online education (offering both distance or blended learning options in the wake of the COVID-19 pandemic) educators of this area face increased difficulties to find appropriate methods and create new content and teaching material. Sharing ideas and experiences regarding what worked and what was less successful could enhance our knowledge of BIS distance education.

W. Abramowicz et al. (Eds.): BIS 2021 Workshops, LNBIP 444, pp. 61–72, 2022.
https://doi.org/10.1007/978-3-031-04216-4_6

Creating third level programs to educate professionals who are able to meet current and emerging expectations drew different answers from different stakeholders [1–3]. In fact, even the name of programs offered show some divergence: depending on native language, history, and culture, BIS-like education is offered under different names including: BIS, Management Information Systems, Business Informatics, Business and IT, Business IT, Business Computing, and so on. Beyond the names there are, of course, curricula offered with different goals and focus ranging from those closer to technology and computer science, through business analytics (or even data science), to more business focused options [3].

Over the last decade two trends may be observed in the demand for BIS graduates. On the one hand some employers, especially SMEs expect graduates who can take on responsibilities almost right away (i.e. having a wide range of specific skills including programming or use of certain tools), while other organizations (mostly large and multinational ones) expect newhires to be flexible, with convertible skills (as they will provide them with customized corporate training) [4, 5]. These demands add to the challenges of an already complex educational setting.

Therefore, to understand the full picture of BIS education of our days, this theoretical discussion offers a systematic overview of various challenges BIS programs need to address and concludes in a proposed integrated approach based on the TOGAF framework [6]. Therefore, this paper first looks at everyday experiences of BIS educators including current best practices. The third section then provides a discussion of challenges along the following dimensions: employer and alumni opinion about required BIS job skills, changes in abilities of incoming high-school students, latest trends in teaching methods, modern assessments techniques, tools and trends of online education, and the special situation highlighted by the recent COVID-19 pandemic. Based on these challenges the fourth section lays down the foundation for a potential curriculum design approach intended to address those issues. The paper closes with summary and further directions.

2 BIS and Its Education

2.1 Typical Characteristics of the BIS Field

The labour market where Information Systems professionals are employed may be characterized by the fast-changing nature of the jobs and the resulting demand for a combination of up-to-date technical knowledge and adaptive skills rooted in the project-oriented and teamwork-based reality of developing, implementing, and managing IT/IS solutions in an organizational context [7]. Beyond the obvious basics of the trade, interpersonal skills, team building and the ability to combine individual efforts with group work are an essential part of BIS professionals [8]. Employers value problem-solving skills and independence with the ability to learn quickly. On top of basic knowledge of IT and business concepts, a broad professional outlook, the right mindset and a systems approach is expected. A good level of English is a must these days.

BIS graduates are typically hired to bridge the gap between IT and business. This gap is especially relevant for large and medium sized companies, or for IT service providers. Typical business IT jobs include: business analysis, system development, digitization, presale activities, logical and physical design of IT services, database management, data analysis and data mining, IT demand management, IT project management, IT services financial controlling, application and service support, IT risk analysis, automation of business processes, and software testing. Regarding specialized IT-IS areas, the list ranges from artificial intelligence and its application, automation, autonomous systems, and process optimisation, to GDPR, cybersecurity, IT security, but industrial modernisation (industry 4.0), databases, BI and data visualisation are also strongly emphasised.

In IT related areas knowledge of the basics changes quickly. Knowledge gained during institutional training can quickly become obsolete, and technologies learnt can get outdated. It is an important goal for students to be able to learn independently, and be capable of self-driven, self-regulated learning.

2.2 A Brief Overview of Some BIS Programs Around the World

A comparative analysis of practices in BIS programs was conducted for 26° programs around the world (see [9] in this volume). They are offered by institutions highly ranked on the Times Higher Education World University Rankings list. These institutions are concentrated in Europe and mainly in the UK, but three are from Asia, two from Australia and one from the United States. 4 programs focus on Information Management, 8 programs are about Information Systems, 5 programs focus on IT, 5 are from the analytics or data science side and 4 covers business administration domain. 35% of these programs aims at focusing on how technology drives business.

Typical career paths at these institutes are business analyst, consultant, project manager and developer in banking, finance, IT sector etc. Companies are involved in not just hiring students as interns, but also in carrying out these courses. Interesting result was that project work was a quite popular methodology applied by these institutions Practice-oriented knowledge transfer is realized in this way. Some programs provide minor program besides major but other ones make their students specialized by elective modules. Their subjects go around e-business, data science, information systems auditing, artificial intelligence, digital transformation domain and so on. The typical length of studies is 6 or 8 semesters, but the total number of credits vary on a broad scale from 120 to 480. Practically, Anglo-Saxon institutes prefer providing programs with high total credits. Their courses usually have 15–20 credits and they put relatively small attention on foundation subjects (approx. 10% of total credits). It is quite common for all peer institutions that Information Technology and Manager Information Systems subject appear in the same degree in the programs.

3 Specific Challenges to Teach BIS

3.1 Challenges of Teaching BIS Arising from the Nature of the Field

BIS is a truly interdisciplinary subject and its education covers several fields – albeit in differing depths – and successfully applying them requires the understanding of their interworking. There are mathematical foundations (analysis, algebra, statistics) but it is also rooted in the basics of economics (e.g. macro and micro economic theories). It requires computer science (hardware, software, and network architecture and programming) foundations too, as well as knowledge of operating systems and various protocols. It also builds on organizational studies (including organizational functions, management, and production processes) [10]. Most importantly it has specific areas involving the application of all of the above, such as functional and enterprise systems, as well as systems development, deployment, and impact analysis. The main challenge for a comprehensive Business Information Systems education curriculum is the pace of-change in its target market and the resulting change in knowledge and skills requirements.

Beyond its multidisciplinary nature, pedagogically it is characterized by a typically high ratio of seminars, the need for project focus, and the requirement of working in groups [8]. BIS education in a classroom context - considering the Bachelor level - may be described by what the literature calls 'active learning' focusing on student interaction. For online options video and audio solutions are usually augmented with less synchronous means such as text messages or sharing files, this still does not make up for lost personal proximity. Using document sharing options and working on the same file together raises new challenges just as much as offering new opportunities. To be successful in this setting of increased complexity and expectations lecturers could use any help they could get - let it be experiences, best practices, successful methodologies, or even ready-made materials [11].

3.2 Global Generational Challenges to 3rd Level Education

Incoming students who arrive to a BIS BSc program show a strong character of digital readiness, even more than their peers in general. The accelerated evolution of our digital world fundamentally determines the life of our youth (even in less developed countries). They may be described being "phygital" [12], as in their world everything physical now has a digital equivalent. From physical reality they have moved to digital communication [13]. For them the real and the virtual is strongly coupled and forms a unity [14]. The continuous technological revolution leading to newer and newer solutions appearing with increasing frequency requires flexibility, creativity, and a fast adaptation to desirable behavioural patterns.

The available information is almost infinite, sources of information are countless and change fast. At the same time, the content of knowledge and the forms of teaching-and-learning (T&L) have also been changing. It seems that traditional forms of knowledge transfer have become less efficient [15]. Students acquire a growing portion of their knowledge from sources outside educational institutions. In the fast expanding informational space stimulus threshold (of attention) is raised, the youth longs for newer and newer impulses and information. The vast opportunity for quick information also makes them less patient. However, when it comes to making decisions they appear more uncertain and tentative despite their apparent confidence in getting information. They need outside assistance, guidance, and help with avoiding the temptation of constant interruptions to focused studying, because there is a steady influx of activity requests such as visiting webpages not related to the learning material, checking emails, visiting social media profiles, joining a chat, or simply playing games [16]. Due to the too high pressure coming from education institutions and teachers coupled with inadequate time-management skills this generation of teenagers has a higher tendency of mental problems than their predecessors had (for example, in the United States youngsters who have experienced some form of depression reached almost 60% in 2017) [16].

Today's high-school students expect relevant, quickly applicable knowledge from education institutions mostly covering technical literacy (such as math, coding, basic technical sciences), as well as data management and interpersonal skills (to be able to connect to others). They value system level thinking, creativity, and knowledge about human behaviour – preferably acquired through experience-based learning [14]. They expect different teaching methods and educational arrangements during their (mandatory) formal training. Their preference is increasingly shifting towards forms of active learning, that are based on gaining experience through practical exercises and require intensive communication [17]. At the same time, they constantly seek feedback, long for reassurance and expect rewards [18].

3.3 Availability of New Advanced 3rd Level Teaching Methods

Over the last two decades or so major changes may be observed in the pedagogical methods available to the university and college teaching community. Some of these are rooted in general new pedagogical approaches while others consider improvements to online options. And, of course, there is a clear drive to integrate them as well.

Problem based learning (PBL) has emerged from constructivist didactics and builds upon students' preliminary knowledge, expectations and interest. For this the starting point of learning is a problem or an issue to solve and students first get familiar with it before learning the information necessary to create a solution. The method is characterized by student-centeredness, work in small group, the presence of the teacher as a facilitator, and the work being organized around the problem [19]. The method incorporates the gaining of knowledge with the development of general skills and attitudes. It also promotes the development of numerous important soft-skills, e.g. communication skills, teamwork, problem-solving, independence, sharing information, and the respect of others [20]. Since one of the starting points of the method is taking the students' individual differences (interest, preliminary knowledge, etc.) into consideration, it is typical that students are motivated to work, spend much time on their studies and intensively take part in course work - especially if they also have an opportunity to have a say in defining the problem [21].

Inquiry-based learning (IBL) is a group of student-centred methods driven by inquiry or research [22]. According to Spronken-Smith et al. [23], IBL is used typically for teaching natural science subjects, where participants experience the process of knowledge creation, and discover the meaning and relevance of information through a sequence of steps. This way learners reach conclusions and reflections related to the newly gained knowledge. The method builds on the curiosity of students about the world surrounding them. Its aim is to develop critical thinking, increase the ability for independent research and raise awareness among students that they are responsible for their own learning, growth, and full maturity [24]. Ernst, Hodge and Yoshinobu [25] who examined the efficiency of the method in relation to the teaching of mathematics, emphasize the deep engagement in rich mathematics (and in general the topic) and the opportunities to collaborate (in some form) during problem solving. The claim is that during the application of this method students' learning performance increases, so does teachers' joy of teaching along with the number of teacher-student and student-student interactions.

The *flipped classroom* educational process model is a form of blended learning. During the application of this instructional strategy preliminary, individual processing of the material of traditional lectures takes place first (typically online), which is then followed by an active classroom work also incorporating problem-based, cooperative methods [26]. In the interpretation of Bishop and Verleger [27], during the preliminary preparation students process multimedia contents. According to Lo, Hew, and Chen [28] this method is based on the use of online technology such that video teaching materials (prepared in advance in short portions of 8–15 min) should be watched by students. Then actual classroom work is composed of short lectures as well as problem solving exercises (individually or in small groups). According to the creators of the model [29], watching the videos just before class is not enough for success. He finds that real information processing and learning should take place at home and students are to arrive to class with notes and questions, which are checked and answered by the teacher. Tucker [30] emphasizes the rethinking of all aspects of teaching and names 'best utilization of the time spent on learning' as the main goal of education.

In case of the so called '*mirrored classroom*' educational process students found their knowledge with their preliminary preparation, which is deepened by conversations during (in-school) classes, complex and cooperative tasks, and teachers' feedback. All these promote the autonomy and cooperation of students while matching students' individual needs much better [31].

Agile Teaching/Learning Methodology (ATLM) is designed for higher education by building upon the best practices and ideas from the field of software engineering. It utilizes concepts from agile software methodologies [32] which is based upon the observation that the processes of software development and learning are in many ways very similar: participants with different (sometimes clashing) goals work together until a certain deadline, based on a very tight schedule, possessing limited resources, and facing many expected/unexpected events. Therefore, both processes require detailed planning/scheduling, follow up and governance, with continuous assessment and feedback from key stakeholders. Building upon these similarities, application of the agile method in education (i.e. during the planning of teaching-learning processes) focuses on three key characteristics: agility, extremes and independence.

Constructivist learning theory assumes that there are no two identical students: everyone has different abilities, preliminary knowledge, ranges of interest and learning needs. Not all students are able to learn at the same pace, along identical methodologies. Consequently, the most important task of teachers is to help students in learning and construing their knowledge. All these have to be taken into consideration when planning the teaching-learning process through emphasizing the interaction and communication required among students/teachers and the requirement of adapting to changing needs. These assume agility: the teacher should be able to adapt quickly to students' skills and needs and modify courses. Adaptation and 'finetuning' of courses can happen properly and in a planned way if students get continuous feedback on their work (in the form of formative evaluation) and they also help teachers with informal (and often anonymous) feedback. This is the 'extreme' characteristics of the ATLM. During the application of this method the central role of teachers continuously fades away, they gradually pull out of the teaching-learning process, while students get more and more self-confident to learn independently and gain the skills, which are important from the aspect of lifelong learning. This is the dimension of independence. For all these to get implemented in classroom practice, the following methodological solutions are proposed: knowledge sharing among students, continuous feedback and teaching learning.

3.4 Challenges to Assessments

While summative methods are very important and they have a clear pedagogical foundation with new methods of teaching come new methods of assessment as well [33]. These methods move beyond the traditional approaches and propagate in-process evaluation of students' comprehension and progress. Formative assessments are formal and informal procedures conducted by teachers during the learning process and are aimed for supporting learning. They are supportive and development focused assessment techniques [34] and include for example diagnostic testing, heterogenic assessment, as well as self- and peer assessment. In addition, to treat students in a holistic manner, it is not irrelevant how students feel about themselves and their education. Consequently, student well-being is considered as a fundamental condition of successful teaching [35]. A clear challenge for BIS is how this philosophy and corresponding techniques may be integrated with the nature of the field as discussed in previous subsections.

3.5 Challenges of Teaching BIS Online

The pace of technological development constantly offers new opportunities and the context of learning is increasingly shaped by digital media including the personal ownership of various (e.g. mobile) devices. But teaching BIS online – or mostly online – is not straightforward and has its – already existing – challenges of its own. This is due to the fact, that in a digitalized world, education, like many other sectors, could not avoid adopting new technologies. A lot has changed over the past two decades or so since the birth of the idea of 'online learning'. Even the terminology has integrated a mushrooming set of new expressions: e-Learning, blended learning, distance education, technology enabled teaching, hybrid education, MOOC, virtual classroom, just to name a few. This is even further magnified by the difficulty of teaching computer soft skills (personal communication, groupwork, project management, etc.) online [36].

3.6 Special Challenges of the COVID Pandemic

While eLearning has been around for over two decades now, the recent global challenge invoked by the COVID-19 pandemic has put online education into the forefront of academic attention – both as a technological opportunity to maintain the continuity of teaching (at all levels of education) and as a challenge to innovate and apply new methodological approaches. The current pandemic put extra strain and challenge on most universities to retain the quality of their education. It has become clear that innovative approaches are needed – and needed fast: approaches that can help to deliver high-quality education from a distance. COVID thus has pushed online, blended, hybrid solutions, but those have their own problems in themselves which this rush to respond just further. As a most recent development our social context – shaped by fear and protective distancing – influences expectations and modes of knowledge exchange as well. "*During the pandemic the learning space has become fully digital including the same learning resources. While learning space is transforming, we also need to rethink about the other qualities of the learning design in IT education and proceed with potential adjustments*" [37] (p. 1).

However, while (aforementioned) modern teaching methods assume a well-organized learning space to be successful, this appeared not to be the case under the changes introduced as a reaction to COVID. While there was a (sudden) move to online or blended education, results may be described being only partial solutions in the sense that while teaching is now technically online, it really only utilizes technology to allow access. Indeed, it does not seem to involve full methodological adjustment to take advantage of technology. In other words, reorganizing teaching did not fully happen along clear methodological guidelines (such as flipped classroom practices, for example), instead, it simply moved more materials online. This is true even for videos, which were prepared out of necessity and their creation was not a result of applying consistent methodological principles (i.e. it has happened more reactively as opposed to being carefully planned). Thus, learning spaces were more ad-hoc than designed.

4 Towards an Advanced BIS Curriculum Framework

The world of information systems and info-communication technologies (ICT) in general are changing fast, sometimes rapidly. Therefore, everything we say about systems design or IS education is rather relative and need to be put into historical context to understand why changes happen and what is the expected lifetime of a paradigm-shift in the field. However, the goal of redesigning BIS education is not only the need to keep up with this pace, indeed, as it was demonstrated so far, there are additional factors that influence the way BIS may be taught.

It was already realized by Zachman ([38], see also [39] for an update) that due to the development in data processing, implementation of IT supported business functions often result in rather isolated solutions, and instead of being an accelerator of adaptivity and supporter of competitiveness, costly IT solutions often freeze the enterprise at the technological and application level applied at a given time. Indeed, because IT system are usually large investment and even when old can still work relatively well, their replacement by the latest technology, platforms, or solutions is a difficult decision, not to mention the costs and risks. Instead, a more flexible view is needed, one that is built on architectural concepts. This approach eventually led to the development of The Open Group Architecture Framework or TOGAF [40]. Indeed, this approach provides a guideline around which an evolutionary BIS curriculum design approach may be organized.

TOGAF differentiates among several architecture domains (called the Enterprise Architecture Model – EAM): business architecture, information systems (data and application) architecture and technology architecture. For enterprises this view may be used to create a process of systematic redesign. In each domain there is a baseline and a target architecture, and a gap analysis can create a roadmap of change. This way the organisation and IT management can follow a well-controlled and coherent development scenario. The suggested architecture development method (ADM) is split into four phases: creating the architecture context; architecture delivery; transition planning; and architecture governance. This creates an opportunity for a customizable framework, repeatable architecture development, which means stepping further towards the advanced, integrated solution; considering re-usability, standardization, interoperability, and portability.

From a different point of view the dimensions and the process of TOGAF may be utilized as a backbone for BIS curriculum design since the EAM is built on the strong correlation between IT technology and business management (which is key to BIS). If a curriculum is considered as set of requirements that need to be met during T&L, it is easy to see that these requirements may change by time, place, type of audience, level of education, and the way of implementation. Requirements might reflect professional viewpoints (such as the AIS guideline) or the short-term interest of the labour market or long-term, future demand of the world of labour (that may be hard to predict). Therefore, similarly to the TOGAF philosophy, separation of requirements (or in the EU, competences) from implementation is a must.

This could be augmented with the latest pedagogical approaches to make it fresh, approachable, and ready for blended learning. In every stage of the architecture development method developers have to contrast the information technology solution with business objectives, processes, and maturity. This and the relatively low level of complexity of EAM fits well to the idea of problem-based T&L. Students may be posed an (organizational IS) problem and seeking solution(s) would force them to explore relevant concepts, information, and techniques while incurring required skills and competencies. One may skip traditional course design, since problem-solving is now placed into the centre of learning. Indeed, one may even start with a very complex, almost unsolvable problem, which then would need to be split into smaller issues first.

Once the original problem was broken down into smaller ones, instead of being 'taught', students will study the business, its environmental and societal context, and its characteristics which, hopefully, will lead to even smaller sub-problems that are more feasible to solve. The expected final outcome is an outline of a working model. During this process – instead of studying material from isolated courses and dedicated lectures – students would need to learn business economics (including firm theory, sociology, regulations, and so on) and at every stage they will need to learn the corresponding IT technology part as well. The problem-solving process under this case-based framework will indicate where and when to introduce system design principles, and procedures, database design, business intelligence methods, or governance issues. It may depend on the timeframe of training, but at least two iterations are necessary. Student audience must confront the barriers of the (suggested) solution. This way they will also understand what the roles of maturity models, transitioning, and audit are – thus getting a full picture of an organization and its information systems. The approach would especially be effective in a dual education (internship, work placement) context.

5 Summary and Future Direction

This proposition paper provided a systematic and comprehensive overview of the challenges BIS programs to face and need to address in our post-pandemic educational context. The review of employer and alumni opinion as well as current BIS education best practices was combined with a landscape of up-to-date teaching methods with focus on teaching computer and organizational soft skills. In light of an increasingly student-centred world augmented with extended online options the paper put forward a BIS education design framework based on TOGAF. The argument for the need and potential success of this approach is that it is capable of addressing the existing set of interrelated issues and challenges in an integrated manner. Admittedly, one limitation of this paper is that the six dimensions investigated in Sect. 3 were concluded from literature. The obvious next step is then to put the suggested approach into practice and create a BIS curriculum organized around TOGAF.

References

1. Topi, H., et al.: Curriculum Guidelines for Undergraduate Degree Programs in IS. ACM (2010)
2. Topi, H., et al.: MSIS 2016 global competency model for graduate degree programs in information systems. Commun. AIS **40**, 1–107 (2017). Article18
3. Boehler, J.A., Larson, B., Peachey, T.A., Shehane, R.F.: Evaluation of information systems curricula. J. Inf. Syst. Educ. **31**(3), 232–243 (2020)
4. Nasir, S.A., Yaacob, W.W., Aziz, W.W.: Analysing online vacancy and skills demand using text mining. J. Phys. Conf. Ser. **1496**, 012011 (2020). https://doi.org/10.1088/1742-6596/1496/1/012011
5. Cummings, J., Janicki, T.N.: What skills do students need? a multi-year study of it/is knowledge and skills in demand by employers. J. Inf. Syst. Educ. **31**(3), 208–217 (2020)
6. Lankhorst, M.: Enterprise architecture at work. Springer Berlin Heidelberg, Berlin, Heidelberg (2017)
7. Pitukhin, E.: Job advertisements analysis for curricula management: the competency approach. In: 9th Annual International Conference of Education, Research and Innovation, Seville, pp. 2026–2035 (2016)
8. Dubey, R.S., Tiwari, V.: Operationalisation of soft skill attributes and determining the existing gap in novice ICT professionals. Int. J. Inf. Manage. **50**, 375–386 (2020). https://doi.org/10.1016/j.ijinfomgt.2019.09.006

9. Szabó, I., Neusch, G.: Comparative analysis of highly ranked BIS degree programs. In: Proceedings of the Business Information Systems Conference Workshops, Hannover, Germany, June 14 (2021). (In this Volume). https://doi.org/10.1007/978-3-030-04849-5
10. Apigian, C.H., Gambill, S.: A descriptive study of graduate information systems curriculums. Rev. Bus. Inf. Syst. (RBIS) **18**(2), 47–52 (2014). https://doi.org/10.19030/rbis.v18i2.8978
11. Landry, J.P., Saulnier, B.M., Wagner, T.A., Longenecker, H.E.: Why is the learner-centered paradigm so profoundly important for information systems education? J. Inf. Syst. Educ. **19** (2), 175–180 (2008)
12. Mamina, R.I., Tolstikova, I.I.: Phygital generation in free global communication. Int. J. Open Inf. Technol. **8**(1), 34–41 (2020)
13. Tolstikova, I., Ignatjeva, O., Kondratenko, K., Pletnev, A.: Generation Z and its value transformations: digital reality vs. physical interaction. In: Alexandrov, D.A., Boukhanovsky, A.V., Chugunov, A.V., Kabanov, Y., Koltsova, O., Musabirov, I. (eds.) DTGS 2020. CCIS, vol. 1242, pp. 47–60. Springer, Cham (2020). https://doi.org/10.1007/978-3-030-65218-0_4
14. Cook, V.S.: Rethinking learning engagement with Gen Z Students. e-mentor **80**(3), 67–70 (2019). https://doi.org/10.15219/em80.1425
15. Davidson, C.N.: The new education: how to revolutionize the university to prepare students for a world in flux. Hachette, UK (2017)
16. Geiger, A.W., Davis, L.: A growing number of American teenagers–particularly girls–are facing depression. Pew Res. Center **12** (2019). https://www.pewresearch.org/facttank/2019/07/12/a-growing-number-of-american-teenagers-particularly-girls-are-facingdepression/
17. Gehlen-Baum, V., Weinberger, A.: Notebook or Facebook? how students actually use mobile devices in large lectures. In: Ravenscroft, A., Lindstaedt, S., Kloos, C.D., Hernández-Leo, D. (eds.) EC-TEL 2012. LNCS, vol. 7563, pp. 103–112. Springer, Heidelberg (2012). https://doi.org/10.1007/978-3-642-33263-0_9
18. Davidovitch, N., Yossel-Eisenbach, Y.: The Learning paradox: the digital generation seeks a personal, human voice. J. Edu. E-Learn. Res. **6**(2), 61–68 (2019)
19. Barrows, H.S.: Problem-based, self-directed learning. JAMA **250**(22), 3077–3080 (1983). https://doi.org/10.1001/jama.1983.03340220045031
20. McComas, W.F.: Problem based learning. In: McComas, W.F. (ed.) The Language of Science Education, pp. 76–76. SensePublishers, Rotterdam (2014). https://doi.org/10.1007/978-94-6209-497-0_66
21. De Graaf, E., Kolmos, A.: Characteristics of problem-based learning. Int. J. Eng. Educ. **19** (5), 657–662 (2003)
22. Levy, P., Little, S., Mckinney, P., Nibbs, A., Wood J.: The Sheffield Companion to Inquiry-Based Learning. The Univ. of Sheffield, Brook Hill, UK (2010). http://www.shef.ac.uk/ibl
23. Spronken-Smith, R., Angelo, T., Matthews, H., O'Steen, B., Robertson, J.: How effective is inquiry-based learning in linking teaching and research. In: An Int. Colloquium on Int. Policies and Practices for Academic Enquiry, 7, 4, pp. 19–21. Marwell, Winchester, UK (2007)
24. Lee, V.S.: Teaching and Learning Through Inquiry: A guidebook for Institutions and Instructors. Stylus Pub. LLC (2004)
25. Ernst, D.C., Hodge, A., Yoshinobu, S.: What is inquiry-based learning. Notices AMS **64**(6), 570–574 (2017). https://doi.org/10.1090/noti1536
26. Lage, M.J., Platt, G.J., Treglia, M.: Inverting the classroom: a gateway to creating an inclusive learning environment. J. Econ. Educ. **31**(1), 30–43 (2000)

27. Bishop, J.L., Verleger, M.A.: The flipped classroom: A survey of the research. In: 120th ASEE National Conference and Exposition, Atlanta, GA (paper ID 6219). American Society for Engineering Education, Washington, DC (2013). https://doi.org/10.18260/1-2–22585
28. Lo, C.K., Hew, K.F., Chen, G.: Toward a set of design principles for mathematics flipped classrooms: a synthesis of research in mathematics education. Educ. Res. Rev. **22**, 50–73 (2017). https://doi.org/10.1016/j.edurev.2017.08.002
29. Bergmann, J., Sams, A.: Flip Your Classroom: Reach Every Student in Every Class Every Day. International society for technology in education (2012)
30. Tucker, B.: The flipped classroom. Educ. Next **12**(1), 82–83 (2012)
31. Yarbro, J., Arfstrom, K.M., McKnight, K., McKnight, P.: Extension of a Review of Flipped Learning (2014). https://flippedlearning.org/wp-content/uploads/2016/07/Extension-of-FLipped-Learning-LIt-Review-June-2014.pdf
32. Chun, A.H.W.: The agile teaching/learning methodology and its e-learning platform. In: Liu, W., Shi, Y., Li, Q. (eds.) ICWL 2004. LNCS, vol. 3143, pp. 11–18. Springer, Heidelberg (2004). https://doi.org/10.1007/978-3-540-27859-7_2
33. Yorke, M.: Formative assessment in higher education: moves towards theory and the enhancement of pedagogic practice. High. Educ. **45**, 477–501 (2003). https://doi.org/10.1023/A:1023967026413
34. Pereira, D., Assunção Flores, M., Niklasson, L.: Assessment revisited: a review of research in assessment and evaluation in higher education. Assess. Eval. High. Educ. **41**(7), 1008–1032 (2016). https://doi.org/10.1080/02602938.2015.1055233
35. Jones, E., Priestley, M., Brewster, L., Wilbraham, S.J., Hughes, G., Spanner, L.: Student wellbeing and assessment in higher education: the balancing act. Assess. Eval. High. Educ. **46**(3), 438–450 (2021). https://doi.org/10.1080/02602938.2020.1782344
36. Tabatabaei, M., Gardiner, A.: Recruiters' perceptions of information systems graduates with traditional and online education. J. Inf. Syst. Educ. **23**(2), 133–142 (2012)
37. Pappas, I.O., Giannakos, M.N.: Rethinking learning design in IT education during a pandemic. Front. Educ (2021). https://doi.org/10.3389/feduc.2021.652856
38. Zachman, J.A.: Framework for information systems architecture. IBM Syst. J. **26**(3), 276–292 (1987). https://doi.org/10.1147/sj.263.0276
39. Sowa, J.F., Zachman, J.A.: Extending and formalizing the framework for information systems architecture. IBM Syst. J. **31**(3), 590–616 (2010). https://doi.org/10.1147/sj.313.0590
40. OpenGroup: The TOGAF® Standard, Version 9.2. (2018). https://publications.opengroup.org/c182?_ga=2.224960986.1921117380.1619948278-1553644343.1619948278

COVID-19-Related Challenges in Business Information Systems Education: Experiences from Slovenia

Marjeta Marolt(✉), Andreja Pucihar, Gregor Lenart, Doroteja Vidmar, Blaž Gašperlin, and Mirjana Kljajić Borštnar

Faculty of Organizational Sciences, University of Maribor, Kranj, Slovenia
{marjeta.marolt,andreja.pucihar,gregor.lenart, doroteja.vidmar,blaz.gasperlinl,mirjana.kljajic}@um.si

Abstract. Universities have encountered numerous difficulties and challenges during the COVID-19 pandemic. They used various approaches to deal with these challenges. Unfortunately, these experiences are not widely discussed. Therefore, this study provides preliminary insights on how the business information systems department at the Faculty of Organizational Sciences, University of Maribor managed to overcome different challenges and executed the study process completely online in the COVID-19 pandemic. Experiences of conducting several courses at the bachelor and master level are reported in the paper. We also provide some suggestions on how to overcome specific challenges faced by students and lecturers. In the future, we wish to conduct a multiple case study including the viewpoints of lecturers, support staff, and students.

Keywords: Business information systems · Study process · Digitalization · COVID-19 pandemic

1 Introduction

The higher education system has undergone numerous changes over the years [1]. Online courses, more diverse and international communities, and more study options are just a few examples of the changes taking place at universities worldwide. Among the most prominent drivers behind these changes has been information and communication technology [2], which also proved useful in the COVID-19 pandemic [3].

The COVID-19 disease broke out in China in December 2019. In Slovenia, the first infected person was detected on March 4, 2020, and the disease began to spread across the country. On March 12, 2020, the pandemic was officially declared and the government began to implement new measures to contain the pandemic, which, as elsewhere in the world, drastically limited public life in the country. The government withdrew the declaration of a pandemic on 31 May 2020. The first wave of the

W. Abramowicz et al. (Eds.): BIS 2021 Workshops, LNBIP 444, pp. 73–78, 2022.
https://doi.org/10.1007/978-3-031-04216-4_7

pandemic in Slovenia lasted 12 weeks. The second wave of the pandemic followed in the fall, when the numbers of infected and dead began to rise sharply.

During the COVID-19 pandemic universities had to conduct lectures and tutorials online, as all educational institutions were closed. Numerous difficulties and challenges have been encountered in this process. There are also various approaches used by faculties to deal with these challenges. Unfortunately, these experiences are not widely discussed. Therefore, to better deal with the difficulties and challenges presented by the COVID-19 pandemic, it is important to share best practices and learn from other experiences.

In this study, the focus is on business information systems education in Slovenia and how the Faculty of Organizational Sciences, University of Maribor managed to overcome different challenges and executed the study process completely online in the COVID-19 pandemic.

2 Materials and Methods

This descriptive study reports the experiences from the Faculty of Organizational Sciences, University of Maribor during the COVID-19 pandemic. This faculty was chosen because the authors are part of the Information Systems department and were involved in the study process during the mentioned period. This department consists of 13 members that cover different fields of information systems and are lecturing different courses in the undergraduate and master programme of Organization and management of information systems (OMIS) study programme. The authors are involved in several courses, for example, Business Model Design and Business Information Systems, Digital Business, Decision Theory, and Data mining to name a few.

Each author provided experiences on the digitalization of teaching and learning during the COVID-19 pandemic. These insights were then combined by one author and then reviewed by all of them. The revised version of combined insights is provided in the following section.

3 Findings

Traditionally, the faculty's study programmes are all accredited as a blended learning, combining in class and online lectures and tutorials. For this purpose, we used Moodle as a common learning platform for all courses for many years. After the pandemic was declared, teaching was switched to MS Teams. As this happened virtually overnight, without any training for faculty and students' different problems occurred, especially concerning the student's access to different program solutions, students' collaboration, and lack of material appropriate for online classrooms. The problems and how it was dealt with are presented in the following subsections. First, the experiences at the undergraduate level are presented, followed by the presentation of the experiences at the master level.

3.1 Undergraduate Level Experiences

At the undergraduate level students become acquainted with different work environments and problem-solving practices related to the development, operation, and maintenance of information systems.

In this section we report observations from three courses in the OMIS study program. The two of them - Business Model Design and Business Information Systems - are taught in the second year and Digital Business in the third year. All three courses are mandatory for the students of the OMIS program. In all three courses, we use teamwork, problem-based learning, and various business modelling tools. There were 12 students enrolled in each of the first two courses, and 25 students in the third course.

In the first course we use a simple business case where students are divided into groups and evaluate different problems and learn different tools while applying them in different phases of Soft System Methodology. Later, they approach a complex case developed based on a real environment - a small manufacturing enterprise. Normally, we often use a blackboard, large posters, markers in different colors, and "post-it" papers during class. Student groups used separate parts of classrooms for discussing and collaborating. This was much more difficult with the MS teams. Sometimes a simple drawing to show how to use a particular tool took a lot of effort to draw using MS Visio or another tool, even if the templates were used. For students to collaborate, we prepared channels in MS Teams, one for each group. However, they used these channels only during lectures. For collaboration on their assignments outside lectures, they used Discord. During the lectures, it was more difficult to engage students in conversation, because it was the first time we met in a virtual environment. They were shy and only participated in the conversation when invited and called by name.

After this course, we met again with the same group of students in the Business Information Systems course. There was less teamwork in the lectures and we did not use any tools, because they were already familiar with the tools from the previous course and they needed these skills only for completing their seminar work later in the exercises. We focused more on the technological perspectives of information systems and discussing the different cases. Student engagement was the same as in the previous course. Only a few students participated in the discussion. The others participated only when called upon by the lecturer.

The Digital Business course is taught in the third, final year of the undergraduate program. We met with this group of students in person before the pandemic. However, half of the students were from the program Business Engineering. And it was with these students that we met for the first time. Surprisingly, these students were also more engaged, even though they were not familiar with the content of Information Systems or Digital Business. This course was more dynamic, with more discussions. This course looks for the innovation potential of digital business models. Students have to apply all the knowledge, skills, and tools they know from previous courses. The result of their assignment and exam must be an innovative digital business model. For this purpose, we use design thinking to stimulate the students in the innovation process. They also presented their ideas to other peers during the course and we had several facilitated discussions as well as brainstorming sessions. In the end, students provided new ideas that were elaborated according to the syllabus.

More technological problems occurred during the tutorials, where students are working with different program solutions (e.g. SAP, Datalab Pantheon X). Before the COVID-19 pandemic access to all the program solutions was provided through computers located in the computer classrooms. As such, students did not need to install program solutions by themselves on their computers. When they needed to install program solutions themselves, many of them had problems because of the different hardware equipment, operating systems, and other settings they are using. When we managed to overcome these first barriers we were able to proceed with tutorials. The next problem that occurred, was related to very slow progress through the assignments. For example, the majority of students did not have an additional screen that would enable them to more easily follow the tutor instructions. Furthermore, some students had issues with the microphone and they reported problems by writing them into chat, which further hindered the progress. To proceed more quickly, tutors had additional sessions with those students who had problems. As the tutors' overload was detected this was just a temporary solution.

3.2 Master Level Experiences

At the master level students develop research, technical, analytical, communication and management (managerial) competencies that enable them to lead effective groups and manage business activities through information and communication technology.

In this section we report educational experiences from two courses on the OMIS program: Decision theory and Data mining. The lecturer was using a PC with two monitors, audio-video equipment for communication, and a graphic tablet (Wacom Cintiq 16). During both courses several programs were used, mostly open source, fully available for students (i.e. Orange data mining, Silver decisions, DEXi, etc.).

The decision theory is the first-year obligatory course for the OMIS students and an elective course for the students of the Enterprise engineering programme. Data mining course is an elective course for both the IS and enterprise engineering programmes in the second year (fourth semester) of master level studies. Usually, in the first year of master studies most students (80%) previously finished the bachelors studies of one of the Faculty of Organizational Sciences programmes. The remaining students are from other faculties and fields of studies (i.e. economics, social sciences, computer science, mathematics, etc.). There are usually some international students enrolled as well. There were 29 students enrolled in this course. This is an educational challenge in itself. Students have different previous knowledge, different levels of understanding of complex management problems, they often have no real-life experiences, and during the Decision theory course the lecturer must bring some of those real-life ideas of complexity to them. In previous years we have developed business simulators to support teaching about complex management problems [4]. The experiments that we have conducted from 2003 to 2006 showed that group supported simulation based learning contributes to a better understanding of the management problems and support learning about decision-making [5]. However, we could not use the business simulator in the online environment, because of the lack of IT support. Also, in the usual classroom environment the lecturer can use the blackboard to present the methods and techniques of decision modelling. The graphic tablet can somewhat replace the blackboard, but this means that the lecturer is working on at least two screens. At the beginning of the course the students were a bit shy, but by the end they started to communicate more freely. They were purposively assigned into working groups, matching those who previously finished the Faculty of Organizational Sciences with those that came from other faculties.

The data mining course is different in the way that the group is smaller (up to 15 students), and by the fourth semester they have already known each other and have worked together before. Based on our experiences from the first wave of the pandemic, we have adjusted the lectures so that the lecturer used the Orange data mining software, while explaining the theory behind the methods and algorithms. This way, the students were more focused on the topic, there was more feedback seeking behaviour present and they were readier for the tutorials and independent work that followed the lectures. However, not all the students have two screens that would enable following lectures and simultaneously working in Orange. The students sought feedback more often compared to the first year's students. There was almost 100% attendance at the lectures, and all the students passed the exam on the first attempt, which was not the case at the Decision Theory course.

When tutorials took place, we encountered a lower level of productivity among students. Although more students were able to attend tutorials, some of them seem to only partly follow them, as they attended them during their work or other obligations. Thus, they were not able to fully focus on assignments and actively participate in the discussion. Consequently, we noticed a decrease in the quality of the submitted assignments.

4 Discussion and Conclusion

During the two waves of the COVID-19 pandemic, we have encountered different challenges. Some of them were quickly solved, while some remain even after the second wave of the COVID-19 pandemic. We did not notice essential differences between the experiences at undergraduate and master levels, but we can make some general conclusions based on our observations.

We noticed that many problems related to technological issues were resolved to the extent that the majority of our students can work in the learning environments that are required for each course. Unfortunately, the internet connection and technical equipment issues that students are facing [6] cannot be resolved by the faculty. Therefore, it is important to inform students before the course, what kind of technological equipment they will need to complete it.

Switching from traditional to computer-based learning/teaching is another problem that we noticed. In the first wave, we encountered difficulties to adapt from both sides, teachers, and students, but in the second wave mostly by students in the first year of undergraduate and master studies. We relate this adaptability struggle to unique circumstances, as our students were not able to meet in person at the beginning of the student year. Furthermore, lack of computer literacy prevented these students to manage their assignments and courseware provided [7]. To overcome this problem, the faculty should offer their students (at least to the first-year students) basic courses in computer literacy.

Teachers have also struggled to motivate students at times. As some of the approaches did not prove successful we tried new ways, including offering them extra credit, adjusting the expectations, and laying out estimated completion times for each assignment. We also tried to motivate them by setting a collaborative environment [8]. This approach proved to be more effective in smaller groups where students already knew each other. In contrast, the larger groups that had never met in person were more difficult to involve in conversation with the teachers and among students.

Overall, even though substantial efforts were made by faculty to implement the distance study process, the disruption caused by the COVID-19 pandemic will have a long-lasting effect. Based on these experiences, we can expect in the future, a variety of new, more hybrid, and flexible teaching approaches that will motivate students to learn and help the teacher to better deal with situations, such as the COVID-19 pandemic [9].

References

1. Teichler, U.: Changing Patterns of the Higher Education System. The Experience of Three Decades. Higher Education Policy Series, vol. 5. Taylor and Francis Group, Bristol (1988)
2. Brennan, J.: Higher education and social change. High. Educ. **56**(3), 381–393 (2008). https://doi.org/10.1007/s10734-008-9126-4
3. Peres, K.G., Reher, P., de Castro, R.D., Vieira, A.R.: Covid-19-related challenges in dental education: Experiences from Brazil, the USA, and Australia. Pesqui. Bras. Odontopediatria Clin. Integr. **20**, 1–10 (2020). https://doi.org/10.1590/pboci.2020.130
4. Škraba, A., Kljajić, M., Leskovar, R.: Group exploration of system dynamics models - is there a place for a feedback loop in the decision process? Syst. Dyn. Rev. **19**(3), 243–263 (2003). https://doi.org/10.1002/sdr.274
5. Borštnar, M.K., Kljajić, M., Škraba, A., Kofjač, D., Rajkovič, V.: The relevance of facilitation in group decision making supported by a simulation model. Syst. Dyn. Rev. **27**(3), 270–293 (2011). https://doi.org/10.1002/sdr.460
6. Gillies, D.: Student perspectives on videoconferencing in teacher education at a distance. Distance Educ. **29**(1), 107–118 (2008)
7. Mishra, L., Gupta, T., Shree, A.: Online teaching-learning in higher education during lockdown period of COVID-19 pandemic. Int. J. Educ. Res. Open **1**, 100012 (2020). https://doi.org/10.1080/01587910802004878
8. Kaup, S., Jain, R., Shivalli, S., Pandey, S., Kaup, S.: Sustaining academics during COVID-19 pandemic: the role of online teaching-learning. Indian J. Ophthalmol. **68**(6), 1220–1221 (2020). https://doi.org/10.4103/ijo.IJO_1241_20
9. Rapanta, C., Botturi, L., Goodyear, P., Guàrdia, L., Koole, M.: Online university teaching during and after the Covid-19 crisis: refocusing teacher presence and learning activity. Postdigital Sci. Educ. **2**(3), 923–945 (2020). https://doi.org/10.1007/s42438-020-00155-y

Successful Project Completion During the COVID-19 Pandemic - A Lesson Learnt

Md Shakil Ahmed[1,2], Marco Gilardi[1(✉)], Keshav Dahal[1], and Dave Finch[2]

[1] University of the West of Scotland, High Street, Paisley, UK
Shakil.Ahmed@goldencasket.co.uk, {Marco.Gilardi,Keshav.Dahal}@uws.ac.uk
[2] Golden Casket Group, Greenock, UK
Dave.Finch@GoldenCasket.co.uk

Abstract. COVID-19 pandemic has taught us how to continue with the day-to-day activities interacting and working from remote locations. In this paper, we have highlighted the positive approach necessary to complete a project with success under this constraint by interacting regularly with the relevant stakeholders keeping focus on the final project deliverables. The salient points with supporting references are chalked out which might be helpful for others to follow if faced with stressful situations that COVID-19 pandemic taught us.

Keywords: Project management · Small and medium size enterprises (SMEs) · Self-reflection · Information technology · Business intelligence

1 Introduction

Involving oneself in a project is one thing and completing the project is another. The project manager (PM) plays a key role in progressing the project, in collaboration with the other members of the project's team, towards the vision for the end-result that has been set out at the very beginning of the project. However, establishing the vision of a project and what the possible project outcome should be it is not easy to set out in the initial days of the project. Following constructive methodologies makes it easier for the PM to develop such vision and identify desirable project outcomes. To establish the outcomes and their requirements it is necessary to embed the interaction with stakeholders related to the project in the project management process from the project ideation stages. The more accurately a PM knows the requirements for the project, the more they can focus on key milestones that can determine the project success and ensure that stakeholders are satisfied with it. In doing so the PM may target too broad goals within the timeframe of the project. During the envisioning stages of the project there is the risk that the PM may cast a wide net that determines goals too broad for the project. To avoid this risk, it would be wise for the PM to narrow down the project's vision to one that is practical and useful for the end-users of the project deliverables. Although the project vision should be the keystone that mandate how the project evolves, problems and requirements set by end users might require that such vision

W. Abramowicz et al. (Eds.): BIS 2021 Workshops, LNBIP 444, pp. 79–85, 2022.
https://doi.org/10.1007/978-3-031-04216-4_8

changes and evolves during the project. This evolving vision might have dire repercussions on the project, both economical and practical. To mitigate this risk, it is important that the PM knows what the real-world situation is, and although there will always be a certain degree of unpredictability, appropriate requirements and constraints are set within the vision.

In this paper, the lessons learnt from a successfully completed recent project within a small and medium size enterprise (SME) will be discussed. The project was funded by a knowledge transfer partnership (KTP), a UK funding scheme aimed at embedding academic knowledge into businesses by promoting collaboration between higher education institutions (HEI) and industry.

The COVID-19 situation imposed considerable stress on the project. In this paper, it will be discussed how such stresses were managed to ensure that the project could continue and be successful. The aim is to share the experience we gained through a case study with the hope that it will be of benefit to academic and businesses that embark on a similar project.

2 Related Work

Project management is an established area of professional expertise and academic research that offers a methodical approach to all stages of a project, ensuring each stage is carefully planned, measured, and monitored [13]. Modern approaches of project management can be tailored to fit the demands for the smaller organizations such as SMEs [8]. By the UK definitions, small firms are those which employ 1 to 49 people, whilst medium-sized firms are those which employ 50 to 200 people [12]. SMEs play a vital role in the economy of any country as they represent the major share of business activities [7]. KTP projects help the SMEs to acquire skills and technology that strategically enhance their competitiveness through collaborations with higher education institutions (HEI). The knowledge from the HEI is embedded into the company through a project undertaken by a qualified graduate known as KTP Associate employed by higher education institutions (HEI) [9].

SMEs need these kinds of projects because they are normally suppliers for larger companies, such as ASDA, TESCO etc. in our case study, or cover a specialized niche in the market. SMEs often create the final products or services for the end-users, but due to their size and limited resources it is difficult for SMEs to compete with larger companies without due consideration to the customer desires [7]. SMEs require to be equipped with people that have mastery of modern technological skills such as digital marketing, or knowledge of business intelligence tools, so that they can sustain their business. However, acquiring such competencies is quite expensive. KTP projects help to create a collaborative environment between various stakeholders so that the SMEs can embed the discussed skills withing their business and endeavor to fulfil their strategies.

The success for this type of projects depends on the systematic approach of the KTP Associate in moving forward with the project, and on his project management skills. Research suggests that a project becomes successful if the project planning laid out by the PM is well developed [8]. Having the clear objectives, with support from top-management is important for the success factors of SMEs' projects [8]. A typical KTP project will be presented as a case study in the reminder of this paper to highlight the importance of project management skills, emphasizing on the vital role the KTP associate plays to make the project a success story.

3 The KTP Project

Our case study is a recently completed two-year long KTP project, which started on 8th April 2019. This collaborative project involved three organizations: Innovate UK, University of the West Scotland, and Golden Casket. The company Golden Casket is located at the north-western side of Scotland whose primary function is to produce chocolates and sweets and distribute them to various customers. This company is growing consistently ever since its foundation about 50 years ago. The project was primarily related to the development of a mobile platform supporting the sales representatives and enhancement of existing business intelligence platforms which are pivotal to the company's day to day decision making in the long run. The main stakeholders from this project are HEI supervisory team, which supports and guide the project from the technical point of view, the KTP adviser, which works as a link between the HEI and the business ensuring that all the supporting elements and decisions are actioned, the company supervisor, which ensures that the company can embed the knowledge and supports the development of the project, and finally the end-users, that drive the projects by establishing the requirements and the constraints, in this case study the end-users were sales managers and salespersons within the company. To monitor the continuous progress of the project, local management committee (LMC) meetings took place quarterly.

3.1 Approach from KTP Associate

The KTP associate works within the KTP as a project manager. At the initial stages of the project, the focus of the associate was on understanding the exact requirements, this was achieved through a contextual study [6] with the sales-reps, in which the KTP Associate shadowed and interviewed the reps, understanding their workflow and the limitations imposed by their context during work. The contextual study helped developing a set of user personae that were then used to develop the project requirements and mock-up the user interfaces of the mobile platform. In addition to the contextual study the KTP associate worked with other company staff, understanding how the company operation works, in particular during the life of the project the KTP associate worked closely with the Director of Operations (company supervisor) which helped the associate envision how the mobile platform will fit within the company operations and what BI tools will be required to enhance the company outputs. Within a couple of months from the start of the project, the project's main requirements were documented describing a systematic approach to reach the project deliverables. This document was then circulated to both the HEI advisory team and the company supervisor for final approval. The documentation enabled HEI supervisory team, associate, and company supervisor to develop the vision of the project and remain focused on developing the deliverables and make the project a success. Once the requirement document was developed, the next stage was to establish the technical feasibility of the requirements and determine the optimum tools necessary to develop the project deliverables. To achieve these goals existing company resources such as the IT infrastructure, the technical documentations already in possession of the company, and what the existing know-how of existing systems are, were studied. This helped scope existing infrastructures and determine whether sufficient resources were in place to support the project. For example, the company already had access to Microsoft Dynamics-AX connected with a Microsoft SQL database server. This system was in operation since 2017 and the company supervisor was the only knowledgeable person to run it smoothly since that time. Once the scoping of the requirement for further development to IT infrastructure was completed, the KTP associate could concentrate on other elements of the project's requirements. At this stage, it was helpful to know how the company benefit from existing systems, aiming at identifying which areas within the company the KTP project could augment and provide a clearer vision to for the project deliverables. To achieve this insight, the KTP associate informally interviewed staff within different departments of the company, such as the office personnel, the warehouse staff, other company directors, and the end-users. The communication skills and the note-taking habits helped a lot at this phase of the project. This preliminary work and understanding of the company operations resulted to be very helpful when the project faced the remote work conditions due to the COVID-19 pandemic starting from March 2020, which coincided with the second year of the project. The data gathered through contextual studies, informal interviews and notes were very helpful to develop and compile business intelligence documents when the need for face-to-face meetings were felt necessary but could not be materialized.

3.2 Importance of LMC Meetings

As part of monitoring the project progress, quarterly LMC meetings were regularly conducted throughout the project time-period. All the stakeholders, comprising of the HEI supervisors, the company supervisor, and the funder representative, took part in these meetings. The KTP associate, in his role of PM, had to deliver a presentation mentioning the project progress, future courses of actions and highlighting the requirements for any technical or training support from different corners. The LMC meetings main advantage was for the KTP associate, which helped keep his motivation and instill a sense of urgency from the team towards the smooth progress of the project. The LMC meetings also helped the company supervisor remain confident that the project would be successful despite the difficulties imposed by the pandemic.

3.3 Crucial Implementation Stages

At the planning stage, it is easy to underestimate the effort needed to complete a project objective and produce the required deliverables. However, a project can only be successful if it fully satisfies the end users, it is therefore important to ensure that the planning and time management allocate sufficient time for the implementation and buffer period that allow for unforeseen delays. Implementation is a crucial stage during the development of software, and it is often the case that during iterative testing end-users ask for more functions or requirements to be added with the deliverables. This is very challenging for the developer (which in this case study is the KTP associate). Because those requirements were not catered for in the initial planning or requirement documents other technical support from the HEI could also not be available as the project moved closer to its end. This challenging situation could be avoided if the KTP associate spend a longer period of time completing the contextual study and gather more data about the end-users during preliminary stages of the project. Although an agile methodology Beck [1] was taken for the implementation of the mobile platform the preliminary contextual study could have captured more requirements if more time were allocated for it, which would have helped during the COVID-19 pandemic.

During the COVID-19 pandemic, the access to the end-users was challenging and often not possible. Efforts were taken to develop prototypes on a PC and test them remotely with the end users. This was not an easy task to do because the end-users were available to do so, had other concerns (the pandemic), or they were sufficiently literate in IT to perform these tests interactively. Best efforts were made by the KTP associate to achieve a timely completion of the project. The success was possible thanks to the strong sense of belongingness with the project developed during the initial stages of the project, the motivation instilled by supervisors and LMC meetings, the personal willpower of the associate, and the vision of success developed by the KTP associate.

4 Lessons Learned and Approach with Self-reflection

While working as a KTP associate a great number of skills could be acquired, namely software development skills, such as how to follow the software development lifecycle (SDLC), and negotiation skills, as the KTP associate had to face a good number of intelligent people who were very helpful and demanding so that the project could be a success in the long run. The confidence by all supervisors and funder representative in a successful project completion motivated the KTP associate to seriously think about how to expand this skill to develop a business model where the same kind of software could be developed for other small companies of similar nature. Practical work experience with the company also provided the associate a great insight on the use of the six techniques and thinking tools for business model generation highlighted in [10], such as: customer insights (customer's perspective), ideation (creative process of generating business ideas), visual thinking (capturing big pictures through visualization of nine building blocks), prototyping (sketches allowing discussion/inquiry on business model), storytelling (facilitates effective communication) and scenarios (reflections on future business model). The 6th thinking tool named as 'scenarios', allows creativity via future contexts on business model designing for varied environments. Highlighting on some models of reflective activities are also important to note as these activities were always encouraged from the Knowledge Transfer Adviser from Innovate UK, who worked as an anchor role in the project.

Different reflective models such as Kolb's Learning Cycle [5], Gibb's model [4], Schön [11] are important to note. Kolb's reflective model suggested this practice as a tool to obtain the conclusions and ideas from the experiences, and the process consists of four phases named as concrete experience, reflective observation, abstract conceptualization, and active experimentation. On the other hand, Gibb's model helps a person to reflect after the experience and is useful for people who is new to reflection. This is formed of six stages such as, description, feelings, evaluation, analysis, conclusion, and action plan. This is easy to understand and aids with sensible judgement [3]. Another approach to reflection was work done by Schön [11] which differentiates between reflection-in-action and reflection-on-action. Reflection-in-action is an efficient method as it allows one to react and change an event while in action. On the other hand, Reflection-on-action requires deeper thought as it encourages one to consider causes and options, which should be informed by a wider network of understanding from research [2]. The KTP associate followed reflection approach close to the concept of Gibb's model. The thinking process was mostly goal directed throughout the KTP project's progress till its implementation and completion stages. The associate always thought first before any action and formulated the chunks of short-term goals in actions to obtain the long-term achievement. This provided the associate a kind of happy feelings when the long-term goal was achieved. The associate always reflected on the approach and activities of the accomplished task so that the next assignments or goals could be done better through important LMC meetings and the face-to-face meetings with different stakeholders of the project.

5 Recommendations to Make a KTP Project Successful

To make a project success, it is wise to visualize the end-result and formulate the activities to reach that goal. The overall activities could be divided into small chunks of small reachable goals. Efforts should be given as much as possible to complete those small chunks of activities as complete as possible so that the requirements for revisions or recapitulation could be avoided. This is important because people forget the things easily and sometimes systematic note taking might be helpful.

Excellent communication skills and collaborative team efforts are very important for the project's smooth progress. The time management is also important.

Gaining the confidence from the stakeholders as quick as possible is also paramount for the projects successful end. If the stakeholders do not have enough faith on the PM, then it would be very difficult for him to progress effectively.

PM should also know the limitations as one might not be equipped with all the necessary know-how. The team effort becomes very important at this stage of the project because people are always there to help but whom to approach and how to approach are the key. In this internet era, there are forums and also plenty of resources where the solutions would be available, only perseverance to search for those solutions are important.

Some type of belongingness to the project is very important as PM is given importance throughout the project period. To value that importance and honor, the PM should own the project so that it moves to a success story.

6 Conclusions

In this paper, a typical project management role on behalf of a project manager is discussed. How PM face the different events of the project stages and continue to give the best effort taking maximum help from all the supporting members of the project team is also described so that the other PMs doing similar type of projects for the SMEs can learn some lessons.

References

1. Beck, K., et al.: Manifesto for Agile Software Development (2001). http://agilemanifesto.org
2. Cambridge International Education: Getting Started with Reflective Practice (2020). https://www.cambridge-community.org.uk/professional-development/gswrp/index.html. Accessed 16 Dec 2020
3. EPM Expert Program Management (2019) Gibbs' Reflective Cycle (2016). https://expertprogrammanagement.com/2019/05/gibbs-reflective-cycle/. Accessed 16 Dec 2020
4. Gibbs, G.: Learning by Doing. Oxford Brookes University (1988)
5. Kolb, D.: Experiential Learning. Prentice Hall, Englewood Cliffs (1984)
6. Lazar, J., Feng, J., Hochheiser, H.: Research Methods in Human-Computer Interaction. Morgan Kauffman (2017)
7. Mohammadjafari, M., Ahmed, S., Dawal, S.Z.M., Zay, H.: The importance of project management in small-and medium-sized enterprises (SMEs) for the development of new products through E-collaboration. Afr. J. Bus. Manage. **5**(30), 11844–11855 (2011). https://doi.org/10.5897/AJBM10.1265
8. Murphy, A., Ledwith, A.: Project management tools and techniques in high-technology SMEs. Manage. Res. News **30**(2), 153–166 (2007). https://doi.org/10.1108/01409170710722973
9. Ogunleye, O.A.: Knowledge transfer partnership: a successful academic-industry relationship in UK with a focus on SMEs. In: 2007 IEEE International Engineering Management Conference, pp. 241–248 (2007). https://doi.org/10.1109/IEMC.2007.5235022
10. Osterwalder, A., Pigneur, Y.: Business Model Generation: A Handbook for Visionaries, Game Changers, and Challengers. Wiley (2011)
11. Schön, D.: The Reflective Practitioner: How Professionals Think in Action. Basic Books, New York (1983)
12. UK Government: Small and Medium-sized Enterprise (SME) Statistics for the UK and Regions 2008 and 2009, Department of Business Innovation and Skills (2010). https://data.gov.uk/dataset/dbaa6871-9750-4b1f-93fa-89c2683f9b73/small-and-medium-sized-enterprise-statistics-sme-for-the-uk-and-regions-2008-and-2009
13. White, D., Fortune, J.: Current Practice in Project Management - an empirical study. Int. J. Project Manage. **20**(1), 1–11 (2002). https://doi.org/10.1016/S0263-7863(00)00029-6

Comparative Analysis of Highly Ranked BIS Degree Programs

Ildikó Szabó(✉) and Gábor Neusch

Corvinus University of Budapest, Budapest, Hungary
ildiko.szabo2@uni-corvinus.hu

Abstract. Student centeredness in curriculum development has pedagogical benefits. Learning outcomes describe what competences will be possessed by students after graduating. Degree programs as well as labour market needs may be compared based on such competences. Developing or redesigning Business Information Systems (BIS) degree programs may utilize offers of competitor institutions as best practices. This was the case at a Hungarian university where the process of reforming the BIS program to meet changing market requirements included a review of peers. This paper presents both competitor and labour market analysis to create a baseline how programs offered by peer institutions ranked on Times Higher Education World University Rankings perform.

Keywords: Business Information Systems · Labour market analysis · Competitor analysis · Student-centred education

1 Introduction

A curriculum development process is either teacher-based that focuses on what teachers know or student-based that supports them during the lifetime of their learning process. As part of a program design, competences including practical skills, knowledge, attitude, autonomy and responsibility are determined to describe what students will be capable of doing and knowing after graduation. Competency based frameworks have been used to describe learning outcomes for some time. However, these frameworks – such as National Qualification Systems [7], European Qualification Framework [4], or Tuning project [1] – are not evolving dynamically so they can only be used as a static input in curriculum development. On the other hand, employers' needs and competitors' behaviour are driving forces that require more constant tracking.

A review of the Business Information Systems (BIS) degree program is under way at our university. During this exercise above frameworks were defined as non-compulsory reference points in the curriculum development and evaluation process. This paper presents how external parties were investigated during the preparation of the new BIS program. A research framework derived from the work of Boehler et al. [2] and supported by text mining and OLAP analysis to highlight labour market needs as well as best practices applied by other institutions. The framework and data collection are presented in Sect. 2. Degree programs offering by 30 universities ranked by Times

W. Abramowicz et al. (Eds.): BIS 2021 Workshops, LNBIP 444, pp. 86–92, 2022.
https://doi.org/10.1007/978-3-031-04216-4_9

Higher Education World University Rankings were examined along 18 dimensions and key findings are summarized in Sect. 3. Conclusions are drawn in Sect. 5.

2 Research Method and Sample Selection

Boehler et al. [2] conducted comparative analysis of practices in Information System curricula development by 89 institutions accredited by AACSB[1]. The analysis was executed twice: 2013 and 2018. Literature review and AACSB accreditation process defined their research questions which were used as an analytical framework. Market demand for IS program graduates and compliance of MIS programs to the 2010 AIS Guidelines were investigated by two questions, while the third one dealt with how peer institutions can adapt to the challenges and opportunities derived from the turbulent environment and were composed of six parts:

a. Are certain IT/IS subjects emphasized?
b. Is there an emphasis on technical skills or strategic management?
c. Does the curriculum emphasis differ by geographic region?
d. Is there an emphasis on practical experience?
e. Do existing IS curricula emphasize security?
f. Is there an emphasis on data analytics?" [2, p. 234].

Our research followed this path to a certain extent. Hungarian labour market needs were discovered by surveys. Alumni and potential employers were questioned about required knowledge and skills, future competencies, relevant job roles or positions [3]. Compliance with 2010 AIS Guidelines was irrelevant in our case. The same research questions (3a, 3c and 3d) were also examined by us.

Data were collected from 40 degree programs provided by 36 institutions from 4 continents (Europe, Asia, Australia and America). This was a manual collection by our team. Erasmus mobility partners and institutions from the Times Higher Education World University Rankings list were selected. 6 institutions were not ranked on this list, so they were excluded from this analysis. Degree programs not having enough data or double degrees were also excluded from data collection leaving 26 degree programs. Data along 18 characteristics[2] were gathered by a team of 12 people. Credits and semesters were measured on numeric scale. The other variables were textual. At the end, all universities were clearly assigned to its program.

Enrolment data were collected from all universities and added to this dataset. Unfortunately, these data were published in different years – on a scale from 2009 to 2021 - per university. Assuming if these numbers can slightly measure the size of the

[1] Association of Advanced Collegiate Schools of Business.

[2] Institution ID, Institution name, Degree program, Number of semesters, Total credits, Language of Education, Description of degree program, Aim of the program, Possible occupations of graduates, Specializations, Corporate relations, Pedagogical methodologies applied, Credits for Foundation (methodology - math, statistics, etc.) Subjects, Credits for Theoretical Economics Subjects, Credits for Business and Management Subjects, Credits for Generic IT Subjects, Credits for Specific BIS / MIS Subjects and Internship.

universities, Fig. 1 presents how these universities are distributed in size, regionally and on the ranking. They are concentrated on Europe and mainly on UK, but three are from Asia, two from Australia and one from United States. 80% of them have at least 10,000 enrolled students. 80% of these institutes are ranked in the first 200 universities.

Fig. 1. Regional distribution of universities (Source: Authors)

3 Analysing Competitors' Degree Programs

This section focuses on analyzing peer institutions and their degree programs to collect best practices to build them into our curriculum. It focuses on investigating questions about IT/IS subjects (3a) and examining geographical distribution of curricula (3c).

The names of these programs revealed that 4 programs focus on Information Management, 8 programs are about Information Systems (actually 1 is from both fields), 5 programs have focus on Information Technology, 5 from analytics or data science side and 2 from business administration domain. There is no correlation between the number of semesters and the total credits. Practically, Anglo-Saxon institutes prefer providing programs with high total credits (see in Fig. 2). Their courses usually have 15–20 credits. Interesting result was that project work was a quite popular methodology applied by these institutions. Companies are involved in not just hiring students as interns, but also in carrying out these courses. Practice-oriented knowledge transfer is realized in this way (Question 3c).

Number of semesters

Country	Abbrev	Insitution (University)	Degree program (Degree ..	Number of semesters	
Australia	USYD	University of Sydney	Bachelor of Advanced Co..	10	240
	UO	Queensland University of ..	Bachelor of Information T..	6	288
Austria	TUW	TU Wien	UE 033 526 - Bachelor's Pr..	6	180
China	CityU	City University of Hong Ko..	Bachelor of Business Adm..	8	180
	HKUST	The Hong Kong University	BBA in Information Syste..	8	288
	Tsinghua SEM	Tsinghua University	Information Management..	8	360
Denmark	CBS	Copenhagen Business Sch..	BSc in Business Administr..	6	180
Hungary	UD	University of Debrecen	Business Informatics	7	210
Netherlands	VU	Vrije Universiteit Amster..	BSc in Business Analytics	6	180
	UvA	University of Amsterdam	Business Analytics	6	180
	UM	Maastricht University	Business Engineering	6	
	TU (Tilburg)	Tilburg University	Joint BSc in Data Science	6	180
Slovenia	UM (Maribor)	Maribor University	E-business	6	180
Switzerland	UZH	University of Zurich	Information Systems	6	120
Thailand	TU	Thammasat University Ba..	BSc in Data Science and In..	8	126
United Kingdom	UWS	University of the West of ..	BSc Business Technology	8	480
	Edin	University of Edinburgh	BSc Computer Science an..	8	440
	Brun	Brunel University	Business Computing (eBu..	6	360
	UKC	University of Kent	Business Information Tec..	6-8	180
	BU	Bournemouth University	Business Information Tec..	4-6	360
	UOA	University of Aberdeen	Business Management an..	8	340
	UWA	Aberystwyth University	Data Science BSc	6	350
	UCL	University College London	Information Management..	6	360
	NorthU	Northumbria University a..	Information Technology ..	6-8	360
	UOM	The University of Manche..	Information Technology ..	8	360
United States	UCI	University of California, Ir..	BS in Business Informatio..	8	

Fig. 2. Overview of degree programs (Source: Authors)

A dashboard (Fig. 3.) was created to investigate whether curricula emphasis differ by geographic region (Question 3c) and to what degree IT/IS subjects are emphasized. Missing data were excluded from the analysis. The proportions of credits of each subject (Foundation, Business Management and Information Systems) were calculated using feature engineering in the Tableau software.

Institutions from UK are highlighted in Fig. 3. Statistical and other mathematical subjects are minor part of programs (in max 10%) at Anglo-Saxon programs except Business Computing at Brunel University. Actually, foundation subjects are part of the programs in only the 10% of 16 cases. Proportion of business management credits varies considerably among degree programs. BSc in Business Administration of Copenhagen Business School stands out, meanwhile BSc Business Technology has a less focus on this subject area. Maximum 20% of their program is related to business subjects in UK institutions. The Data Science (DS) program of the Tilburg University deals with MIS subjects in more detail than IS subjects. The Business Information Technology (BIT) program of the University of Kent goes the opposite way. It is worth investigating that the reason behind this phenomenon. Is Data science approached from practical point of view? Why are MIS trends not emphasized in BIT curricula?

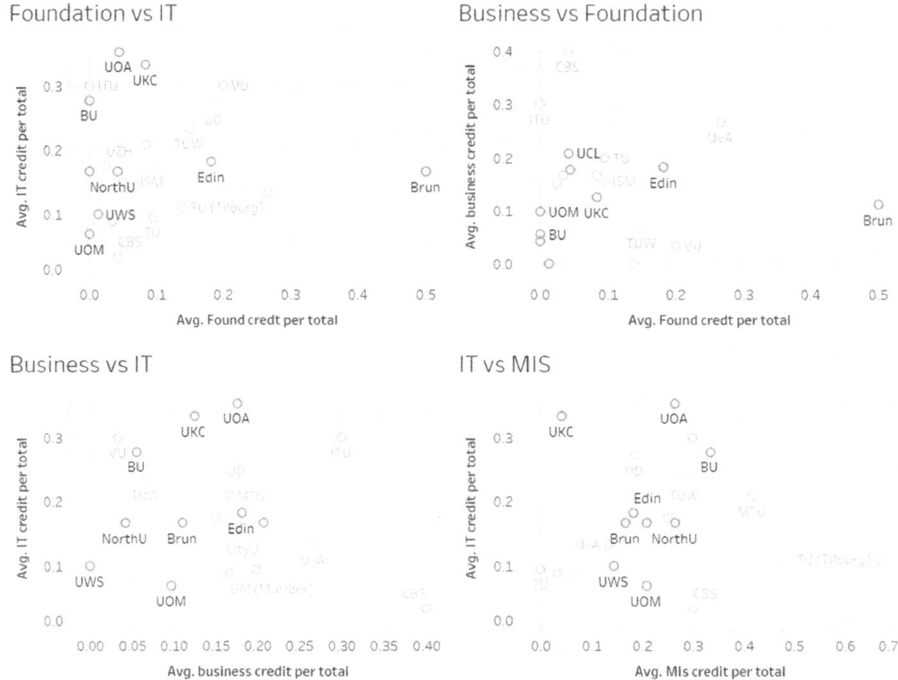

Fig. 3. Analysis of UK universities (Source: Authors)

In other cases, Anglo-Saxon degree programs provide IT and MIS curricula in the same proportion except two (BSc Business Technology from the University of the West Scotland and Information Technology from the University of Manchester).

We have seen that degree programs offered from Anglo-Saxon area have special features such as not emphasizing foundation. The Anglo-Saxon programmes show a quasi-uniform behaviour compared to the other programmes, so we have focused on this region for the analysis of practice orientation (Question 3d). We looked at what needs are emerging here and whether and how each training programme is trying to meet them. A labour market analysis in 2019 in pre Covid-19 period was executed to see what skills required by the UK labour market that should be provided by universities. This dataset was selected due to fact its data are not influenced by the pandemic.

4 UK Labor Market Analysis

Pitukhin et al. found that professional requirements for a candidates appear in job listings [6]. Wowczko also highlights that along with the emergence of online recruiting a huge amount of potentially useful information is available for researchers about competencies [8]. Researchers of CareerBuilder.com have found upon examining a sample taken from half billion job listings is English that ninety percent of them contain expressions that can be accepted as competencies [9]. Nasir et al. [5] mention that job listings processed by them emphasized vacancy information mainly than required skills.

Investigating competency needs in the UK labour market for the IT sector was carried out on dataset from March 2019. This period was not affected by the pandemic, so it was useful to compare positions from the labour market with positions for which students are prepared. Job ads were selected to find these positions. Indeed.uk as the No.1 job portal was chosen to provide them. Positions, their locations and descriptions were collected by a Python crawler. After data cleansing, text mining process was executed to highlight the most needed positions and skills.

The positions were varied among different categories. Development positions (software engineer, dev ops engineer, web applications developer, senior integration developer), consultants (risk assurance manager, senior business consultant, O365 consultant, business development consultant) and data analyst jobs (customer data analyst, bi support analyst) were needed especially among others. BIS programs train to fill these positions. The skills required primarily were written and oral communication, analytical and problem-solving skills, time management, the ability to work independently and in teamwork. Expert advice, project management and knowledge in technology were searched especially as knowledge.

Comparing these needs with career options claimed by universities it was found that 11 degree programs (4 in UK) train analysts, 7 (2 in UK) prepare for consultant positions, 5 (3 in UK) train web developer and 2 prepares for software engineering jobs. This section was about investigating practical side of these programs (Question 3d.). It was noted that internship and project works are usually to get practical experiences required by each position aimed by the programs.

5 Conclusion

This paper looked out at what best practices are at external parties (competitor higher education institutes or employers). Three research questions from the analytical framework of Boehler et al. [2] were used in our analysis. It was revealed that the proportion of providing IT and MIS courses or subjects are quite the same in UK (Question 3a). Geographical analysis pinpointed that foundation subjects are underrepresented in degree programs and courses with high number of credits are general in Anglo-Saxon areas (Question 3c). Practice-oriented teaching is realized by project work and internship. Companies are involved in teaching process. Examined institutes fit to labor market needs because required positions (consultant, software engineer, web developer) from pre Covid-19 period are mentioned as job career at them. These are practical positions. (Question 3b).

References

1. Adam, J.-M., Mercerone, A., Milosz, M..: Using the tuning methodology in the TEMPUS Eramis project. In: Proceedings of the International Scientific and Education Congress, Science of the Future, pp. 22–23 (2010)
2. Boehler, J.A., Larson, B., Peachey, T.A., Shehane, R.F.: Evaluation of information systems curricula. J. Inf. Syst. Educ. **31**(3), 232–243 (2020)
3. Csáki, C., Szabó, I., Szabó, Z., Csillik, O., Gábor, A.: Reconsidering the challenges of BIS education in light of the COVID pandemic. In: Abramowicz, W., et al. (eds.) Proceedings of the Business Information Systems Conference Workshops, BIS 2021. LNBIP, Hannover, Germany, 14 June 2021, vol. 444, pp. 61–72. Springer, Heidelberg (2022). https://doi.org/10.1007/978-3-031-04216-4_1
4. Méhaut, P., Winch, C.: The European qualification framework: skills, competences or knowledge? Eur. Educ. Res. J. **11**(3), 369–381 (2012). https://doi.org/10.2304/eerj.2012.11.3.369
5. Nasir, S., Yaacob, W.F.W., Aziz, W.A.H.W.: Analysing online vacancy and skills demand using text mining. J. Phys. Conf. Ser. **1496**, 012011 (2020). https://doi.org/10.1088/1742-6596/1496/1/012011
6. Pitukhin, E., Varfolomeyev, A., Tulaeva, A.: Job advertisements analysis for curricula management: the competency approach. In: 9th Annual International Conference of Education, Research and Innovation Proceedings, Seville, pp. 2026–2035 (2016)
7. Ure, O.B.: Learning outcomes between learner centredness and institutionalisation of qualification frameworks. Policy Fut. Educ. **17**(2), 172–188 (2019). https://doi.org/10.1177/1478210318774689
8. Wowczko, I.A.: Skills and vacancy analysis with data mining techniques. Informatics **2**(4), 31–49 (2015). https://doi.org/10.3390/informatics2040031
9. Zhao, M., Javed, F., Jacob, F., McNair, M.: SKILL: a system for skill identification and normalization. In: Proceedings of the 29th AAAI Conference on Artificial Intelligence, Austin, Texas, pp. 4012–4017. AAAI Press (2015)

DigBD Workshop

DigBD 2021 Workshop Chairs' Message

Evoked by recent trends, such as big data, data science or cloud computing, the digitization of organizations necessitates new approaches for the planning and engineering of IS in today's data-driven world. In many cases, sophisticated approaches are required to overcome the data-intensive nature of such endeavors. At this point, established technologies, as they have been used for many years, are reaching their limits. In the flux of digitization novel and innovative technologies and concepts, such as automation, cloudification, continuous integration, micro services, model-driven engineering, or decision support systems appear to be promising "enablers" to meet the current demands. To overcome their realization shortcomings and to create competitive advantages in data-intensive scenarios, a plethora of facets must be handled.

Hence, the first installment of the Workshop on Digitization in the Flux of Big Data Scenarios was created to allow researchers from diverse backgrounds to present and discuss their work, facilitating the understanding of the domain's underlying issues and challenges, as well as to present potential approaches on how to bridge the identified gaps.

In this workshop, we invited a variety of research approaches including, but not limited to, theoretical articles, reviews, and use case studies that are related to the generation of competitive advantages through big data as well as the use of innovative technologies for planning, engineering, deploying, testing, benchmarking, and operating data-intensive systems, with the aim of facilitating their dissemination and increasing their effectiveness.

The submitted papers were each assessed by three or four reviewers, with four papers being accepted. Those were authored by researchers from four countries and three continents. This regional diversity was also present in the Program Committee, with its members representing institutions from seven countries and three continents. We want to thank all of them for their support in helping us realize this workshop. Furthermore, we would like to thank all the authors that submitted their work as well as everyone who engaged in the discussions in the course of the session.

We would also like to express our gratitude to everyone who made the BIS 2021 conference, as well as the DigBD 2021 workshop, possible. While we, as probably everyone else, would have preferred to have the event in-person instead of online, the organizers still did an exceptional job in assuring an enjoyable experience. We also want to use the opportunity to especially thank Elżbieta Lewańska who was always helpful and committed whenever we had an issue concerning the organizational aspects of the workshop.

Danielc Staegemann
Matthias Volk
Naoum Jamous

Organization

Chairs

Daniel Staegemann	Otto-von-Guericke University Magdeburg, Germany
Matthias Volk	Otto-von-Guericke University Magdeburg, Germany
Naoum Jamous	Otto-von-Guericke University Magdeburg, Germany

Program Committee

Mohammad Abdallah	Al-Zaytoonah University of Jordan, Jordan
Darlan Arruda	SSENSE, Canada
Samuel Fosso Wamba	Toulouse Business School, France
Bernhard Heiden	FH Kärnten, Austria
Johannes Hintsch	T-Systems International GmbH, Germany
Matthias Pohl	Otto-von-Guericke University Magdeburg, Germany
Hosseinzadeh Hariri Reihaneh	Oakland University, USA
Zaher Al-Sai	Universiti Sains Malaysia, Malaysia
Marcus Thiel	Otto-von-Guericke University Magdeburg, Germany

Multi-agent System for Weather Forecasting in India

A. G. Sreedevi[1], S. Palaniappan[1], P. Shankar[1], and Vijayan Sugumaran[2(✉)]

[1] Amrita School of Engineering, Amrita Vishwa Vidyapeetham, Chennai 601103, India
{ag_sreedevi,p_shankar}@ch.amrita.edu, ch.en.u4cse19023@ch.students.amrita.edu
[2] Oakland University, Rochester, MI 48309, USA
sugumara@oakland.edu

Abstract. Accurate weather prediction is a challenging task. It involves large amount of data and computation, which vary dynamically. This paper discusses a novel idea of using Multi Agent System (MAS) for weather forecasting, particularly in the Indian context. The proposed approach incorporates a deep neural network model within MAS with a hybrid ANN algorithm to recognize the static and dynamic weather conditions. The approach uses ensemble prediction to account for indeterminism in weather conditions. Predictions and alerts given by MAS can help the government and local authorities to plan precautions in a timely manner. The paper discusses the implementation challenges and advantages of a MAS Model compared to Numerical Weather Prediction method.

Keywords: Artificial Neural Network · Multi Agent Systems · Numerical Weather Prediction · Weather forecasting

1 Introduction

Weather forecasting is considered as one of the challenging operational tasks carried out by meteorological stations throughout the globe [1]. The major attributes that affect weather conditions of a particular region includes: humidity, wind speed, atmospheric pressure, rain and temperature. Accurate weather prediction in India is challenging because of climate change and global warming. Traditionally, meteorologists use Numerical Weather Prediction (NWP) Models for weather prediction. This classical approach attempts to model the fluid and thermal dynamic systems for grid-point time series prediction based on boundary meteorological data. NWP solves a large system of nonlinear mathematical equations with the help of super computer algorithms to provide better forecasts.

Use of Multi Agent Systems (MAS) in weather prediction is a novel idea as it would reduce the human intervention and human errors. The multi agent systems have interconnected agents that receive input from the environment and make their own decisions. Indeterminism from the environment can be handled by the multi agent system. Even if one agent fails, the other agents are capable of tolerating it. It is also

W. Abramowicz et al. (Eds.): BIS 2021 Workshops, LNBIP 444, pp. 97–102, 2022.
https://doi.org/10.1007/978-3-031-04216-4_10

easier to add new agents to the system. Using MAS in weather prediction makes the network scalable, robust and cost effective.

Indian weather prediction still relies on numerical methods which is intrinsically associated with errors due to large deterministic chaos. A new architecture involving MAS for weather forecasts using grid model will be more precise and faster. This fast prediction of weather with minimized error will help to deliver precise early alerts and warnings to people in the wake of major disasters. This research in progress paper uses MAS for data collection, prediction and monitoring the environment. Each agent in this MAS system has specific roles to play. Since weather data is collected using a large number of sensors scattered on a vast area through sensor networks, data acquisition, ensuring integrity, data storage and management are complex tasks. Current big data tools and technologies are required to handle these tasks efficiently.

The remainder of this paper is structured as follows. The next section discusses the related works, followed by the Indian weather prediction method. Subsequent section explains the proposed model for better prediction and forecast. Then discussion of the expected outcomes of the model implementation and final conclusion are provided.

2 Related Works

The use of numerical weather prediction techniques was started in early 1990s by Bjerknes and Richardson. Early approaches suffered from the initial value problem of statistics based on linear regression. Later, it was improved to include Multiple regression, Map Reduce Framework and Support Vector Machine (SVM). Initial analysis was over the statistics. However, ANN technique has been drawing considerable attention of researchers, as it can handle the complex non-linearity problems better than the conventional statistical techniques. Deep learning-based weather prediction models use long short-term memory (LSTM), temporal convolutional networks (TCN) [2], back-propagation networks (BPN) [3], and Naïve Bayesian networks (NBN) [4] for prediction. However, these approaches have some limitations. For example, LSTM is unable to handle the dynamic sudden change of initial parameters of weather prediction where as TCN can capture only two levels of information. The BPN and NBN methods could achieve only 65–70% accuracy in weather prediction. In [5], an intelligent agent-based weather prediction system, iJADE Weather MAN is used for multi station weather prediction. The model used MAS along with humans for data collection and was trained to predict only the rainfall.

Using MAS for weather prediction can deal with the non-linearity and indeterminism in the data. This paper proposes a MAS for data collection, prediction and monitoring. The hybrid ANN algorithm in MAS considers the static and dynamic properties of the environment, and will be capable of predicting the weather.

3 Numerical Weather Prediction in India

Many of today's weather forecasting systems depend on meteorologists' observations and interpretation based on traditional concepts and models. India, uses Weather Research and Forecasting (WRF) model as an operational model for making its weather forecasts. The first step is the Initial guess and data assimilation. NWP being an initial value problem, the more accurate the estimation of the Initial conditions, the better is the quality of forecasts. The initial guess in India, is made using the observation data gathered from the earth system, which is the main observation platform. It includes Automated Weather Station, GPS Enabled Balloons measuring wind, temperature, pressure, humidity, rainfall and solar radiation, and Satellite Observations. This observed data along with predicted data is used to calculate errors. Based on these calculated values the prediction is made. The process is termed as data assimilation. These observations are used to solve a set of equations governing conservation of momentum (Newton's 2nd law), conservation of mass, conservation of energy, and relationship among pressure (p), volume (V), and temperature (T). These equations are expressed in NWP as a system of nonlinear partial differential equations.

Worldwide, there exists two types of NWP models: Grid Point and Spectral Models. Grid point models consider data at discrete fixed grid points. Spectral models rep-resent data by using continuous wave functions. Since the data involved is non-linear, probabilistic forecasting, the uncertainty of a forecast, is used in many forecast stations for better prediction. In physical NWP this is achieved by running an ensemble of forecasts using different initial conditions and different stochastic model [6].

4 Multi-agent System for Indian Weather Forecasting

As discussed in previous sections, NWP method involves human effort, formulation, complex calculations with increased chance of human error. Here, we propose a new system model using MAS and ANN for Indian Weather Forecasting which is scalable, robust and flexible to accommodate sudden climate changes. This will also reduce the human error in the predictions. The study of Indian weather showed two main characteristics. Firstly, the climate may remain static for a period of time without much change in initial conditions and weather. We call it as static weather. Secondly, the climate changes drastically with time, which is termed as dynamic weather condition. The probability of NWP to be true in dynamic weather situation is less.

We propose a novel model for weather prediction taking into account the static and dynamic weather conditions. This proposed model shown in Fig. 1(a) is applied over the grid model of NWP. It deploys MAS in two layers: First and the Second layer. MAS plays different roles in each of these layers. In the First Layer, each grid of 13 km $\times$ 13 km consists of one or more of lower-level MAS to collect and process the local weather data. The Architecture of this multi-agent system is shown in Fig. 1(b). First layer agents include: Acquisition, Prediction, Notification and Archive Agents. Acquisition agents collect environmental data from the sensor networks deployed in the region of interest. The collected data are fed to the proposed prediction layer of MAS. Detailed architecture of the prediction layer is shown in Fig. 2. The Prediction Layer consists of several modules, which are briefly described below.

Module 1 is the preparatory module which calculates the initial conditions based on collected parameters. This module uses a set of equations governing conservation of momentum (Newton's 2nd law) [7], conservation of mass, conservation of energy, and relationship among p, V, and T to calculate the initial weather conditions.

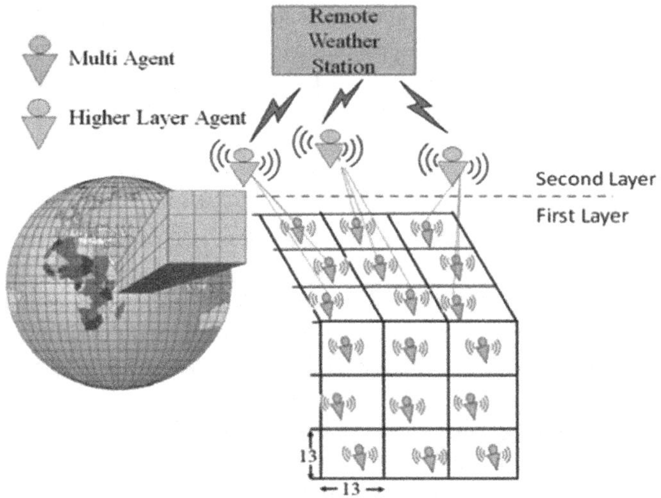

Fig. 1a. System model for weather forecasting

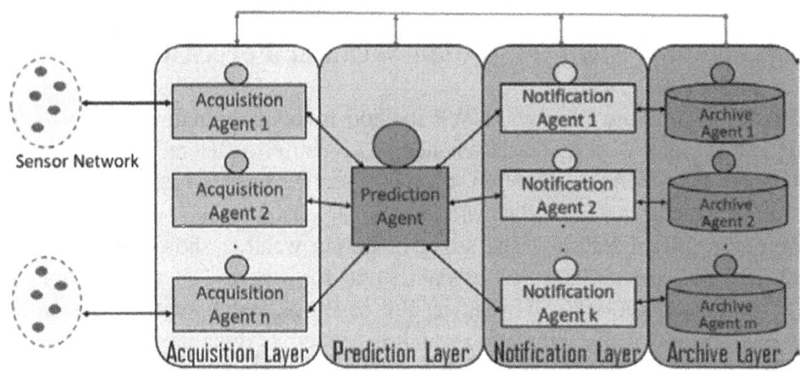

Fig. 1b. Architecture of multi-agent system.

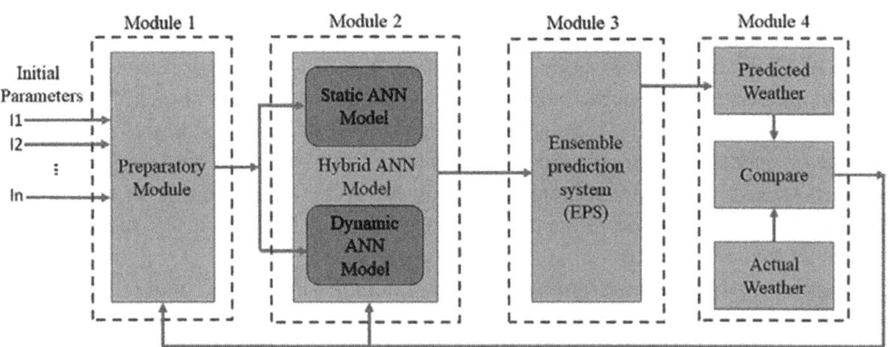

Fig. 2. Proposed MAS prediction layer for weather forecasting

Module 2 uses hybrid ANN models to predict the weather based on this initial condition. It analyzes the non-linear time variant data based on static and dynamism in it. If the condition is found to be static, ANN model uses static algorithm to predict the weather. Any drastic variation in the parameters makes ANN use the dynamic model for prediction. The weather output is fed to the ensemble prediction system (EPS) (Module 3) which deals with the indeterminism in the weather. It adjusts the predicted values based on the achieved uncertainty. The predicted weather is then compared with actual weather and fed back to preparatory module and Hybrid ANN to adjust the model and error for upcoming predictions. The predicted weather information is collected by the notification agent to give timely alerts and warnings. The Archive Agent saves all the data related to predictions in a repository for later use.

MAS in the second layer of Fig. 1(a) acts as higher-level agents which collect data from first level agents over many regions, process them to predict the weather across a larger area. These agents interact directly with remote weather stations to predict cyclones and monsoons.

4.1 Implementation

The intelligent agents capable of processing, predicting and monitoring the weather is designed using the JADE platform. For better quality and efficiency, it is assumed that these agents have efficient communication network available. The system is designed for a region of 39 km × 39 km with the dimension of grid model as 13 km × 13 km (Fig. 1a and 1b). The First layer agents are capable of communicating to nearby agents and also to the Second layer agents within its communication range. The agents in the Second layer are implemented to interact with the first layer agents and also the remote weather stations. The prediction agent is developed with the modules shown in Fig. 2 to prepare, model, and predict the weather. The initial parameters, namely, temperature, humidity, pressure, solar radiation, and wind constitute the dataset. The available set of data is divided into two categories (Training Set and Testing Set) and fed to the JADE MAS model. The training dataset is fed to the preparatory module and then to the proposed Hybrid ANN Model to train the model for future predictions. Hidden layers are adjusted to meet the requirements. It is then fed to the ensemble system for the actual weather prediction. Testing set is used to check the MAS model prediction accuracy. Testing model calculates the error between the actual and predicted values and is fed back to ANN and preparatory module for better future predictions. The model learning is performed through the adjustment of errors that are regularly backpropagated all the way through to the initial time for forecasting prediction.

5 Discussion

The proposed MAS for weather forecasting is under implementation. The preparatory module shown in Fig. 1a and 1b has been completed and the other modules comprising the multi-agent system are currently being implemented. We are also refining the static and dynamic ANN algorithm to improve the forecast accuracy. We expect to produce a working proof-of-concept prototype in the near future and would be able to present it during the workshop. The model will be validated with weather datasets from India for prediction accuracy, convergence time and processing efficiency. It will also be compared to other approaches to demonstrate its efficacy, scalability and robustness.

Weather specific challenges, which prevail in different regions make this model complex to handle. Firstly, the atmosphere being 3-dimensional, the dynamics and spacings of grid tend to change in a vertical direction. Secondly, the training data available is limited. 350,000 samples constitute 40 years of hourly data. These samples correlate in time. Assuming new weather prevailing every day, the sample count will get reduced to 15,000. With lack of empirical support, it is difficult to estimate. The Technical challenges include the nonlinearity of the data we deal with. Data for a single variable with 10 levels may take up to 30 GB of data. For a network with several variables or even at higher resolution, the data might not fit into the CPU RAM.

6 Conclusion

We have presented a MAS model for gathering real-time weather information for processing, forecasting, alerting and monitoring. The novelty and the proposed approach have been highlighted in the paper. The overall system for implementation and the MAS architecture has been discussed. The implementation of the prototype is currently underway using JADE where the model is trained and validated for accuracy of the prediction. The challenges of implementation include technical and weather specific constraints. We have implemented first level MAS with module 1 and are in the process of refining the hybrid ANN algorithm which accounts for the static and dynamic nature of weather conditions. Weather forecasting and monitoring using MAS model can facilitate effective resource management and disaster preparedness.

References

1. Hayati, M., Mohebi, Z.: Application of artificial neural networks for temperature forecasting. World Acad. Sci. Eng. Tech. **28**(2), 275–279 (2007)
2. Hewage, P., Trovati, M., Pereira, E., Behera, A.: Deep learning-based effective fine-grained weather forecasting model. Pattern Anal. Appl. **24**(1), 343–366 (2020)
3. Chung, C.Y.C., Kumar, V.R.: Knowledge acquisition using a neural network for a weather forecasting knowledge-based system. Neural Comput. Appl. **1**(3), 215–223 (1993)
4. Li, B., Liu, J., Dai, H.: Forecasting from low quality data with applications in weather forecasting. Int. J. Comput. Inf. **22**(3), 351–357 (1998)
5. Lee, R., Liu, J.: iJADE weather MAN: a weather forecasting system using intelligent multiagent-based fuzzy neuro network. IEEE Trans. Syst. Man Cybern. Part C (Appl. Rev.) **34**(3), 369–377 (2004)
6. Palmer, T.: The ECMWF ensemble prediction system: Looking back (more than) 25 years and projecting forward 25 years. Q. J. R. Meteorol. Soc. **145**, 12–24 (2019)
7. Holmstrom, M., Liu, D., Vo, C.: Machine Learning Applied to Weather Forecasting, Meteorological Applications. Wiley (2016)

Towards a Data Collection Quality Model for Big Data Applications

Mohammad Abdallah(✉), Alaa Hammad, and Wael AlZyadat

Faculty of Science and Information Technology, Al-Zaytoonah University of Jordan, Amman, Jordan
{m.abdallah, wael.alzyadat}@zuj.edu.jo

Abstract. Big Data and its uses are widely used in many applications and fields; artificial information, medical care, business, and much more. Big Data sources are widely distributed and diverse. Therefore, it is essential to guarantee that the data collected and processed is of the highest quality, to deal with this large volume of data from different sources with caution and attention. Consequently, the quality of Big Data must be fulfilled starting from the beginning; data collection. This paper provides a viewpoint on the key Big Data collection Quality Factors that need to be considered every time the data are captured, generated, or created. This study proposes a quality model that can help create and measure data collection methods and techniques. However, the quality model is still introductory and needs to be further investigated.

Keywords: Big Data · Data collection · Data quality · Data collection quality · Data collection quality model

1 Introduction

Collecting a huge amount of data about the business, health, or anything else. Then analyze this huge amount of data will give a better understanding of the field and will help in taking the right decisions and predictions. Therefore, Big Data plays a significant role in life.

Big Data (BD) is a technological word for the large quantities of heterogeneous data that are quickly generated and distributed, and for which traditional methods for processing, analyzing, retrieving, storing, and visualizing such huge sets of data are now unsuitable and insufficient. This can be seen in a variety of fields, including sensor-generated data, social networking, and digital media uploading and downloading. Big Data can be structured, semi-structured, or unstructured.

The data in any system or application is a vital part. It can be collected from different sources or created in many ways. So, the data must be correct, accurate, and complete to get the maximum benefits out of it [1].

Data collection is the procedure of collecting, measuring, and analyzing accurate insights for research using standard validated techniques. So, Data collection quality becomes an essential part of any data processor data management. In the Big Data period, data quality considerations and problems have already been addressed which

W. Abramowicz et al. (Eds.): BIS 2021 Workshops, LNBIP 444, pp. 103–108, 2022.
https://doi.org/10.1007/978-3-031-04216-4_11

need to be evaluated. The Big Data quality variables also include the factors focused on the data itself and the management of Big Data and customer needs [2, 3].

However, the data source or data collection quality has not been questioned, discussed, and measured deeply in literature. In this research, we have proposed a quality model to measure the data collection process.

In Sect. 2, some related work about data collection quality factors and models are explored. In Sect. 3, a data collection quality models was proposed and finally in Sect. 4 the conclusions and future direction of the research are highlighted.

2 Related Works

The data is a set of qualitative or quantitative variables, it can be in many different forms and formats. Big Data is distinguished from any other data type by the 3V's dimensions; Volume, Velocity, and Variety [4]. The 3V's of Big Data was also the start of Big Data quality factors. Volk et al. [5] have summarized the quality issues that can be raised by the 3 V's of Big Data as: "*Handling huge volumes of data in different formats at high speeds while maintaining resiliency and data security, can be very challenging*".

Big Data and apps are a problem of the consistency of Big Data. Any program that uses it must ensure that data have high-quality requirements to deliver a good quality system. Especially the factors of quality that take into account Big Data are the same as for conventional sets of data. Furthermore, certain additional quality factors deal with many of the facts, such as data management and repair [6].

The consistency factors of data have been debated by most researchers in past studies. However, the data and Big Data have several common indicators of consistency and vary in quality and calculation applications [7].

Data collection is one of the processes that are essentials in any data sets or Big Data application. Big Data Accuracy, Completeness, Redundancy, Readability, Accessibility Consistency, Trusts are the primary data and Big Data Quality Factors. To fulfill the quality factor, every factor is connected to one or more quality parameters or criteria. However, these factors have discussed the data quality from all perspectives; the Data perspective, Management perspective, Processing, and Service perspective [8].

The Big Data quality measurement is not only focused on data quality. Data Quality Management (DQM) is also a quality challenge that must be considered [9], which intersects with data collection in its aspects. DQM has five main aspects [10, 11]: People, Data Profiling, Defining Data Quality, Data Reporting, Data Repair.

Data creation may introduce a group of mistakes. This may be caused by human influence such as typos, misunderstanding, or misrepresentation [12]. As well as data collection tools and techniques [13].

To decrease their influence, several strategies have been applied. Those include procedures to reduce possible causes of mistakes, such as better instructions or the simplification of forms [12]. Heinrich et al. [14] have an application of rule sets and statistical analyses. There metric makes it possible to regard laws with a unique probability that are likely to be followed. The resulting metric values are likely to free the evaluated dataset from internal inconsistencies about the ambiguous rules and thus provide a consistent understanding.

Savosin et al. [15] raised issues that edit Big Data professional sources without authorization might cause them to shut down, although the information itself can be used freely. Ordinary users will, on the other hand, offer a great deal of information with a little pause in different fields of expertise. However, depending on the subject and external factors such as the precise usage of sensor instruments, the distribution of users in that region, etc., their efficiency and precision can vary.

Liu et al. [16] proposed a non-linear optimization programming model with resource management constraints by creating a data-quality Petri net to catch the mechanism by which the information system produces, spreads, and builds problems in the data quality.

As seen from the literature there is a lack of a general model that deals with data collection quality issues. The researchers only try to avoid or tolerate the data collection or creation quality issues. In the next section, a proposed quality model for data collection will be introduced. This model should help in preventing the data from being collected in the wrong way and can help to produce clean, reusable Big Data sets.

3 The Proposed Data Collection Quality Model

The process of data collection is one of the basic stages in improving and developing the quality of the data and the resulting information. The process of data collection is not limited to specific areas, but it is used in many sectors such as Technology, Health care, Engineering, and many others. Therefore, it is become an important step to make sure that the collected data is not only correct but also of good quality. So, this stage requires a lot of accuracy, time, effort, and sometimes cost but to obtain the highest benefit and quality of data, there is a set of proposed conditions for application in the data collection stage.

The proposed quality model for collecting data to be used in Big Data systems is accumulative different quality models for data collection for different purposes. Therefore, we hope this quality model can help any data collection applications that collect data for any reason. The proposed quality model has 7 quality factors, as shown in Table 1.

Reliability of the data source: the data must be correct, complete, and consistent, and coherent with the 4c's of data. A dependable source presents a well-researched and well-supported hypothesis, claim, or discussion based on strong evidence [17]. Reliability is a cornerstone in any quality model, since it is related to the correctness and dependability.

Trustworthy: The data providers will not give the right data if they do not trust the data recipient and they are not convinced about the purpose of data collection. Therefore, it is important to work to build trust between the data providers (the first part) and the data recipient (the second part). As a result of building trust; the effort, time, and cost will be reduced [18]. It also considers the data legitimate and the data provider must be aware of why the data is collected and how it will be used [19].

Table 1. Data collection quality factors

Quality factor	Quality dimensions	References
Reliability of the data source	Correctness, completeness, consistent, coherent, dependability	[17]
Trustworthy	Accuracy, legitimate, intrinsic	[18, 19]
Data suitability	relevance, fitting, usefulness	[20, 21]
Data preservation	Availability, validity, accessibility	[22]
Data integrity	Security, privacy	[23]
Rapid data collection	Timelines	[24]
Data reusability	Understandable, renewable and composable	[25–27]

Data Suitability: The data collected should be related to the purpose of the data use. Collecting data from unrelated sources will produce a huge amount of unfitting data, which increases the time, effort, and cost of data cleansing and analysis [20, 21]. If the data is irrelevant then for sure the decisions that are made depending on it will be faulty and misleading.

Data Preservation: The data should be ensured to be available, valid, and accessible in the long run. Preserving efforts should ensure that the content is accurate, secure, and accessible while maintaining its dignity, such as authentication, signing metadata for the protection, assigning representation records, and ensuring appropriate data structures and file formats [22]. However, the data provider and the data recipient should have an agreement about how long and where the data will be preserved. However, the data can become outdated if it was kept for too long time. Therefore, the data preservation should be correctly applied.

Data Integrity: The data must be secured and encrypted during data collection and transmission. In some cases, the data is collected by a third party. Thus, the data must be not accessible or readable by the third party and only the data recipient can decrypt the data and read it. This means the data must keep its integrity from the source till the last destination [23]. Data leak can cause many problems. Therefore, the data integrity is one of the important quality factors that must be in any data quality model.

Rapid Data Collection: Real-time data (RTD) is the data that is delivered immediately after its collection. For navigation and tracking, real-time data is often used. Real-time computation is usually used to process such data, but it can also be saved for review later or offline [24]. Collecting huge amount of data needs to be fast to get the most of the data needed in a short time. Otherwise, the system will take ages to collect the data needs and that may effects its accuracy.

Data Reusability: The collected data should be understandable, renewable, and composable. The process of cleaning and converting raw data before processing and analysis are known as data preparation. It's a crucial step before processing that usually entails reformatting data, making data corrections, and merging data sets to enrich data. For data professionals or business users, data preparation may be time-consuming, but it is essential to place data in perspective to transform it into information and remove prejudice caused by poor data quality. Standardizing data formats, enriching source data, and/or eliminating outliers are all common steps in the data preparation phase. Also, Data reuse refers to the practice of repurposing data for a different task or purpose than it was created for. The use of suitable metadata schemas will help to describe datasets and enable them to be reused over time [25].

4 Conclusions and Future Directions

Big Data is one of the quickest tools used in many applications nowadays. Data inspires agriculture, healthy production, and a variety of market choices. Data quality must be taken into account and assessed because bad decisions are based on low data quality. In this research, a quality model for the data collection process was introduced. The model has 7 quality factors that measure the data collection process. The factors start with the reliability of the data source then the regulations of data collecting and then the nature of the targeted data.

In the future, the model may be expanded and evaluated against other quality models. It also may be applied in some experiments as well. Also, the quality model can be used to produce Big Data area-specific data collection quality models.

References

1. Staegemann, D., Volk, M., Nahhas, A., Abdallah, M., Turowski, K.: Exploring the specificities and challenges of testing big data systems. In: 15th International Conference on Signal-Image Technology & Internet-Based Systems (SITIS 2019), Sorrento, Italy, pp. 289–295. IEEE (2019)
2. Batini, C., Rula, A., Scannapieco, M., Viscusi, G.: From data quality to big data quality. J Database Manage. **26**(1), 60–82 (2015). https://doi.org/10.4018/jdm.2015010103
3. Staegemann, D., et al.: A preliminary overview of the situation in big data testing. In: In 6th International Conference on Internet of Things, Big Data and Cyber Security (IoTBDS 2021), pp. 296–302. SciTePress (2021)
4. Laney, D.: 3D management: controlling data volume, velocity, and variety. Application delivery strategies. META Group Res. Note **6**(70), 1 (2001)
5. Volk, M., Staegemann, D., Pohl, M., Turowski, K.: Challenging big data engineering: positioning of current and future development. In: In 3rd International Conference on Internet of Things, Big Data and Cyber Security (IoTBDS 2019), pp. 351–358 (2019)
6. Staegemann, D., et al.: Challenges in data acquisition and management in big data environments. In: 6th International Conference on Internet of Things, Big Data and Security, pp. 193–204. SciTePress (2021)
7. Abdallah, M.: Big data quality challenges. In: The International Conference on Big Data and Computational Intelligence (ICBDCI 2019), Pointe aux Piments, Mauritius, pp. 1–3. IEEE (2019)
8. Abdallah, M., Muhairat, M., Althunibat, A., Abdalla, A.: Big Data quality: factors, frameworks, and challenges. COMPUSOFT Int. J. Adv. Comput. Technol. **9**(8), 3785–3790 (2020)

9. Wang, R.Y.: A product perspective on total data quality management. Commun. ACM **41** (2), 58–65 (1998). https://doi.org/10.1145/269012.269022
10. Lebied, M.: The Ultimate Guide to Modern Data Quality Management (DQM) For an Effective Data Quality Control Driven by the Right Metrics. https://www.datapine.com/blog/data-quality-management-and-metrics/. Accessed 18 July 2021
11. Galin, D.: Software Quality: Concepts and Practice, 1 edn. Wiley-IEEE Computer Society (2018)
12. Janssen, M., van der Voort, H., Wahyudi, A.: Factors influencing big data decision-making quality. J. Bus. Res. **70**, 338–345 (2017)
13. Izadi, D., Abawajy, J.H., Ghanavati, S., Herawan, T.: A data fusion method in wireless sensor networks. Sensors **15**(2), 2964–2979 (2015)
14. Heinrich, B., Klier, M., Schiller, A., Wagner, G.: Assessing data quality – a probability-based metric for semantic consistency. Decis. Support Syst. **110**, 95–106 (2018)
15. Savosin, S.V., Mikhailov, S.A., Teslya, N.N.: Systematization of approaches to assessing the quality of spatio-temporal knowledge sources. In: Journal of Physics: Conference Series, vol. 1801, no. 1, p. 012006 (2021)
16. Liu, Q., Feng, G., Zhao, X., Wang, W.: Minimizing the data quality problem of information systems: a process-based method. Decis. Support Syst. **137**, 113381 (2020)
17. Abowitz, D.A., Toole, T.M.: Mixed method research: fundamental issues of design, validity, and reliability in construction research. J. Constr. Eng. Manag. **136**(1), 108–116 (2010)
18. Byabazaire, J., O'Hare, G., Delaney, D.: Using trust as a measure to derive data quality in data shared IoT deployments. In: 29th International Conference on Computer Communications and Networks (ICCCN), Honolulu, HI, USA, pp. 1–9 (2020)
19. Berger, C., Stefani, P.D., Oriola, T.: Legal implications of using social media data in emergency response. In: 11th International Conference on Availability, Reliability and Security (ARES 2016), Salzburg, Austria, pp. 798–799 (2016)
20. Khayyat, Z., et al.: BigDansing: a system for big data cleansing. In: Proceedings of the 2015 ACM SIGMOD International Conference on Management of Data, Melbourne, Victoria, Australia, pp. 1215–1230 (2015)
21. Maletic, J., Marcus, A.: Data cleansing: beyond integrity analysis. Iq: Citeseer, pp. 200–209 (2000)
22. Berman, F.: Got data? A guide to data preservation in the information age. Commun. ACM **51**(12), 50–56 (2008)
23. Lebdaoui, I., Hajji, S., Orhanou, G.: Managing big data integrity. In: International Conference on Engineering & MIS (ICEMIS), Agadir, Morocco, pp. 1–6 (2016)
24. Zeng, Y., Jiang, W., Wang, F., Zheng, X.: Real-time data collection and management system for emergency spatial data based on cross-platform development framework. In: IEEE 4th Advanced Information Technology, Electronic and Automation Control Conference (IAEAC 2019), Chengdu, China, pp. 260–265 (2019)
25. Mehboob, B., Chong, C.Y., Lee, S.P., Lim, J.: Reusability affecting factors and software metrics for reusability: a systematic literature review. Softw. Pract. Exp. **51**, 1416–1458 (2021)
26. Alzyadat, W., AlHroob, A., Almukahel, I.H., Muhairat, M., Abdallah, M., Althunibat, A.: Big data, classification, clustering and generate rules: an inevitably intertwined for prediction. In: The International Conference on Information Technology (ICIT 2021), Amman, Jordan, pp. 149–155. IEEE (2021)
27. Staegemann, D., Volk, M., Lautenschläger, E., Pohl, M., Abdallah, M., Turowski, M.: Applying test driven development in the big data domain–lessons from the literature. In: The International Conference on Information Technology (ICIT 2021), Amman, Jordan, pp. 511–516. IEEE (2021)

Investigating the Incorporation of Big Data in Management Information Systems

Daniel Staegemann(✉), Hannes Feuersenger, Matthias Volk, Patrick Liedtke, Hans-Knud Arndt, and Klaus Turowski

Otto von Guericke University Magdeburg, Universitätsplatz 2, 39106 Magdeburg, Germany
{daniel.staegemann,hannes.feuersenger,matthias.volk, patrick.liedtke,klaus.turowski}@ovgu.de, hans-knud.arndt@iti.cs.uni-magdeburg.de

Abstract. In a world that is more and more driven by data, decision makers are provided with a huge amount of information. However, while this appears to be a good development, they also face the challenge of getting through those masses to get to the actually important insights. To ease this task for managers that oversee highly complex situations, management information systems (MIS) provide valuable support. While those systems were usually drawing from internal data sources that were rather structured, for several years the paradigm of big data has been on the rise. This however brings not only new possibilities for gaining insights, but also additional challenges. To help in dealing with those, the publication at hand presents a review of the literature that considers the incorporation of big data in MIS and reflects on current trends as well as challenges for future researchers.

Keywords: Management information system · Big data · Literature review · Decision support system · Knowledge management system

1 Introduction

One of the most important factors determining the success of an organization's decision making is constituted by its ability to incorporate relevant and high-quality information into the corresponding process [1]. One widely used measure [2] to meet these needs is the use of management information systems (MIS). With an ever increasing amount of more and more complex data being produced [3, 4], the traditional ways and tools for dealing with data are no longer sufficient. This led to new approaches, techniques and paradigms being developed. Those are amalgamated under the terms big data, respectively big data analytics [5] and received a lot of attention since their emergence more than a decade ago [6]. To further increase the value provided by MIS it appears to be an obvious decision to expand them into the big data realm.

However, as of now, the incorporation of big data in MIS is still in a rather immature state with numerous independent endeavors and strategies, but no unified agenda. To obtain an overview of the current situation and provide researchers as well as practitioners with insights that help them to improve the integration of big data in

W. Abramowicz et al. (Eds.): BIS 2021 Workshops, LNBIP 444, pp. 109–120, 2022.
https://doi.org/10.1007/978-3-031-04216-4_12

MIS, the following research question (RQ) that is partitioned in four sub research questions (SRQs) shall be answered:

RQ: What is the current situation regarding the application of big data in management information systems?

SRQ1: In which way do current management information systems incorporate big data?

SRQ2: How can management information systems benefit from big data analytics?

SRQ3: Are there any trends regarding the development of the publications dealing with the application of big data in management information systems?

SRQ4: What are the most promising avenues for future research?

To answer each of those SRQs and consequently also the superordinate RQ, the publication at hand is structured as follows: After this introduction, the general concept of MIS is introduced, succeeded by an analogous section regarding the topic of big data. Ensuing, a review of the relevant scientific literature is presented that is subsequently analyzed and discussed. Finally, a conclusion is drawn and possible directions for future research outlined.

2 Management Information Systems

The concept of MIS originates from the 1960s as well as early 1970s, when computers were supposed to support management [7]. A MIS corresponds to the information and communication view of management systems, where a management system maps all the requirements in an organization that are necessary to achieve its goals. Starting from the origin, there is a long MIS history in which a variety of different terms have been created and also discarded. The early approach to MIS is considered to be the Total System Approach, which was intended to include a monitoring and control function in addition to information tasks [7]. However, these early holistic approaches to MIS failed [7], which is why the associated terminology has also changed again and again over the years. More specific terms such as Decision Support System (DSS) or Knowledge Management System (KMS) have been introduced. In a DSS, the focus is on supporting the decision-making process and decision-making itself. For both strategic and operational decisions, such a system ensures that all relevant information is identified, processed, analyzed and its insights are presented in a structured manner [8]. Meanwhile, a KMS attempts to structure and contextualize existing knowledge so that it is available as broadly as possible throughout an organization and not just to specific individuals [9]. In more simple terms, such a system attempts to extract the tacit knowledge of employees from their heads and capture it in digital form into explicit knowledge. In addition to the creation of more specific terms, there has been a contradictory development of terms that, like the earlier ones, has aimed for holism and generality. In this context, the concept of Business Intelligence (BI), has become established. BI is understood to mean the systematic application of information and communication technology and the corresponding techniques not only to collect data, but also to analyze it and to present the resulting insights visually [10]. Based on this abstract definition, not only the two concepts of DSS and KMS described above, but many more can be subsumed under the generic term of BI.

As it is the case for the majority of information and communication technology, the concept of big data, which has been getting traction since around 2011 [11], has significantly impacted the domain of MIS. When big data is integrated into MIS, a whole new wealth of potential insights can be tapped into, which however comes at the cost of massively growing complexity and a need to sift through the deluge of available information to identify those that are actually beneficial.

3 Big Data

While there is no unified definition of the term big data [12], it generally refers to data that are especially demanding in terms of their volume, velocity, variety and variability, when trying to analyze them, resulting in a need for highly sophisticated solutions that surpass the abilities of traditional techniques [5]. Those characteristics are also referred to as the four *Vs* of big data. *Volume* refers to the huge amount of data, in size as well as the number of files that have to be handled and is probably the most prominent one. While the definition of "big" is somewhat fuzzy, today's workloads can be in the petabyte-region [11]. *Velocity* corresponds to two aspects. On the one hand the speed with which data are incoming and on the other hand how fast they have to be processed [13]. *Variety* corresponds to the heterogeneity of data types, ways of formatting, languages and similar properties that can oftentimes be found in an analysis' underlying set of data [14], but also the diversity in the data's origins [5]. *Variability* stands for the change of the previously mentioned characteristics [13], which can occur temporarily [15] or long-term, in the latter case potentially necessitating a corresponding adaption of the big data analytics solution [16].

However, even though those characteristics are presumably the four most common ones, there are also many others that can be found in the literature [17]. Examples comprise the likes of veracity [18], validity [19] or value [20]. While the implementation of big data into an organization's workflow is a challenging task with many aspects that have to be considered [13, 21–25], it also promises substantial gains [24, 26, 27], which are in many cases realized through an improved decision-making by the respective management. Subsequently, to assure their usage and maximize the benefits, it is necessary to make sure that decision makers are provided the available information in an easily accessible and comprehensible way.

4 Literature Review

To determine the current status, but possibly also trends concerning the incorporation of big data analytics in MIS, a structured literature review is conducted. To ensure its comprehensiveness and comprehensibility [28], the methodologies of Levy and Ellis [29] as well as Webster and Watson [30] were followed, resulting in a multi-stepped procedure that is further described in the following.

4.1 Protocol

For the purpose of identifying the relevant literature, instead of using a publisher-bound database like IEEE Xplore, the source-neutral alternative Scopus was used. Due to the fact that it incorporates the contents of numerous scientific databases like ACM Digital Library, IEEE Xplore, ScienceDirect and additionally publications from many other publishers such as Springer, Wiley, Taylor & Francis or Sage. This allows for a more comprehensive search, which likely improves the results of the review.

The used search term was constructed as follows. To assure a strong connection of the research's content, the phrase "Big Data" had to appear in title, abstract or the paper's keywords. Furthermore, to also have the connection to MIS, at least one of the terms "Management Information System" or "Management Support System" or "Enterprise Information System" or "Executive Support System" or "Decision Support System" or "Data Support System" or "Management Reporting System" or "Knowledge Management System" was required in title, abstract or keywords. While those are not necessarily synonyms for MIS, they are potentially related and were therefore included for the first phase of the review process, assuring a more comprehensive coverage. Additionally, only conference papers or journal articles were considered, since those are usually peer reviewed, which promises a certain degree of quality. Moreover, the contributions had to be written in English and literature reviews were excluded to only get primary insights, avoiding to inadvertently reiterating the same findings repeatedly. The further inclusion and exclusion criteria generally aimed at contributions that provide business-centered insights instead of mere unproven concepts or pure technical implementations. Whenever one inclusion criterion was not valid or at least one exclusion criterion was applicable, the paper was rejected. A list of all applied criteria is depicted in Table 1.

Table 1. Inclusion and exclusion criteria

Inclusion Criteria	Exclusion Criteria
Published in conference or journal	Paper is a literature review
Strong attention to MIS	Not written in English
Strong attention to big data	Only focused on technical implementation
Strong focus on business aspects	Focused on the automatic triggering of actions based on information
Focus on the provision of information to managers	Paper only discusses theoretical concepts, methods, models or algorithms without in-practice evaluation

Using the previously described search parameters, 1315 publications were found and subsequently examined regarding their relevance. An overview of this process is depicted in Fig. 1. In a first step, the titles were read to assess their general relevancy for answering the RQs. After this procedure, 478 papers were remaining. For those, the abstracts and keywords were examined to further reduce the list, leaving 174 contributions for further examination. Reading their introduction and conclusion, which usually gives a good grasp of a paper's essence, decreased the number to 45. As the fourth step, to further determine their pertinence, they were skimmed in total, resulting in a set of 14 relevant publications that were subsequently thoroughly read in total and are further discussed in the findings section. The whole procedure was conducted by four researchers. In earlier stages, each person investigated one fourth of the respective body of literature. This split was conducted to help in overcoming the high number of publications. After each stage, the remaining papers were permutated between the authors. In doing so, it was assured that the publications that reached later stages were each evaluated by multiple researchers.

In the end, for the skimming in the last step of this process, the approach was slightly modified. Whereas in the previous phases, in cases of doubt, the respective contribution was carried over to the next step, here, in case of uncertainty concerning the classification, the paper was also examined by the other researchers. Those contributions were then assessed by each one. If at least three of the four voted for rather including the paper or rather dropping it, the respective action was taken. For the remaining cases, the considerations were more detailed, using the quality criteria depicted in Table 2.

Table 2. Relevancy criteria

Category	ID	Criterium	Answer
Thematical Fit	F1	Explicitly Covering MIS	Yes/No/Partially
	F2	Strong Focus on Big Data	Yes/No/Partially
	F3	Focus on MIS/BD in General	Yes/No/Partially
	F4	Explicitly Answers at Least One RQ	Yes/No/Partially
Quality	Q1	Clear Description of the Paper's Objectives	Yes/No/Partially
	Q2	Research Method is Adequately Explained	Yes/No/Partially
	Q3	Paper is Supported by Primary Data	Yes/No/Partially
	Q4	High Quality of References	Yes/No/Partially

For each criterium, each of the four reviewers individually noted if he considered the criterium fully, partially, or not sufficiently fulfilled. Those answers were converted into the numbers 1, 0.5, respectively 0 and subsequently added for each paper and category. The ones that had a combined score of at least 4 and for both categories (Thematical Fit and Quality) not lower than 1.5 respectively were then added to the literature set, while the others were omitted. An overview of the whole search and filter process, showing the conducted steps (first level) as well as the respectively remaining numbers of publications (second level) is depicted in Fig. 1.

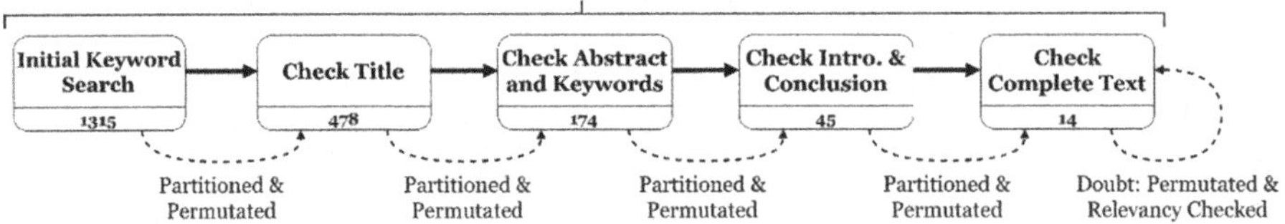

Fig. 1. The review process

4.2 Findings

By following the review protocol, a set of 14 relevant publications was identified, which is listed in Table 3. The following examination of these publications provides the answers for SRQ 1 and SRQ 2.

Generally speaking, the papers emerging from the literature review can basically be divided into two thematic areas, the ones that cover the use of big data for decision support and those that deal with the use of big data for knowledge generation.

Table 3. The final set of literature

Ref	Title	Type
[31]	Big data analytics for knowledge generation in tourism destinations – A case from Sweden	Journal
[32]	A Solution for Information Management in Logistics Operations of Modern Manufacturing Chains	Journal
[33]	The Roles of Big Data in the Decision-Support Process: An Empirical Investigation	Conference
[34]	Business process improvement by means of Big Data based Decision Support Systems: a case study on Call Centers	Journal
[35]	Study on analysis and decision support system of power grid operation considering meteorological environment based on big data and GIS	Conference
[36]	Decision Support in Big Data Contexts: A Business Intelligence Solution	Conference
[37]	Analysing a large amount of data as a decision support systems tool in Nigeria organisation	Journal
[38]	Schema on read modeling approach as a basis of big data analytics integration in EIS	Journal
[39]	Big data creating new knowledge as support in decision-making: practical examples of big data use and consequences of using big data as decision support	Journal
[40]	Integrated Understanding of Big Data, Big Data Analysis, and Business Intelligence: A Case Study of Logistics	Journal
[41]	Big data analytics for operations management in engineer-to-order manufacturing	Journal
[42]	A decision support system based on ontology and data mining to improve design using warranty data	Journal
[43]	Methodology for the Implementation of Knowledge Management Systems 2.0	Journal
[44]	An analytic infrastructure for harvesting big data to enhance supply chain performance	Journal

In [40], the authors attempt to bridge the gap between the research areas of big data and BI by doing a literature analysis. As a result, they come to the hypothesis that the two methods can no longer coexist independently, but rather represent an integrated decision support system. The raw data previously used by the BI system is now big data. A subsequent single case study in the largest logistics company in Korea proves that such an integrated DSS can significantly increase both productivity and cost-effectiveness in the area of loading and unloading of docks. Staying in the realm of logistics, Ilie-Zudor et al. [32] developed a predictive-analysis-based decision support platform, which is based on machine learning and can be used in several logistics scenarios. Due to this application area, the authors highlight the relevance of the following data sources: historical data, semi-static hub data, dynamic depot data, adjacency data, travel time data, economic data and other external data. Pereira and Costa [36] also recognized great potential in the use of a DSS, trained with big amounts of data, for retail. They developed a BI solution in a Portuguese company (i.e. in a sporting goods) store, that analyzes the movement data of people inside closed spaces using cell phone data. By using the recorded data, decision makers get well-founded support in deciding how the store should be set up. For supporting business process improvement, Vera-Baquero et al. [34] developed and evaluated a five-step-approach for building a big data based DSS: First, the process structure should be defined (definition phase) and the DSS should be configured (configuration phase). In the execution phase the implementation of the system and monitoring of defined events takes place. The resulting outcomes are analyzed by business users in the subsequent control phase. Finally, in the diagnosis phase, an as-is comparison of the expected improvements and the generation of new optimization potentials is performed.

The use of big data techniques for optimizing decision-making has also made its way into the public sector [39]. Based on a case study with the Finnish city government of Turku, a number of use cases were identified in which entirely new insights could be gathered through the use of big data techniques. Basically, the increase in resource efficiency and the resulting decrease in costs can be seen in all areas of application. Many employees not only understand the potential benefits of using big data but are even demanding increased use in the future. This shows the positive attitude that is widespread in large parts of the workforce. Zhan and Tan [44] described the fact that supply chain managers should be able to make better decisions through competence set analyses on big data. Therefore, they developed an integrated analytic infrastructure which is divided in five stages: 1) data capture and management, 2) data cleaning and integration, 3) data analytics, 4) competence set analysis and deduction graph, 5) information interpretation and decision making. In Li et al. [35] the influence of the meteorological environment on the power grid should be analyzed in the context of a case study of the so-called East China grid. Because of the fact that big data analytics enables users to make data driven decisions, an approach of an analysis and decision support system for the power grid operation taking into account the meteorological environment is provided. The analysis of traffic data represents another interesting use case for the integration of big data in Enterprise Information Systems (EIS). Based on the three levels of an EIS - corporate data model level, data warehouse level and report level - there are three different approaches for integrating big data analytics. According to [38] integration must take place at least at the corporate data model level and one other level. This can be extended to integration at all three levels. For solving recurring problems, Alkahtani et al. [42] used warranty data for product design in the automobile manufacturing. The authors present a framework for a DSS which uses ontology-based text mining to facilitate the knowledge search. The DSS also uses Self Organizing Maps (SOM) to detect the specific defective component" [42].

The research from Fadiya [37] was based on an analysis of the integration of big data and decision support applications (BDDSA) in Nigerian MIS user organizations. By conducting a study with 400 clients the author gives answers regarding the use of BDDSA in different points (e.g. the perceived benefits, the complexity and the compatibility). The expectations of using a BDDSA show that on the one hand there are many advantages (e.g. cost reduction and the understanding of the data by the senior management). On the other hand, there are some challenges of using BDDSA, like the complexity and the confusion of the system. The paper of Kozjek et al. [41] deals with the use of big data for an engineer-to-order manufacturing. They present a case study which uses real data from a manufacturing execution system (MES). Before, there were problems with the production planning and scheduling. To eliminate those, the goal of the study was to develop a simulation tool which is able to forecast the production process. Poleto et al. [33] proposed a model for the integration of big data, BI and decision making. Their model is divided into six elements: 1) The first element is the content which comes from external and internal data sources and is divided in public content and private content. 2) With the intelligence element the data is separated in different contexts. 3) Opportunities and alternatives have to be created which is done by another element. 4) A decision support system is intended to support the user in decision-making and thus facilitate work. 5) The implementation of a decision describes the fact that the decision was made. 6) Complementary to this, organizational learning should be included, which refers to the knowledge that has been built up through decisions. Furthermore, in [31] the authors developed a BI-based destination management information system (DMIS) in the context of a case study at a large Swedish tourism destination. With the help of this system, tourism companies should not only generate new knowledge, but also apply and learn from it. In this context the authors divide between the generation and application of knowledge and thus also offer a clear separation of the systems used: While the data warehouse and big data instruments are primarily required for the generation of knowledge, the DMIS serves the application. For the knowledge generation phase, the authors identified the following indicators relevant for the tourism sector: economic performance indicators, customer behavior indicators and customer perception and experience indicators. In order to be able to apply the newly created knowledge, a dashboard was developed, with the help of which the most important insights are presented visually. Finally, Orenga-Roglá and Chalmeta [43] developed the so-called Web 2.0 Knowledge Management (W2KM) methodology, which guides the entire process of developing and implementing a KMS 2.0 (i.e. a KMS that also uses Web 2.0 and big data technologies). Due to the high complexity of such a project, the authors recommend going through the following seven phases in a structured manner: draft, planning, analysis, design, development, implementation and control. In a case study the methodology was applied on a big oil and gas company. Besides the realization of an improvement of the methodology it was noted that the methodology offered great potentials to improve the overall business processes within the company and that especially the combination of Web 2.0 and big data tools had a good impact.

4.3 Discussion

When examining the relevant literature, it becomes apparent that only relatively few publications sufficed the search's requirements but those are highly diverse. They span across different countries and continents, are situated in different industries and follow vastly different general approaches to provide managers with the information they need to strengthen their decision making. At the same time, they all emphasize the benefits provided by the integration of big data and MIS. The general sentiment of the studies can therefore be described as favorable. In [39] this positive attitude is furthermore explicitly extended to the point of view of the organization's employees who even demanded increased future use, highlighting the huge potential of such solutions. The observation stands in contrast to, for example, [45] where still a rather strong skepticism of users towards heavily data focused decision making approaches was ascertained, which apparently faded over time. This could be an indication that the use of big data and acting on the basis of data are becoming more accepted, especially since Fredriksson is talking about the public sector, where technical acceptance is usually even more difficult to achieve than in the private sector. However, no conclusive statement can be made here due to it being only an individual case analysis and not a large-scale study.

With regard to SRQ 3, it can be observed that from 2014 onwards, each year one or more publications dealing with the topic have been created, showing a continuous interest. The types however changed. Whereas in the beginning conference publications and journal articles were present in equal proportions, for the time since 2017 only the latter made it into the list. This could be an indicator for growing importance of the field, leading to journals having a bigger interest in it. Yet, even though the general maturity of the research seems to be increasing, in the papers presenting specific technical solutions, the aspect of quality assurance is not further elaborated, despite being a highly important issue when striving to obtain competitive advantages through the use of data driven decision making [21]. Furthermore, it is also noticeable that the regarded publications, despite them fitting the criterium on referring to MIS, in several cases emphasize other terminology as for example *decision support system*. This shows how intertwined the corresponding terms oftentimes are. While it is of no greater significance in practice, it constitutes a challenge for researchers who have a hard time compiling all the relevant contributions.

As a result, review papers, as the publication at hand, are of great importance. They provide an overview of the corresponding literature and are therefore a valuable foundation for succeeding research endeavors [28]. Those, however, are necessary to further advance the domain, as it has also become apparent in the literature. Additionally, it was found that several of the publications dealt with issues on how to define workflows and process stages for harnessing big data in MIS or propose certain models, which shows that there are no standard procedures that are univocally applied. Yet, those might be a big step in the direction of making data driven decision making even more accessible, allowing more organizations to profit from the huge possible advantages [26]. Those two directions can be therefore seen as promising avenues, constituting the answer to

SRQ 4. However, the time horizon for both likely differs, with the latter probably being a harder to come by challenge.

5 Conclusion

With data production continuously rising, decision makers are faced with a data deluge that they have to overcome to obtain that information that is beneficial for steering their respective organizations in the right direction. For this task, MIS constitute a valuable aide that helps in turning pure data into actual insights that can be acted upon. However, with the rise of big data, those MIS are faced with the challenge of its integration. Within this publication it was explored, by conducting a structured literature review, what is the current situation regarding the application of big data in MIS. In doing so, the contribution is twofold. On the one hand, future researchers and practitioners are provided with an overview of the domain and a list of relevant publications for further enquiry, which they can use as a foundation for their own endeavors and projects. This is especially valuable since the heterogeneity and diversity of the used terminology complicates the search for literature, making it even more cumbersome than in other research streams. The thus saved time can consequently be used to benefit the actual advancement of the cause instead of spending it sifting through a myriad of contributions to find the small fraction that might be helpful. On the other hand, this publication also highlighted avenues for future research, which appear to be especially promising. In doing so, it provides prospective authors with potential directions for their own work, helping to advance the domain as a whole. However, in the future, an extension of the search concerning databases and search terms might yield even more comprehensive findings. Furthermore, as for any review paper and despite the authors' best efforts, a certain degree of subjectivity will always remain in the decisions concerning the inclusion or exclusion of the evaluated papers. This might in turn influence the obtained results. Though, with the findings showing rather strong tendencies and unanimity, this effect can be considered insignificant.

References

1. Ghasemaghaei, M.: Understanding the impact of big data on firm performance: The necessity of conceptually differentiating among big data characteristics. Int. J. Inf. Manag. **57**, 102055 (2021)
2. Burgard, M.: Empirische Untersuchung der Wettbewerbsrelevanz von Business-Intelligence-Konfigurationen auf der Basis des Resource-based View. Universaar, Saarbrücken (2011)
3. Dobre, C., Xhafa, F.: Intelligent services for big data science. Futur. Gener. Comput. Syst. **37**, 267–281 (2014)
4. Yin, S., Kaynak, O.: Big data for modern industry: challenges and trends [point of view]. Proc. IEEE **103**, 143–146 (2015)
5. NIST: NIST Big Data Interoperability Framework: Volume 1, Definitions, Version 3. National Institute of Standards and Technology, Gaithersburg, MD (2019)
6. Parlina, A., Ramli, K., Murfi, H.: Theme mapping and bibliometrics analysis of one decade of big data research in the scopus database. Information **11**, 69–94 (2020)

7. Oppelt, R.U.: Computerunterstützung für das Management: Neue Möglichkeiten der computerbasierten Informationsunterstützung oberster Führungskräfte auf dem Weg von MIS zu EIS? Oldenbourg, München (1995)
8. Sprague, R.H.: A framework for the development of decision support systems. MISQ **4**, 1–26 (1980)
9. Maier, R.: Knowledge Management Systems: Information and Communication Technologies for Knowledge Management; with 91 Tables. Springer, Berlin (2007). https://doi.org/10.1007/978-3-540-71408-8
10. Dedić, N., Stanier, C.: Measuring the success of changes to existing business intelligence solutions to improve business intelligence reporting. In: Tjoa, A.M., Xu, L.D., Raffai, M., Novak, N.M. (eds.) CONFENIS 2016. LNBIP, vol. 268, pp. 225–236. Springer, Cham (2016). https://doi.org/10.1007/978-3-319-49944-4_17
11. Gandomi, A., Haider, M.: Beyond the hype: big data concepts, methods, and analytics. Int. J. Inf. Manage. **35**, 137–144 (2015)
12. Volk, M., Staegemann, D., Turowski, K.: Big Data. In: Kollmann, T. (ed.) Handbuch Digitale Wirtschaft, vol. 58, pp. 1–18. Springer Fachmedien Wiesbaden, Wiesbaden (2020) https://doi.org/10.1007/978-3-658-17291-6
13. Volk, M., Staegemann, D., Pohl, M., Turowski, K.: Challenging big data engineering: positioning of current and future development. In: Proceedings of the 4th IoTBDS (2019)
14. Kaisler, S., Armour, F., Espinosa, J.A., Money, W.: Big data: issues and challenges moving forward. In: Proceedings of the 46th HICSS, pp. 995–1004. IEEE, Piscataway, NJ (2013)
15. Katal, A., Wazid, M., Goudar, R.H.: Big data: Issues, challenges, tools and Good practices. In: Parashar (Hg.) – 2013 sixth International Conference, pp. 404–409 (2013)
16. Staegemann, D., Volk, M., Daase, C., Turowski, K.: Discussing relations between dynamic business environments and big data analytics. CSIMQ **7**, 58–82 (2020)
17. Emmanuel, I., Stanier, C.: Defining big data. In: Proceedings of the BDAW 2016, pp. 1–6. ACM Press, New York, New York, USA (2016)
18. Alaoui, I.E., Gahi, Y., Messoussi, R.: Full consideration of big data characteristics in sentiment analysis context. In: Proceedings of the 4th ICCCBDA, pp. 126–130. IEEE (2019)
19. Khan, M.A.-u.-d., Uddin, M.F., Gupta, N.: Seven V's of big data understanding big data to extract value. In: Proceedings of the 2014 Zone 1 Conference of the American Society for Engineering Education, pp. 1–5. IEEE (2014)
20. Younas, M.: Research challenges of big data. SOCA **13**, 105–107 (2019)
21. Günther, W.A., Rezazade Mehrizi, M.H., Huysman, M., Feldberg, F.: Debating big data: a literature review on realizing value from big data. J. Strat. Inf. Syst. **26**, 191–209 (2017)
22. Abdallah, M., Muhairat, M., Althunibat, A., Abdalla, A.: Big data quality: factors, frameworks, and challenges. Compusoft: Int. J. Adv. Comput. Technol. **9**, 3785–3790 (2020)
23. Al-Sai, Z.A., Abdullah, R., Husin, M.H.: Critical success factors for big data: a systematic literature review. IEEE Access **8**, 118940–118956 (2020)
24. Maroufkhani, P., Tseng, M.-L., Iranmanesh, M., Ismail, W.K.W., Khalid, H.: Big data analytics adoption: determinants and performances among small to medium-sized enterprises. Int. J. Inf. Manag. **54**, 102190 (2020)
25. Staegemann, D., Volk, M., Jamous, N., Turowski, K.: Understanding issues in big data applications - a multidimensional endeavor. In: Proceedings of the 25th AMCIS (2019)
26. Müller, O., Fay, M., Vom Brocke, J.: The effect of big data and analytics on firm performance: an econometric analysis considering industry characteristics. J. Manag. Inf. Syst. **35**, 488–509 (2018)
27. Ghasemaghaei, M.: Improving organizational performance through the use of big data. J. Comput. Inf. Syst. **60**, 395–408 (2020)

28. Vom Brocke, J., Simons, A., Niehaves, B., Reimer, K., Plattfaut, R., Cleven, A.: Reconstructing the giant: on the importance of rigour in documenting the literature search process. In: Proceedings of the ECIS 2009 (2009)
29. Levy, Y., Ellis, T.J.: A systems approach to conduct an effective literature review in support of information systems research. Inf. Sci. Int. J. Emerg. Transdisc. **9**, 181–212 (2006). https://doi.org/10.28945/479
30. Webster, J., Watson, R.T.: Analyzing the past to prepare for the future: writing a literature review. MISQ, **26**, xiii–xxiii (2002)
31. Fuchs, M., Höpken, W., Lexhagen, M.: Big data analytics for knowledge generation in tourism destinations – a case from Sweden. J. Destin. Mark. Manag. **3**, 198–209 (2014)
32. Ilie-Zudor, E., Kemény, Z., Ekárt, A., Buckingham, C.D., Monostori, L.: A solution for information management in logistics operations of modern manufacturing chains. Procedia CIRP **25**, 337–344 (2014)
33. Poleto, T., de Carvalho, V.D.H., Costa, A.P.C.S.: The roles of big data in the decision-support process: an empirical investigation. In: Delibašić, B., et al. (eds.) ICDSST 2015. LNBIP, vol. 216, pp. 10–21. Springer, Cham (2015). https://doi.org/10.1007/978-3-319-18533-0_2
34. Vera-Baquero, A., Colomo-Palacios, R., Molloy, O., Elbattah, M.: Business process improvement by means of big data based decision support systems: a case study on call centers. Int. J. Inf. Syst. Proj. Manag. **3**, 5–26 (2015)
35. Li, L., Yao-qiang, X., Xing-zhi, W., Kai, W.: Study on analysis and decision support system of power grid operation considering meteorological environment based on big data and GIS. In: 2016 China International Conference on Electricity Distribution, pp. 1–6. IEEE (2016)
36. Pereira, J.L., Costa, M.: Decision support in big data contexts: a business intelligence solution. In: Rocha, Á., Correia, A.M., Adeli, H., Reis, L.P., Mendonça Teixeira, M. (eds.) New Advances in Information Systems and Technologies, vol. 444, pp. 983–992. Springer International Publishing, Cham (2016). https://doi.org/10.1007/978-3-319-31232-3
37. Fadiya, S.O.: Analysing a large amount of data as a decision support systems tool in Nigeria organisation. IJCRSEE **5**, 121–129 (2017)
38. Janković, S., Mladenović, S., Mladenović, D., Vesković, S., Glavić, D.: Schema on read modeling approach as a basis of big data analytics integration in EIS. Enter. Inf. Syst. **12**, 1180–1201 (2018)
39. Fredriksson, C.: Big data creating new knowledge as support in decision-making: practical examples of big data use and consequences of using big data as decision support. J. Decis. Syst. **27**, 1–18 (2018)
40. Jin, D.-H., Kim, H.-J.: Integrated understanding of big data, big data analysis, and business intelligence: a case study of logistics. Sustainability **10**, 3778 (2018)
41. Kozjek, D., Vrabič, R., Rihtaršič, B., Butala, P.: Big data analytics for operations management in engineer-to-order manufacturing. Procedia CIRP **72**, 209–214 (2018)
42. Alkahtani, M., Choudhary, A., De, A., Harding, J.A.: A decision support system based on ontology and data mining to improve design using warranty data. Comput. Ind. Eng. **128**, 1027–1039 (2019)
43. Orenga-Roglá, S., Chalmeta, R.: Methodology for the implementation of knowledge management systems 2.0: a case study in an oil and gas company. Bus. Inf. Syst. Eng. **61**, 195–213 (2019)
44. Zhan, Y., Tan, K.H.: An analytic infrastructure for harvesting big data to enhance supply chain performance. Eur. J. Oper. Res. **281**, 559–574 (2020)
45. Namvar, M., Cybulski, J.L.: BI-based organizations: a sensemaking perspective. In: Proceedings of the Thirty Fifth International Conference on Information Systems (2014)

The Perception of Test Driven Development in Computer Science – Outline for a Structured Literature Review

Erik Lautenschläger(✉)

FernUniversität in Hagen, 58084 Hagen, Germany
erik.laute@gmail.com

Abstract. Test driven development (TDD) is a practice that aims to improve product quality and maintainability by interweaving the design and implementation with its testing. It is most prominent in the software development domain. However, its usefulness is not undisputed, making it a somewhat controversial topic. Besides giving a short introduction regarding the principles of TDD, the publication at hand motivates and outlines a structured literature review to obtain new insights regarding the perception of TDD in computer science, hopefully contributing to the corresponding discourse. Furthermore, by already conducting the first steps of the review, it provides a first impression regarding the vastness of the potentially relevant literature base and gives a rough indication regarding the extend that is to be expected for the completed work.

Keywords: Test driven development · TDD · Structured literature review · Computer science · Software development · Digitization

1 Introduction

In the course of the ongoing digitization, the importance and ubiquity of information technology for our daily lives has been growing for many years [1] without any indication that this trend will stop in the foreseeable future [2]. Subsequently, also the significance of software products and their development has risen [3] and concepts like IoT and big data emerged [4]. However, with the increasing technical possibilities and corresponding opportunities, also the complexity of the software itself has reached new levels [5]. Thus, depending on their purpose, the development of new applications as a driving force of the ongoing digitization can be very expensive, time-consuming, and is also prone to error [6, 7]. As a result, approaches which promise to solve or alleviate those challenges are highly relevant for practitioners and thence also for researchers [8]. One of these strategies is test driven development (TDD), the application of which in the big data domain has moreover recently been brought into discussion [9]. However, the opinion concerning its actual value regarding software quality and required expenditure is not uniform [10–13]. Therefore, the publication at hand seeks to answer the following research question (RQ).

W. Abramowicz et al. (Eds.): BIS 2021 Workshops, LNBIP 444, pp. 121–126, 2022.
https://doi.org/10.1007/978-3-031-04216-4_13

RQ: How is the perception of test driven development in the computer science literature?

Furthermore, to facilitate a clearer structure, the RQ is partitioned into three sub research questions (SRQ), with each covering a particular aspect of the issue, leading to a holistic assessment, when regarded in conjunction.

SRQ1: Which are perceived advantages and benefits of TDD?

SRQ2: Which are perceived disadvantages and challenges of TDD?

SRQ3: What is the general sentiment regarding TDD?

The remainder of the text is structured as follows. After the introduction, an overview regarding the topic of TDD is given. Then, the envisaged methodology for the proposed study is outlined. Lastly, it is presented, which steps are still to be conducted in the future, which results are aspired and how they will contribute to the body of knowledge.

2 Test Driven Development

TDD is an approach that is mostly used in software development and aims at increasing the quality of the implemented product by not only regulating its testing, but also the way it is designed [14, 15]. However, due to the focus of the publication at hand, the following considerations are explicitly only geared towards the development of software, even though most of it might also be applicable to other domains like process modeling [16] or the development of ontologies [17, 18]. When developing software, the traditional way of doing so is to first think of a change in functionality that ought to be integrated, implement it, and then test it. However, when leveraging the process of TDD, the last two steps are interchanged. That is, a change in functionality is envisioned first, subsequently one or multiple tests for the change are written and executed, but supposed to fail without a corresponding software counterpart to run against [19]. Only then, the implementation of the change is conducted. To be valid, it must pass the previously written tests [15]. Subsequently, a refactoring of the written code is conducted, so not only the tests are passed, but an adherence to the environment's standards and best practices is warranted [19]. This approach also severely affects the software design, since an underlying principle of TDD is that a change is planned as a smallest possible unit, instead of a big working package [20]. This allows for small, incremental modifications [21], facilitating the interlocking of testing and development to enable the developer to leverage short testing cycles [22] and therefore achieve a high test frequency. Commonly, most of the tests are specifically written for those units. However, a variety of other tests can be leveraged in TDD, such as integration, system, and acceptance tests [23]. Another method often used in the context of TDD is continuous integration (CI) in combination with test automation [24, 25]. A CI server automatically starts all available tests when new code is committed to a versioning system, thus assuring that newly implemented changes have no negative effect on already existing parts of the software. This in turn facilitates faster adjustments to changing requirements as they can be observed for some applications as for example in some big data use cases [26].

3 Methodology

The publication at hand aims to explore the perception of TDD in the literature. Consequently, the literature review is the method of choice. While an unstructured approach might be suitable to obtain a first impression of a topic, it lacks the thoroughness and reliability that is provided by following a rigorous search protocol. Since a comprehensive and trustworthy analysis is needed to meaningfully assess the concept of TDD concerning its perceived value, instead a structured literature review appears to be necessary. If carried out diligently, it also allows for future researches to retrace the results and potentially build upon them, therefore providing additional benefit [27]. For meeting those requirements, the chosen approach is closely related to the procedures described by Webster and Watson [28] and Levy and Ellis [29].

To assure a broad coverage of the relevant literature, the search is conducted using *Scopus*, which comprises the contents of most relevant search engines, and *SpringerLink*. For both, the search period was set to the time from 2010 to 2020, assuring that the obtained results are at least somewhat recent. To increase relevancy, only contributions from the *computer science* domain were considered. Furthermore, publications hat to be written in English or German and needed to be published in a journal or as part of conference proceedings. While those inclusion criteria are the same for both search engines, due to their different available input parameters, the search terms are not identical. In SpringerLink, the phrases *"Test Driven Development"*, *"Test-Driven Development"* or *"Test-Driven-Development"* had to appear anywhere in the text for a contribution to be further considered. Since Scopus does not allow for a full-text search, instead one of the abovementioned phrases or *"TDD"* had to be part of title, abstract or keywords.

Following this protocol, in Scopus, 1327 conference papers and 460 articles were found, resulting in a total of 1787 publications. SpringerLink added another 436 conference papers and 146 articles, totaling to 582 contributions. Therefore, in total, the initial set of literature consists of 2369 items, comprising 1763 conference papers and 606 journal articles. Removing duplicates led to the deletion of 17 journal articles, leaving 589. For the conference publications, the number was reduced by 64, resulting in 1699 remainders. Combined, 81 duplicates were deleted, leaving 2288 unique items. To find the contributions that are actually relevant for answering the RQ and SRQs, for the future, a multi-stepped refinement process is intended. In a first phase, the titles, abstracts and keywords of all remaining publications will be examined to exclude those, which are clearly not relevant. The second phase comprises the reading of the introduction and the conclusion to further narrow the collection down. Subsequently, in phase three, the publications deemed fitting by now will be skimmed in total. The remaining contributions will be read in total. Furthermore, for those findings, a backward search will be conducted. Additionally, also a forward search regarding references to the particular work as well as its authors is to be carried out. This process is intended to proceed until it appears to yield no new results. However, due to the relatively clear delimitation of the topic, making it unlikely to find relevant literature without the initially defined search terms, it is not expected to find many more additional publications by this method. Yet, to increase validity and significance of the results, it is still necessary. An outline of the described review protocol is depicted in Fig. 1. Once the relevant set of literature is determined, the actual analysis can take place.

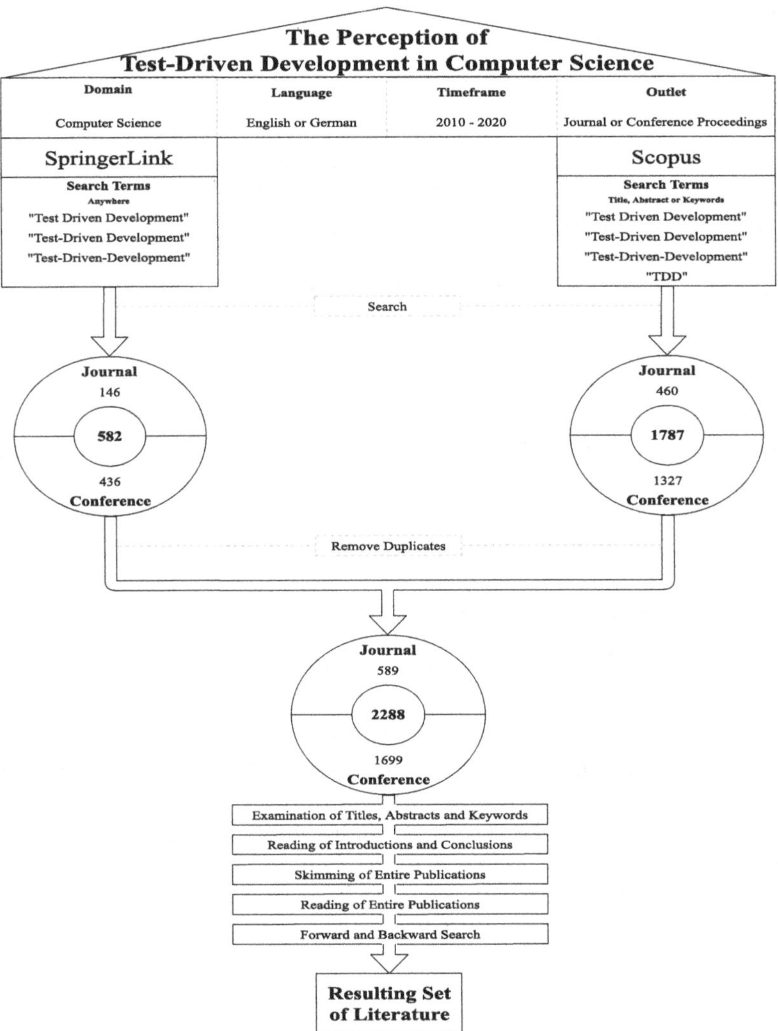

Fig. 1. The review protocol.

4 Future Work and Expected Results

While the concept for the proposed research is already in an advanced state, it is still a work in progress. As mentioned above and indicated in Fig. 1, the steps following the removal of the duplicates are still to be conducted. Once the relevant literature is identified, its thorough analysis will provide answers to the RQ and the SRQs, allowing to obtain a comprehensive summary of the perception of TDD in the scientific literature. This will hopefully allow to derive new insights regarding its usefulness, strengths, weaknesses, and preferred application contexts, providing valuable input for researchers, as well as practitioners, and enriching the corresponding discourse.

References

1. Deb, S.: Information technology, its impact on society and its future. Adv. Comput. **4**, 25–29 (2014)
2. Theis, T.N., Wong, H.-S.P.: The end of Moore's law: a new beginning for information technology. Comput. Sci. Eng. **19**, 41–50 (2017)
3. Sommerville, I.: Software Engineering. Pearson, Boston, Columbus, Indianapolis, New York, San Francisco, Hoboken, Amsterdam, Cape Town, Dubai, London, Madrid, Milan, Munich, Paris, Montreal, Toronto, Delhi, Mexico City, São Paulo, Sydney, Hong Kong, Seoul, Singapore, Taipei, Tokyo (2016)
4. Volk, M., Staegemann, D., Trifonova, I., Bosse, S., Turowski, K.: Identifying similarities of big data projects–a use case driven approach. IEEE Access **8**, 186599–186619 (2020)
5. Damasiotis, V., Fitsilis, P., O'Kane, J.F.: Modeling software development process complexity. Int. J. Inf. Technol. Project Manage. **9**, 17–40 (2018)
6. Staegemann, D., Volk, M., Jamous, N., Turowski, K.: Understanding issues in big data applications - a multidimensional endeavor. In: Proceedings of the Twenty-Fifth Americas Conference on Information Systems (2019)
7. Volk, M., Staegemann, D., Pohl, M., Turowski, K.: Challenging big data engineering: positioning of current and future development. In: Proceedings of the IoTBDS 2019, pp. 351–358 (2019)
8. Khan, A.A., Keung, J.W., Fazal-E-Amin, Abdullah-Al-Wadud, M.: SPIIMM: toward a model for software process improvement implementation and management in global software development. IEEE Access **5**, 13720–13741 (2017)
9. Staegemann, D., Volk, M., Jamous, N., Turowski, K.: Exploring the applicability of test driven development in the big data domain. In: Proceedings of the ACIS 2020 (2020)
10. Tosun, A., et al.: An industry experiment on the effects of test-driven development on external quality and productivity. Empir. Softw. Eng. **22**(6), 2763–2805 (2016). https://doi.org/10.1007/s10664-016-9490-0
11. Bissi, W., Serra Seca Neto, A.G., Emer, M.C.F.P.: The effects of test driven development on internal quality, external quality and productivity: a systematic review. Inf. Softw. Technol. **74**, 45–54 (2016)
12. Janzen, D.S., Saiedian, H.: On the influence of test-driven development on software design. In: 19th Conference on Software Engineering Education & Training (CSEET 2006), pp. 141–148. IEEE (2006)

13. Choma, J., Guerra, E.M., da Silva, T.S.: Developers' initial perceptions on TDD practice: a thematic analysis with distinct domains and languages. In: Garbajosa, J., Wang, X., Aguiar, A. (eds.) XP 2018. LNBIP, vol. 314, pp. 68–85. Springer, Cham (2018). https://doi.org/10.1007/978-3-319-91602-6_5
14. Shull, F., Melnik, G., Turhan, B., Layman, L., Diep, M., Erdogmus, H.: What do we know about test-driven development? IEEE Softw. **27**, 16–19 (2010)
15. Crispin, L.: Driving software quality: how test-driven development impacts software quality. IEEE Softw. **23**, 70–71 (2006)
16. Slaats, T., Debois, S., Hildebrandt, T.: Open to change: a theory for iterative test-driven modelling. In: Weske, M., Montali, M., Weber, I., vom Brocke, J. (eds.) BPM 2018. LNCS, vol. 11080, pp. 31–47. Springer, Cham (2018). https://doi.org/10.1007/978-3-319-98648-7_3
17. Keet, C.M., Ławrynowicz, A.: Test-driven development of ontologies. In: Sack, H., Blomqvist, E., d'Aquin, M., Ghidini, C., Ponzetto, S.P., Lange, C. (eds.) ESWC 2016. LNCS, vol. 9678, pp. 642–657. Springer, Cham (2016). https://doi.org/10.1007/978-3-319-34129-3_39
18. Davies, K., Keet, C.M., Lawrynowicz, A.: More effective ontology authoring with test-driven development and the TDDonto2 tool. Int. J. Artif. Intell. Tools **28**, 1950023 (2019)
19. Beck, K.: Test-Driven Development: By Example. Addison-Wesley, Boston (2015)
20. Fucci, D., Erdogmus, H., Turhan, B., Oivo, M., Juristo, N.: A dissection of the test-driven development process: does it really matter to test-first or to test-last? IIEEE Trans. Software Eng. **43**, 597–614 (2017)
21. Williams, L., Maximilien, E.M., Vouk, M.: Test-driven development as a defect-reduction practice. In: Proceedings of the 14th ISSRE, pp. 34–45. IEEE (2003)
22. Janzen, D., Saiedian, H.: Test-driven development concepts, taxonomy, and future direction. Computer **38**, 43–50 (2005)
23. Sangwan, R.S., Laplante, P.A.: Test-driven development in large projects. IT Prof. **8**, 25–29 (2006)
24. Shahin, M., Ali Babar, M., Zhu, L.: Continuous integration, delivery and deployment: a systematic review on approaches, tools, challenges and practices. IEEE Access **5**, 3909–3943 (2017)
25. Karlesky, M., Williams, G., Bereza, W., Fletcher, M.: Mocking the embedded world: test-driven development, continuous integration, and design patterns. In: Embedded Systems Conference. UBM Electronics (2007)
26. Staegemann, D., Volk, M., Daase, C., Turowski, K.: Discussing relations between dynamic business environments and big data analytics. CSIMQ **23**, 58–82 (2020)
27. Vom Brocke, J., Simons, A., Niehaves, B., Reimer, K., Plattfaut, R., Cleven, A.: Reconstructing the giant: on the importance of Rigour in documenting the literature search process. In: Proceedings of the ECIS 2009 (2009)
28. Webster, J., Watson, R.T.: Analyzing the past to prepare for the future: writing a literature review. MISQ **26**, 13–23 (2002)
29. Levy, Y., Ellis, T.J.: A systems approach to conduct an effective literature review in support of information systems research. Informing Sci. J. **9**, 181–212 (2006)

DigEx Workshop

DIGEX 2021 Workshop Chairs' Message

It is our pleasure to present the post-proceeding papers of the 3rd International Workshop on Transforming the Digital Customer Experience (DigEx-2021) that was held in conjunction with the 24th International Conference on Business Information Systems (BIS 2021). The workshop was initiated in response to an increasing need from the scientific and business communities to find a space to exchange ideas and knowledge in this area. As with the previous workshop, DigEx 2021 took place online due to the COVID-19 pandemic; however, it did not stop the scientific community from participating in the workshop and exchanging ideas.

In this edition, after the reviewing process, two papers were accepted from all the submissions for presentation. Each article was reviewed by at least three reviewers. The authors were allowed to revise the papers, taking into account the reviewers' comments and the discussions of the presentation, before including them in the proceedings.

This year the Program Committee (PC) included researchers from 10 universities (Universidad de los Andes, Universidad Nacional de Colombia, INSA Strasbourg, University of Jyväskylä, University of the Republic, Eindhoven University of Technology, TU Dortmund, University of Auckland, Poznan University of Economics and Business, and Thammasat Business School). The effort of the PC in the revision of the articles enabled us to ensure the workshop quality in terms of the value of the scientific contributions.

We would like to thank the authors of the submitted papers and the members of the Program Committee who participated and helped in any way to promote the workshop. We are also grateful to the organizers of the BIS 2021 conference from the Poznan University of Economics and the Leibniz Information Centre for Science and Technology (TIB) for their help in the organization of the conference and associated workshops online. Without their efforts, it would have been impossible to organize this academic event.

Oscar Avila
Virginie Goepp
Beatriz Helena Diaz

Organization

Chairs

Oscar Avila	Universidad de los Andes, Colombia
Virginie Goepp	INSA Strasbourg, France
Beatriz Diaz	Universidad Nacional de Colombia, Colombia

Program Committee

Oscar Avila	Universidad de los Andes, Colombia
Sonia Camacho	Universidad de los Andes, Colombia
Beatriz Díaz	Universidad Nacional de Colombia, Colombia
Lauri Frank	University of Jyväskylä, Finland
Virginie Goepp	INSA Strasbourg, France
Laura Gonzalez	Universidad de la República, Uruguay
Paul Grefen	Eindhoven University of Technology, The Netherlands
Markus Siepermann	TU Dortmund, Germany
David Sundaram	University of Auckland, New Zealand
Marcin Szmydt	Poznan University of Economics and Business, Poland
Sagar Sen	Simula Research Lab, Norway
Mathupayas Thongmak	Thammasat Business School, Thailand

Influence of Augmented Reality on Consumer Behaviour in Online Retailing

Jan Schmidt[1], Christopher Reichstein[2(✉)], and Ralf-Christian Härting[1]

[1] Aalen University of Applied Sciences, Business Administration, Aalen, Germany
jan44.schmidt@gmx.de, ralf.haerting@hs-aalen.de

[2] Baden-Wuerttemberg Cooperative State University, Heidenheim, Germany
christopher.reichstein@dhbw-heidenheim.de

Abstract. This study examines the influence of augmented reality on consumer behaviour in online retailing based on the stimulus-organism-response model. In this context, the affective and cognitive response, and the effect on purchasing behaviour are investigated in more detail. For this purpose, a quantitative study was carried out and analysed using structural equation modelling. The results show a positive influence of the perceived augmentation both on emotions during the use of AR and on the perceived amount of information. The attitude towards the use of AR has the greatest impact on purchasing behaviour, followed by the perceived amount of information. In addition, emotions indirectly effect the purchasing behaviour through its attitude as a mediator.

Keywords: Augmented reality · Consumer behaviour · Online retailing · Stimulus-organism-response model · Empirical study

1 Introduction

Consumers are increasingly shopping online. In Germany, online retail sales almost tripled between 2010 and 2019 from EUR 20.2 billion to EUR 59.2 billion [1]. A current accelerator of this trend is the Corona crisis, in which online sales rose at a record-breaking pace. In the crisis month of August 2020, online sales in Germany were 22.9% higher than in August 2019 [2]. Hereby, Augmented Reality (AR) technology could be a possible driver for even greater future growth in online retail. AR can be defined as a technology that "allows the user to see the real world, with virtual objects superimposed upon or composited with the real world" [3]. In online retailing, AR represents a new opportunity for product presentation [4, 39]. Until a few years ago, trying out products before buying them was the exclusive preserve of stationary retailers. With AR, virtual try-ons of glasses, make-up and clothes or setting up a new piece of furniture in the living room are also possible online. Some companies such as Maybelline [5], Mister Spex [6] and IKEA [7] already use AR in their online shops. With the spread of AR technology in online retailing, companies and market researchers are faced with the questions of how consumers react to AR and whether the use of AR in online shops has a positive influence on purchasing behaviour. Insights

W. Abramowicz et al. (Eds.): BIS 2021 Workshops, LNBIP 444, pp. 131–143, 2022.
https://doi.org/10.1007/978-3-031-04216-4_14

into the behaviour of consumers in shopping situations with AR help companies to derive measures for action to optimize their online shops. How consumers react to AR in online shops has not yet been sufficiently investigated in consumer behaviour research [8]. This topic has increasingly become the focus of research in the last few years. However, previous studies have mostly focused on the investigation of the acceptance of AR applications by consumers using the technology acceptance model (TAM) [34–36]. In this technology-oriented view, the actual core characteristic of AR, the augmentation of products, is not yet investigated in research. In contrast, the stimulus-organism-response (SOR) model considers augmentation as a stimulus for internal psychological processes in the organism and subsequent behaviour. Thereby, the SOR model has been largely neglected in the investigation of the impact of AR on consumer behaviour in online retailing. The aim of this paper is to reduce this research gap using the SOR model to gain insights into the extent to which AR applications in online shops influence consumers' purchasing behaviour. Moreover, the study focuses on the role of cognitive and affective responses. Thus, emotions during the use of AR and the attitude towards the use of AR are considered as affective components. The perceived amount of information represents the cognitive component.

The paper is structured as follows. Section 2 introduces the research design and method, followed by the documentation of the results in Sect. 3 with a subsequent discussion in Sect. 4. The paper ends with a conclusion in Sect. 5.

2 Research Design and Method

2.1 Hypothesis Development

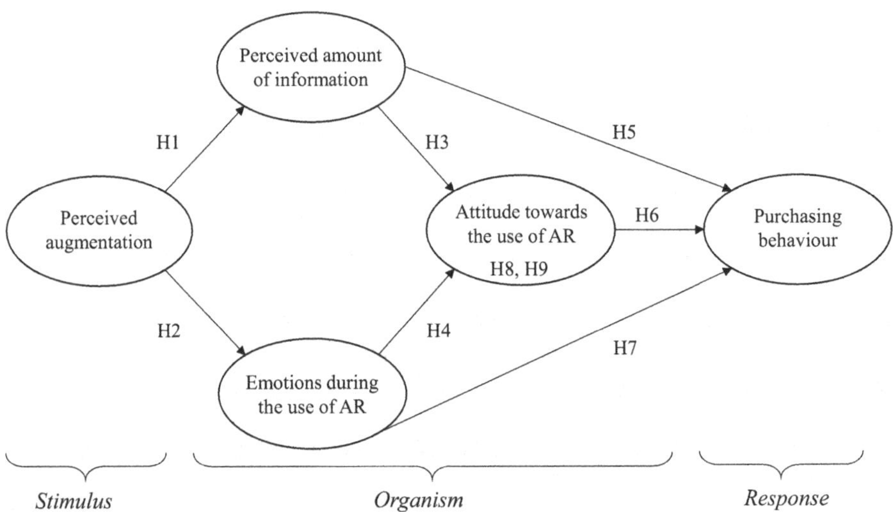

Fig. 1. Research model

The research model of the empirical study is based on the SOR framework [10]. In contrast to the TAM, the SOR model is better suited to depict the causal link between psychological processes and human behaviour in the context of a purchase decision situation. The TAM, on the other hand, focuses more on technological aspects of the application and mainly uses cognitive factors to explain behaviour. According to the SOR theory, stimuli lead to affective and cognitive response, which then influence an individual's behaviour [11]. In this study, the construct of perceived augmentation acts as a stimulus. The cognitive component of the organism forms the construct perceived amount of information. From the field of affective processes within the organism, positive emotions such as enthusiasm, joy and happiness during the use of AR and the attitude towards the use of AR were measures. Purchasing behaviour is the response variable in this study. Based on the interrelationships of the stimulus, organism and response variables, the authors derived nine hypotheses. Previous studies show that virtual product experiences lead to a higher perceived amount of product related information and to a better consumer information than is the case with static product presentations like text or images [12, 13, 38]. Virtual product experiences were stimulated in these studies with 360-degree rotations of products. Due to the characteristics of AR technology, AR product presentations can also be classified as a virtual product experience. Therefore, the authors assume that AR can contribute to a higher diversity of information, which leads to the first hypothesis:

H1: The perceived augmentation of AR applications in online shops has a positive effect on the amount of information perceived by consumers.

In online retailing, AR is used as a stimulus to evoke positive emotions during online shopping and thus to create a customer experience [14]. A study by Javornik has already shown that augmentation as a characteristic of AR can lead to positive affective response [15]. Hence, the authors formulate the following hypothesis:

H2: The perceived augmentation of AR applications has a positive effect on the emotions of consumers in online shops.

Consumers who feel informed by a product presentation tend to have a more positive attitude towards the online shop [16]. The perceived amount of information in a 3D product presentation also influences the attitude of consumers [13]. Based on the above understanding, the following hypothesis is set forth:

H3: The perceived amount of information has a positive effect on consumers' attitude towards using AR.

According to several studies, attitudes towards virtual product presentations are positively influenced by their entertainment value and the emotions they evoke [17, 18]. In this context, the following hypothesis is developed:

H4: Emotions during the use of AR applications have a positive influence on the attitude of consumers towards AR.

According to research findings, the perceived amount of product-related information influences the purchase intention of consumers [19, 20]. In the case of 360-degree product presentations, which create a three-dimensional experience like AR, a study has already found a positive influence of the perceived amount of information on the purchase intention [13]. The authors therefore derive the following hypothesis:

H5: The perceived amount of information provided by AR applications has a positive effect on the purchasing behaviour of consumers in online shops.

There are numerous studies regarding attitudes as an influencing factor on purchasing behaviour. The work by Hausmann and Skiepe shows that the consumers' intention to buy something on a website is positively related to their attitude towards this website [21]. Another study was able to identify a positive relationship between attitudes towards virtual product presentations and the purchase intention of consumers [13]. These findings lead to the following hypothesis:

H6: A positive attitude towards the use of AR has a positive effect on the purchasing behaviour of consumers in online shops.

In the context of online shopping, Wu et al. were able to determine a positive connection between the emotions regarding a website and the purchase intention of consumers [22]. With regard to AR, which is used in online retail to create a customer experience and should contribute to an increase in the conversion rate [14], an experiment by Watson et al. has already found that the positive emotional response when using an AR makeup app positively influences the purchase intention [26]. Thus, the authors postulate:

H7: Emotions during the use of AR applications have a positive influence on the purchasing behaviour of consumers in online shops.

According to Rosenberg and Hovland's theory, attitudes not only include affective and cognitive components, but are also closely related to a behavioural component [23]. Based on this, the authors assume that the attitude towards the use of AR plays a mediator role in the research model with respect to the relationship between the perceived amount of information (H8) as well as the emotions during the use of AR (H9) and the purchasing behaviour of consumers in online shops (Fig. 1).

2.2 Research Method and Data Collection

To analyse the hypotheses, the authors conducted an online survey with a standardized questionnaire using the software LimeSurvey. The questionnaire is structured in an introduction, personal questions, general questions about previous experience with AR in online shops and the survey of the five variables. A filter question is also part of the questionnaire. People who state that they have not yet used any AR applications in online shops cannot take part in the survey of the variables and have to end the survey. The authors used existing multi-item scales from the literature to measure the reflective constructs perceived augmentation [24], the perceived amount of information [20], emotions during the use of AR [25, 26] and attitudes towards the use of AR [9]. To identify possible inconsistencies in the response behaviour of the respondents, these constructs were also measured with a global single item. Since purchasing behaviour in this study only refers to whether AR influences the intention to purchase, the authors decided to measure the endogenous construct with a single item on a ratio scale based on Munyon et al. [37]. The measurement models are shown in Table 3 in the appendix. All items of the constructs were queried on a five-point Likert scale from agree, somewhat agree, neutral, somewhat disagree to disagree.

A pre-test ensured the quality and comprehensibility of the questionnaire. The study started on October 16, 2020 and ended on November 29, 2020. The authors shared the survey via Facebook, Instagram, XING and WhatsApp. Participants could access and take part in the online survey via a link. A total of 399 people took part in the online survey. The data sets of 105 people who stated that they had not yet used AR applications for online shopping were not considered in the analysis. Furthermore, the authors excluded the data sets of 55 other people who cancelled the online survey. The remaining 239 data sets were checked for inconsistent response behaviour and response patterns, such as straight lining [27]. Four data sets indicated that the questions were answered inappropriately and were eliminated from the sample. After data cleaning, the final sample consists of 235 participants.

57.9% of the survey participants are male and 42.1% female. The majority of respondents (56.6%) are between 18 and 24 years old. Less than half of the respondents are 25 or older (41.7%) and 1.7% are under 18. In line with the age structure, a large proportion of participants (54%) are pupils, students or trainees. In contrast, 43.8% are employed. In terms of previous use of AR, about one-third of respondents (32.3%) said they had used an AR application only once, and just under half of respondents (47.2%) used them rarely. Only 18.7% used AR in online shops occasionally. The proportion of those who used AR frequently or very often (1.8%) is noticeably low. Glasses have already been tried on virtually by 75.3% of the sample. 38.7% have also tested furniture with AR in online shops. The participants had comparatively little AR experience with cosmetics (10.2%), clothing (9.4%), jewellery and watches (6%), household appliances (4.3%) and shoes (3.4%).

The statistical method of structural equation modeling is used to examine the relationships of the constructs in the research model [27]. The authors decided to use the variance-based partial least squares estimation method, which is commonly used in marketing and is applicable even with a relatively small sample [27]. In this method, the significance of the paths is determined by bootstrapping. With the support of the SmartPLS software [28], the structural model and the measurement models were evaluated. To evaluate the measurement models, the authors checked internal consistency using Cronbach's alpha (CA) and composite reliability (CR). Convergence validity is ensured by checking the outer loadings of the items, also known as item reliability (IR), and the average variance extracted (AVE) [27]. The authors chose the heterotrait-monotrait (HTMT) ratio to test discriminant validity [27]. The measurement models in this study exceed the recommended thresholds from the literature of 0.7 for CA, 0.6 for CR, and 0.5 for AVE [27]. The outer loadings of items should be >0.7. Items with a value below 0.4 should be removed from the measurement model. Items with values between 0.4 and 0.7 do not necessarily have to be eliminated but can be retained for reasons of content validity. However, eliminating these indicators may be useful if it leads to an increase in CA, CR or AVE beyond the threshold required for each criterion [27]. In the perceived augmentation construct, one item (The applications place virtual objects on your body or in your space) was removed from the measurement model because of an outer loading below 0.4. To reach the requirement level for AVE, the authors removed three items (excited, interested and inspired) from the measurement model of the emotions construct. Furthermore, there is no lack of discriminant validity in the model, since the HTMT values for all construct combinations are below 0.9 [27].

In addition, the structural model was tested for multicollinearity using the variance inflation factor (VIF) [27]. In this study, there is no problem of multicollinearity as all VIF values are less than five. Moreover, the authors use R^2 as a central quality criterion for the explanatory power of the structural model and its constructs [27]. Following Chin [29], the R^2 values for the constructs emotions during the use of AR (0.194) and perceived amount of information (0.266) can be rated as weak and the R^2 values for the constructs purchasing behaviour (0.376) and attitude towards the use of AR (0.413) as moderate [29]. However, in the research discipline of consumer behaviour, an R^2 value of >0.2 is considered high [27]. According to this, all R^2 values for the constructs are to be considered high except for the R^2 value for the emotions construct.

3 Results

The results of the structural equation model are shown in Fig. 2.

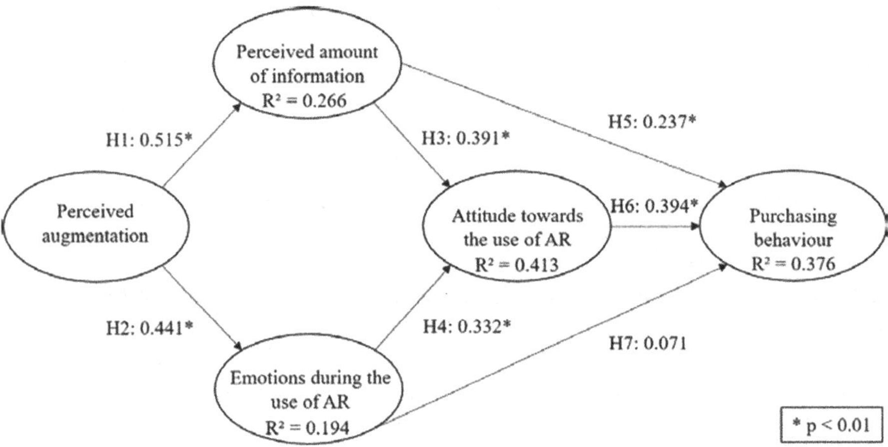

Fig. 2. Structural equation model with coefficients

The results show a significant positive effect (ß = 0.515, $p < 0.01$) of the perceived augmentation on the perceived amount of information, and thus also on the cognitive processes in the organism. H1 can be confirmed. Regarding the affective component emotions, there is also a significant positive relationship (ß = 0.441, $p < 0.01$) based on perceived augmentation. H2 can thus also be confirmed. The results for the processes within the organism show a significantly positive influence of the perceived amount of information on the attitude towards AR (ß = 0.391, $p < 0.01$). This leads to confirmation of H3. There is also a significant positive relationship (ß = 0.332, $p < 0.01$) between emotions when using AR applications and the attitude towards using AR. The authors can confirm H4. Moreover, the results of the empirical analysis show that the purchasing behaviour of consumers is significantly positively influenced by the amount of information they perceive (ß = 0.237, $p < 0.01$). These findings support H5. H6 assumes that a positive attitude towards AR leads to a positive purchasing behaviour of the consumer. The results show a significant and positive relationship (ß = 0.394, $p < 0.01$). Thus, H6 can be confirmed, too. However, the results do not reveal a direct significant influence of emotions when using AR (ß = 0.071, $p > 0.01$) on purchasing behaviour. H7 must be rejected.

Table 1. Direct effects

SEM path	Path coefficient	M	SD	t statistics	p values
Augmentation → Information	0.515	0.518	0.051	10.054	0.000
Augmentation → Emotions	0.441	0.447	0.051	8.559	0.000
Information → Attitude	0.391	0.392	0.067	5.837	0.000
Emotions → Attitude	0.332	0.337	0.066	5.014	0.000
Information → Purchasing behaviour	0.237	0.233	0.076	3.123	0.002
Attitude → Purchasing behaviour	0.394	0.398	0.067	5.900	0.000
Emotions → Purchasing behaviour	0.071	0.072	0.071	1.001	0.317

In addition to the direct effects (Table 1), the indirect effects of the model can also be considered. Dependencies between two model variables can be mediated through indirect effects via third party variables in the model by calculating the product of the regression coefficients along the paths [27]. Based on the indirect effects (Table 2), H8 and H9 are tested below. The indirect effect between the amount of information perceived and purchasing behaviour are weakly significant (ß = 0.154, $p < 0.01$). Since the direct effects are also significant, the conditions for a complementary mediation are met [27]. Accordingly, a part of the effect of the perceived amount of information on purchasing behaviour is explained by the attitude towards AR. Thus, H8 is supported. Since the direct effect between emotions when using AR and purchasing behaviour is not significant, but the indirect effect is significant ($\beta = 0.131$, $p < 0.01$), the conditions for exclusively indirect mediation are met [27]. That means: Positive emotions when using AR lead to a positive attitude towards AR, which then has a positive influence on

Table 2. Indirect effects

SEM path	Path coefficient	M	SD	t statistics	p values
Augmentation → Information → Attitude	0.201	0.204	0.044	4.615	0.000
Augmentation → Emotions → Attitude	0.146	0.151	0.034	4.264	0.000
Augmentation → Information → Purchasing behaviour	0.122	0.121	0.042	2.944	0.003
Augmentation → Emotions → Purchasing behaviour	0.031	0.032	0.032	0.979	0.328
Emotions → Attitude → Purchasing behaviour	0.131	0.134	0.033	3.922	0.000
Information → Attitude → Purchasing behaviour	0.154	0.157	0.040	3.841	0.000

purchasing behaviour. H9 can be confirmed, too. Further indirect effects in the research model are presented below. Via the organism variables, the perceived augmentation has an indirect, significantly positive effect on attitudes towards the use of AR ($\beta = 0.348$, $p < 0.01$) and on the purchasing behaviour of consumers ($\beta = 0.290$, $p < 0.01$).

4 Discussion

This study uses the SOR framework from consumer behaviour theory to examine the impact of AR in online shops on consumers. Based on the interrelationships between the SOR components, the results of this study can be discussed. The results show that perceived augmentation as a stimulus has a positive effect on the organism. Positive effects of AR can be found for both cognitive and affective response. In the cognitive area, the perceived augmentation of AR applications leads to a positive influence on information processing. This finding can be derived from the result of the study that consumers perceive a lot of information about the product through AR product presentations. Compared to the classic product presentation in online shops with text and pictures, a more direct product experience is possible through the virtual trying on of products respectively the virtual placement of products in 3D in the consumer's environment [30]. As a further finding, the present research provides that high perceived augmentation in online shopping leads to positive emotions. This result is supported by Javornik's study, which found a positive effect of perceived augmentation on the affective response with regard to virtual try-ons of glasses [15]. So, the joy of using AR applications can create a shopping experience for consumers.

In this study, the attitude of consumers towards the use of AR applications plays a central role in investigating the influence of AR on consumer behaviour. On the one hand, the additional perceived information about the products and, on the other hand, the positive emotions that a consumer experiences while using AR contribute to a positive attitude formation. These study results show that both cognitive processes of information acquisition and processing as well as affective components, such as emotions, lead to a positive attitude towards AR in online shops. Earlier studies came to similar conclusions, which found a positive effect of perceived usefulness and enjoyment on the attitude towards AR [31, 32].

Another result of the present work is that the attitude towards AR has the greatest positive influence on the purchasing behaviour compared to the perceived amount of information and the emotions when using AR. However, the influence of the perceived amount of information on the purchase decision is also considerable. The additional information provided by AR seems to make it easier for consumers to assess the products they are looking at. This in turn leads to a reduction of uncertainties in consumer decision-making, which was already proven in a previous study by Dacko in the context of mobile AR shopping apps [33]. Surprisingly, the results of the present study show that the positive emotions of consumers while using AR have no direct influence on the purchase decision. In contrast, other studies found that positive emotional response of a virtual product presentation leads to an increased purchase intention [13, 26]. In the present study, it should be pointed out the special role of the attitude in mediating the relationship between emotions and purchasing behaviour. The positive emotions evoked during online shopping with AR have an indirect effect on the intention to purchase via the attitude towards AR.

In summary, it can be derived from the results of this study that the use of AR in online shops is an effective form of product presentation to positively influence consumers' purchase decisions. It is noteworthy that, according to the results of this study, the amount of information about the products is more important for consumers' purchase decisions than the entertainment value of AR applications.

5 Conclusion

The aim of this paper was to gain insights into the extent to which AR applications in online shops influence the purchasing behaviour of consumers. The SOR model was used as a theoretical basis to determine the causal relationships between AR and the cognitive as well as affective processes within the consumer and their influence on purchasing behaviour. The results show that product presentations with AR have a positive influence on intrapsychic processes in the consumer. Both cognitive and affective components in the organism have a positive effect on the purchasing behaviour of consumers in online shops.

In terms of theoretical implications, the study provides essential methodological contributions regarding the influence of AR on consumer behaviour in online retailing. In contrast to previous studies, this paper focuses on internal psychological processes in the organism of the consumer by using the SOR model (user centricity) instead of concentrating on the technology acceptance by users (technology centricity). By doing so, the authors developed a model to directly measure how the attitude towards the use of AR, the perceived amount of information and emotions during the use of AR impacts their purchasing behaviour. Moreover, it is now possible to indirectly measure the attitude towards the use of AR through the perceived augmentation and to indirectly measure the purchasing behaviour. Here, the construct "emotions during the use of AR" only indirectly effects the purchasing behaviour by mediating its influence through the attitude towards the use of AR.

With respect to practical implications, the study provides insights for online retailers. The disadvantage of online retail compared to stationary retail of not being able to try out and check products in real life is considerably reduced by the use of AR technology. Due to its positive effects on emotions, online retailers can use AR as a tool to bind customers emotionally to their store. The supportive effect of AR in decision-making can also be expected to reduce the risk of mispurchases and the associated returns. This in turn can lead to cost savings on the part of online retailers. Furthermore, AR can increase conversion rates and consequently sales revenues for online retailers.

However, the study has some limitations. Differences between the online shops, e.g. with regard to the quality of the AR technology used, were not collected. This also applies to the point in time at which survey participants gained AR experience. Due to the rapidly advancing development of AR technology and the quality of AR applications, the authors assume that more recent AR experiences tended to be rated more positively. When evaluating the results, there was also no breakdown of the product groups viewed by consumers. Assuming that AR applications are differently suitable for different product groups, research on specific products can provide more differentiated results. In this context, further studies can examine the influence of the attitude towards the product. In addition, this study did not measure the real purchases made after viewing an AR product presentation, but only the potential that consumers see in AR to positively influence their purchase decisions. Future research could come up with more precise results by measuring real purchases made through AR. Since purchasing behaviour was measured with a single item, there are limitations regarding the reliability of this construct, too. Due to the complexity of consumer behaviour, the authors did not examine all influencing factors in this study. In another study, users can be differentiated according to the phase in the purchase decision process. It can be assumed that the influence of AR on purchasing behaviour depends on the phase in which the customer is in the purchasing process [37]. Furthermore, the purchasing motivation of consumers can be investigated. When using the AR applications, the participants had different technical circumstances, such as the internet connection or the end devices used. A comparable study with a laboratory experiment could avoid these inequalities.

With increasingly powerful end devices and the ongoing development of AR technology, an even more realistic AR representation of products in online shops will be possible in future. This will presumably lead to consumers perceiving augmentation more intensively and thus shopping increasingly become an experience in which more product-related information can be perceived. It remains to be seen whether these developments will also lead to an even stronger impact on the purchasing behaviour of consumers. However, as this study shows, online retailers should consider AR as a way of presenting their products.

Appendix

Table 3. Measurement models and quality criteria

Constructs	IR	CR	CA	AVE
Perceived augmentation [24]		0.824	0.715	0.540
The virtual objects on the devices seem real	0.737			
The virtual objects seem part of the real environment	0.745			
The virtual objects still seem to be part of the real environment when you turn your body or your device	0.797			
The virtual objects seem to exist in real time	0.653			

(continued)

Table 3. (*continued*)

Constructs	IR	CR	CA	AVE
Perceived amount of information [20]		0.852	0.783	0.535
The AR applications make it possible to learn a lot about the products viewed	0.715			
The product presentations with AR are very informative	0.757			
After viewing the products through the AR applications, you know enough to make an informed purchase decision	0.730			
You can fully trust the information provided by the AR applications	0.724			
The AR applications provide you with a lot of information about the products	0.729			
Emotions during the use of AR [25, 26]		0.824	0.728	0.541
During the use of AR applications…				
…you are enthusiastic	0.788			
…you feel joyful	0.743			
…you feel entertained	0.649			
…you are happy	0.754			
Attitude towards the use of AR [9]		0.890	0.843	0.621
The AR applications are so interesting that you will want to learn more about them	0.612			
It makes sense to use AR applications when shopping online	0.809			
Using AR applications when shopping online is a good idea	0.832			
Other people should also use AR applications in online shops	0.811			
Using AR applications is helpful	0.853			
Purchasing behaviour [37]		1.000	1.000	1.000
AR applications have great potential to positively influence your purchase decision in online shops	1.000			

References

1. Handelsverband Deutschland (HDE): Online Monitor (2020). https://einzelhandel.de/index.php?option=com_attachments&task=download&id=10433. Accessed 21 Jan 2021
2. Statistisches Bundesamt (Destatis): Kaufhäuser in der Krise (2020). https://www.destatis.de/DE/Presse/Pressemitteilungen/2020/10/PD20_N063_45212.html. Accessed 18 Jan 2021
3. Azuma, R.T.: A survey of augmented reality. Presence Teleoperators Virtual Environ. **6**(4), 355–385 (1997)
4. Javornik, A.: Classifications of augmented reality uses in marketing. In: Proceedings of the 13th IEEE International Symposium on Mixed and Augmented Reality, vol. 13, pp. 67–68 (2014)

5. Maybelline: Try It On (n.d.). https://www.maybelline.com/virtual-try-on-makeup-tools. Accessed 6 Mar 2021
6. Mister Spex: Virtual try on (n.d.). https://www.misterspex.co.uk/l/pg/100508. Accessed 6 Mar 2021
7. IKEA: Say hej to IKEA Place (n.d.). https://www.ikea.com/au/en/customer-service/mobile-apps/say-hej-to-ikea-place-pub1f8af050. Accessed 6 Mar 2021
8. Riar, M., Korbel, J., Xi, N., Zarnekow, R., Hamari, J.: The use of augmented reality in retail: a review of literature. In: Proceedings of the 54th Hawaii International Conference on System Sciences, vol. 54, pp. 638–647 (2021)
9. Schreiber, S.: Die Akzeptanz von Augmented-Reality-Anwendungen im Handel. Springer Gabler, Wiesbaden (2020). https://doi.org/10.1007/978-3-658-29163-1
10. Woodworth, R.S.: Psychology: A Study of Mental Life. H. Holt, New York (1929)
11. Mehrabian, A., Russell, J.A.: An Approach to Environmental Psychology. MIT Press, Cambridge (1974)
12. Li, T., Meshkova, Z.: Examining the impact of rich media on consumer willingness to pay in online stores. Electron. Commerce Res. Appl. **12**(6), 449–461 (2013)
13. Park, J., Stoel, L., Lennon, S.J.: Cognitive, affective and conative responses to visual simulation: the effects of rotation in online product presentation. J. Consum. Behav. **7**, 72–87 (2008)
14. Poushneh, A., Vasquez-Parraga, A.Z.: Discernible impact of augmented reality on retail customer's experience, satisfaction and willingness to buy. J. Retail. Consum. Serv. **34**, 229–234 (2017)
15. Javornik, A.: 'It's an illusion, but it looks real!' Consumer affective, cognitive and behavioural responses to augmented reality applications. J. Mark. Manage. **32**(9–10), 987–1011 (2016)
16. Smith, S.P., Johnston, R.B., Howard, S.: Putting yourself in the picture: an evaluation of virtual model technology as an online shopping tool. Inf. Syst. Res. **22**(3), 640–659 (2011)
17. Kim, J., Forsythe, S.: Factors affecting adoption of product virtualization technology for online consumer electronics shopping. Int. J. Retail Distrib. Manage. **38**(3), 190–204 (2010)
18. Kim, J., Forsythe, S.: Adoption of sensory enabling technology in online apparel shopping. Eur. J. Mark. **43**, 1101–1120 (2009)
19. Kim, H., Lennon, S.J.: E-atmosphere, emotional, cognitive, and behavioral responses. J. Fashion Mark. Manage. **14**(3), 412–428 (2010)
20. Kim, M., Lennon, S.J.: Television shopping for apparel in the United States: effects of perceived amount of information on perceived risks and purchase intention. Family Consum. Sci. Res. J. **28**(3), 301–330 (2000)
21. Hausman, A.V., Siekpe, J.S.: The effect of web interface features on consumer online purchase intentions. J. Bus. Res. **62**(1), 5–13 (2009)
22. Wu, W.-Y., Lee, C.-L., Fu, C.-S., Wang, H.-C.: How can online store layout design and atmosphere influence consumer shopping intention on a website? Int. J. Retail Distrib. Manage. **42**, 4–24 (2013)
23. Rosenberg, M.J., Hovland, C.I.: Cognitive, affective, and behavioral components of attitudes. In: Rosenberg, M., Hovland, C., McGuire, W., Abelson, R., Brehm, J. (eds.) Attitude organization and change, pp. 1–14. Yale University Press, New Haven (1960)
24. Javornik, A., Rogers, Y., Moutinho, A.M., Freeman, R.: Revealing the shopper experience of using a "Magic Mirror" augmented reality make-up application. In: Proceedings of the 2016 ACM Conference on Designing Interactive Systems (DIS 2016), vol. 16, pp. 871–882. Association for Computing Machinery (2016)

25. Chang, H.-J., Eckman, M., Yan, R.-N.: Application of the stimulus-organism-response model to the retail environment: the role of hedonic motivation in impulse buying behavior. Int. Rev. Retail Distrib. Consum. Res. **21**(3), 233–249 (2011)
26. Watson, A., Alexander, B., Salavati, L.: The impact of experiential augmented reality applications on fashion purchase intention. Int. J. Retail Distrib. Manage. **48**(5), 433–451 (2018)
27. Hair, J.F., Jr., Hult, G.T.M., Ringle, C., Sarstedt, M.: A Primer on Partial Least Squares Structural Equation Modeling (PLS-SEM). Sage Publications, Thousand Oaks (2016)
28. Ringle, C.M., Wende, S., Becker, J.-M.: SmartPLS 3, Bönningsted (2015). http://www.smartpls.com. Accessed 20 Jan 2021
29. Chin, W.W.: The partial least squares approach to structural equation modelling. In: Marcoulides, G.A. (ed.) Modern Methods for Business Research, pp. 295–358. Lawrence Erlbaum, Mahwah (1998)
30. Verhagen, T., Vonkeman, C., Feldberg, F., Verhagen, P.: Present it like it is here: creating local presence to improve online product experiences. Comput. Hum. Behav. **39**, 270–280 (2014)
31. Pantano, E., Rese, A., Baier, D.: Enhancing the online decision-making process by using augmented reality: a two country comparison of youth markets. J. Retail. Consum. Serv. **38**, 81–95 (2017)
32. Yim, M.Y.-C., Chu, S.-C., Sauer, P.: Is augmented reality technology an effective tool for e-commerce? An interactivity and vividness perspective. J. Interact. Mark. **39**, 89–103 (2017)
33. Dacko, S.G.: Enabling smart retail settings via mobile augmented reality shopping apps. Technol. Forecasting Soc. Change **124**, 243–256 (2017)
34. Alves, C., Luís Reis, J.: The intention to use E-commerce using augmented reality - the case of IKEA place. In: Rocha, Á., Ferrás, C., Montenegro Marin, C.E., Medina García, V.H. (eds.) ICITS 2020. AISC, vol. 1137, pp. 114–123. Springer, Cham (2020). https://doi.org/10.1007/978-3-030-40690-5_12
35. Alam, S.S., Susmit, S., Lin, C.-Y., Masukujjaman, M., Ho, Y.-H.: Factors affecting augmented reality adoption in the retail industry. J. Open Innov. Technol. Market Complexity **7**(2), 1–24 (2021)
36. Park, M., Yoo, J.: Effects of perceived interactivity of augmented reality on consumer responses: a mental imagery perspective. J. Retail. Consum. Serv. **52**, 1–9 (2020)
37. Munyon, T., Matthew, J., Crook, T., Harvey, P., Edwards, J.: Consequential cognition: exploring how attribution theory sheds new light on the firm-level consequences of product recalls. J. Organ. Behav. **40**(5), 587–602 (2018)
38. Wintermann, O.: Perspektivische Auswirkungen der Corona-Pandemie auf die Wirtschaft und die Art des Arbeitens. Wirtschaftsdienst **100**(9), 657–661 (2020). https://doi.org/10.1007/s10273-020-2733-0
39. Härting, R.C., Reichstein, C., Härtle, N., Stiefl, J.: Potentials of digitization in the tourism industry – empirical results from German experts. In: Abramowicz, W. (ed.) BIS 2017. LNBIP, vol. 288, pp. 165–178. Springer, Cham (2017). https://doi.org/10.1007/978-3-319-59336-4_12

Personality Based Data-Driven Personalization as an Integral Part of the Mobile Application

Izabella Krzeminska[1,2(✉)] and Marcin Szmydt[1]

[1] Poznan University of Economics and Business, al. Niepodległosci 10, 61-875 Poznan, Poland
[2] Orange Polska, R&D Labs, Obrzezna 7, Warszawa, Poland
Izabella.Krzeminska@orange.com

Abstract. The article presents the results of the work on the method of intuitive UI and UX personalization of mobile applications. The method is based on the user's personality profile (Big 5) inferred from the available data on the user's phone at the time of installation. The user's personality model was created based on machine learning performed on data from 2,202 people. The proposed method enables personalization from the first contact of the customer with the application. Therefore, it is a significant advantage of the study. Moreover, the method ensures complete data privacy protection since no data about the user is uploaded outside the mobile phone.

Keywords: User · Service personalizing · Data based personalisation · Human-centred services · Detecting personality based on digital data

1 Introduction and Research Objectives

The digital revolution has made the smartphone the most used personal device, and a natural source of information about the user, which is confirmed by numerous studies [4]. The deepening dependence and coexistence with technology mean that profiling based on demographic characteristics is insufficient. Currently, digital data about users named "digital fingerprints" are commonly collected and used for profiling and classification. However, gathering users data is time-consuming and resulted in delaying product adjustment to the user preferences.

Therefore, the first motivation was to look for good classifiers that can be used from the first moment of using the application, right after installation (previous publications related to this research program are [12–14]). Then the insight about the user can be used for dynamic and automatic adaptation of services, e.g. smartphone application. The above motivation defines the main research problem analyzed in this publication, which is: how to define the personality profile of a mobile application user and personalize it to their needs from the moment the application is installed. To solve this problem, the following research

W. Abramowicz et al. (Eds.): BIS 2021 Workshops, LNBIP 444, pp. 144–155, 2022.
https://doi.org/10.1007/978-3-031-04216-4_15

questions were defined: (RQ1) Is it possible to create an automatic method to determine the user's personality based on the available mobile phone data during the app installation? (RQ2) Can this method be used in a mobile application for automatic application personalization?

The article presents the possibility of using telephone data for active profiling and automatic personalization of UX and UI. The first part will present research objectives, key findings from the review of related research, chosen research methodology and scheme. Then the results of research on the data personality model and the concept of personalizing a mobile application based on this profile will be discussed. In the end, conclusions will be presented, and the limitations and further plans for research and development of the concept will be discussed.

2 Research Methodology

The overall methodology chosen to carry out the required research is Design Science [6]. Following the Design Science framework, the presented research consists of the following steps presented in Fig. 1 The first step is identifying the existing problems in data-driven personalised personalisation. A literature review was conducted to search the possible existing solutions with a detailed analysis of available data. The summary of this stage is in section *Related Works*. Next, the research procedure consists of pre-research stages: interviews with customers, the psychometric procedure for creating required personality tools, preparing tools for data collecting and data collection. A summary of these pre-research studies is in section *Pre-research*. Then the primary research for the creation of artefact was conducted. The Hevner's Design Cycle: data processing and artefact development is presented in section *Research*.

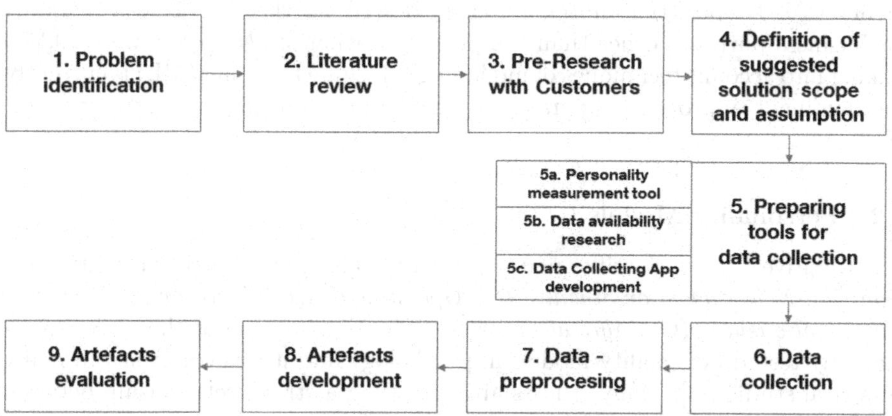

Fig. 1. Research steps defined according Design Science Research (own source)

3 Summary from Literature Review

The systematic literature review was conducted based on the method proposed by [15]. The primary keys used for the input search were: *determining personality based on (digital) data, personalization of services based on the user's personality, user's profiling based on data, customer-oriented services*; *mobile app personalisation*; *personality as a base of human-robot interaction.*

3.1 Prediction Personality from Data

Over the past ten years, many attempts have been made to define personality based on the digital footprint. The citations in this article are not exhaustive, and only examples are given to illustrate the conclusions from the more comprehensive literature review. Most of the research is concerned with determining personality from Social Media (SM) data, mainly through large text data-sets, available as open data on Twitter or MyPersonality App for training purposes (e.g. [2,10,11].

There are attempts of predicting personality from SM profile picture [10,17]. The single studies in this topics, explore other than SM data, e.g. call logs [21], registry from mobile applications [26], eye movements [1], data from devices such as socio-badge [9]. It is worth noting that some of these studies are currently impossible to repeat due to the lack of availability of this data, but still they are valuable from a research perspective. This research relied on a large amount of data, either collected or possessed from service use history, with a few exceptions.

Based on a review of more than 35 studies on this subject, it can be concluded that the basic data for personality detection are text data from posts and tweets on SM platforms. It also seems that some researchers focus mainly on improving the ML methods themselves, without the specific purpose of using these models in practice. Despite the efforts, no cases describing the use of such a model for business purposes other than the personalization of SM were found [20,22]. Taking into account techniques of modelling, Machine Learning (ML) is currently the dominant approach for personality prediction from digital footprint [5,11, 16,23,24].

3.2 Personality Models

The Big Five Theory classifies personality traits along five dimensions: *Extraversion* (E), *Neuroticism or Stability (S)*, *Openness to Experience or Intellect* (O), *Conscientiousness* (C), *Agreeableness*(A). The Big 5 is one of the best experimentally tested personality models in psychology and it was confirmed in many empirical studies [3]. Many studies are indicating a strong relationship between Big 5 and behavior and preferences [8].

There is also another personality typology: the Myers-Briggs Type Indicator (MBTI) [7]. It is an array of 16 personality types, resulting from a combination of 4 binary dimensions: Introvert-Extravert, Intuition-Sensing, Thinking-Feeling, Judging-Perceiving.

Considering the attempts of data-driven personality, The Big 5 is used the most often (23 from 35 found out) while the MBTI is used about less frequently (7 from 35). The remaining 5 are single uses of other approaches. Regardless of the model used (Big 5 or MBTI), it can be stated that the personality model is treated as a set of discrete binary variables in most cases. This approach is coherent only with the Myer-Brigs Theory, a typology composed of a combination of 2-pole classes. Although using personality traits as a binary variable is convenient for building models (avoiding imbalance sample issue), it does not seem justified for personalizing products. For personalization, traits are for identifying those who differ significantly from the typical, average level of the trait. In the Big 5 model, traits are dimensions, and it is possible to define the typical users and the cut-off points of extreme groups. This fundamental difference affects both the interpretation and possible use of the result. Finally, the prediction of binary typology is less likely to differentiate behaviour [25].

3.3 Differentiation of Users Experience Based on Personality

There are some examples of using the personality for personalising advertisement execution [20] and recommendation systems [18]. Considering the creation of personality-aware service, valuable insight about the Big 5 personality impact comes from human-AI and human-robots interaction surveys. For example, the service adaptation process is more straightforward in the case of High E, High C, and High S [19]. However, highly neurotic (Low S) are not resistant to stress, accompanied by a higher level of anxiety and a lower ability to adapt to what can be crucial in brand new services based on advanced technology using Virtual Reality or Augmented Reality. In contrast, High O, when learning about and discovering new things feel satisfying, and such activity is beneficial for them. An additional incentive for people with High Openness is their intellectual involvement, so they have different adaptation paths. The research [27] confirms that High E prefers robots with extroverted behaviours and introverts with introverted ones. Therefore, it can be assumed that the inclusion of personality in the profiling of less advanced but interactive services like a virtual agent or other mobile application will bring benefits for users.

3.4 Identified Gap

The Big 5 approach seems to be adequate for service personalization purposes. Determining the personality is most often based on data collected while using a specific service (usually social media). Creating a profile requires time to record behaviours relevant to the model. Therefore, it is not practiced to calculate the user's personality profile at the time of service installation.

Little attention has also been paid to personalization based on the personality profile and its use in user-beneficial activities (not just tailoring marketing communication, ads, or content recommendation). Thus, there is a need to develop technology to enable much more reliable and people-friendly solutions and automatic personalization of UX and UI based on personality.

Determining the personality in most existing research does not ensure complete user privacy, i.e.the, the profile is calculated using sensitive data and outside the user's end device. Moreover, data, storage, and security were not usually discussed in the literature on determining the user's personality based on digital traces. Therefore, the question arises is it possible to design the counting of the personality profile with better privacy protection, for example, on the end device (e.g. smartphone).

4 Designing and Developing of Artefacts

The primary objective is to develop a novel method of personalizing interactive electronic services like smartphone applications. The design science stage of design and developing of the artefact is presented in Fig. 2.

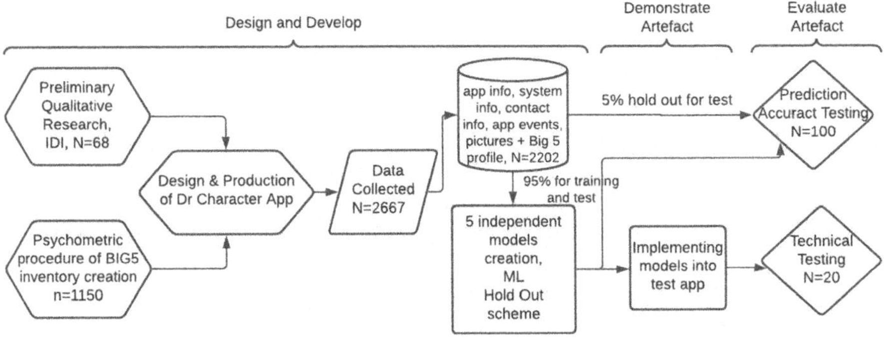

Fig. 2. Developing and evaluation of artefacts.

First, the preliminary qualitative study was carried out to identify needs and collect descriptions of discriminatory behaviour. Additionally, at this stage, users needs and expectations towards the personalization were defined, respectively, to each personality trait extreme groups. Consecutively, the psychometric procedure dedicated to creating the personality electronic assessment tool was executed.[1]). Then, the mobile application (named Dr Charakter) was developed, dedicated to collecting anonymous statistics from mobile phone together with the Big 5 assessment. The data categories processed on mobile phone: telco data (contact list statistics, call logs statistics, text messages log statistics, etc.), applications basic info, system info data, photos (structure of directories, number of photos, photos with faces, etc.), phone settings and statistics e.g. battery consumption, kind of security level. 2666 people were recruited. The participants installed app on the phone, answered Big 5 questions and received the

[1] The 25 items tool was created. Reliability, N = 3331: Alfa-Cronbach coefficient: E:.76, A:.58, O:.59, S:.72, C:.64). Accuracy: r-Pearson coefficients with IPIP-BFM-50: E:.85, A:.55, O:.62, S:.81, C:.76).

personality profile and basic statistics from mobile phone. During the filling the questionnaire, application calculates required statistics from the data available on the phone (with full transparency of what is done and an acceptance of the set of required by law consents). The only anonymous statistics were transferred from the respondent's mobile phone to analytic lab servers. Finally, the personality model was created (artefact 1) based on smartphone data, the same kind which can be available at the moment of any app installation. Finally, the model was implemented in the service prototype (artefact 2) and proceeding with the final validation. A novel method of personalizing smartphone applications.

5 Developing the Data Driven Model

5.1 Data Processing

2,667 people decided to participate in the study, including 1,303 men and 1,364 women. The average age was 31. Therefore, the final age distribution is similar to the characteristics of Internet users in Poland. In addition, the participants must possess skills sufficient to install the application on their own, agree with five consents and carry out the procedure by performing commands on the screen. Finally, the age distribution was: 44% aged 18–29, 32% aged 30–39, 24% aged 40 and higher. There were 250 raw data types, from which 143 were input for presented in this article modelling (Table 1). In the case of data collected by the *Dr Character* app, quality testing and data control have been performed.

Table 1. Descriptions of the data categories taken from the smartphone by Dr Character application

Data category	Examples of data	Definition and description
Standard android information (99)	device security, screen layout, color mode, font scale, keyboard parameters, battery level, rotation, alarm alert, tone/mute/vibrate ring	Current phone settings (during the test)
Contacts (33)	numbers with COE, with address, with e-mail, contacted last month, with name included family names from the list	List of contacts described in statistics. Number of contacts grouped according to 33 different criteria
Applications list (5)	package name, names, categories, found URL	Application general information such as the application list
Applications statistics (6)	package name, first timestamp, battery consumption, last timestamp, last time used, total time foreground	Application statistics - general information such as the date of instalment

The following cases were excluded: people who misplaced a personality questionnaire, installed applications on unused phones, cases with data incomplete or doubled due to the events of interrupted procedures (data transmission errors). After a pre-processing and filtering out records with errors, the model was built on data from 2202 unique participants (82,63% of the initial sample). Personality scores were normalized into the Sten scale. For the selected Sten scale, the unit of the standardized Sten scale is one sten. The number of units is 10. Thus, one sten covers 0.5 SD of the population (reference groups), and the mean of this scale is 5.5.

5.2 User's Initial Smartphone Personality Profile Model

User's Initial Smartphone Personality Profile (UISP) model was created in Python. All machine learning was conducting using *scikit-learn* library. In general, the model creation consists of the following stages: (1) Data set processing, (2) Creating Set of Additional Features, (3) Finding the best solution for the unbalanced sample problem, (4) Finding the best in the class model predicting personality (5) Creating the model on training sample (6) Validating model on the test sample.

For model creation, ensemble techniques were used. This technique is a combination of multiple machine learning algorithms or models. They are used because of the best controlling of the bias-variance trade-off, increasing model performance, and providing good model stability. There are 12 Machine Learning Methods chosen among others for tests: Random Forest Classifier (RF), k-Nearest Neighbor (KNN), Support Vector Machine (SVM), Logistic Regression (LR), Linear Discriminant Analysis (LDA), Decision Tree Classifier- CART, Gaussian Naive Bayes (NB), ETC = Extra Trees Classifier, Bagging Classifier (BC), AdaBoost Classifier (ADA), LGBM Classifier (LGBM), XGB Classifier (XGB). All of them were used to find out the best model fit for each dimension. The model was created using the standard hold out procedure, in which the data are divided into a training set (95%) the set for validation (5%) which is separate and not used for training. The training procedure is iterative and consists of a cycle of training (80% of the sample) and test (20%).

To evaluate the model quality is worth determining the "baseline" level of the prediction, i.e. the result to which the model's predictions will be compared. For regression problems, a random variable is often used, e.g., from 0 to 100, for comparisons. Considering the skewed distribution of 3 classes of trait level (low, medium, high), it will be approximately 33%-34% for the average baseline. The baseline for each of the models was defined in the context of the model usage (business perspective). So the assumed baseline is not using personality for personalization, which caused every user to receive the not-personalized service version (for average target). In this situation, the middle class is personalized with 100% precision and, for those with a High or Low level, precision is 0%. Based on that assumption, the baseline presented in the Table 3 was calculated. So far, a personality model is built based on 3 data categories of an Android mobile phone: Application Info, System Info, and Contacts. Of the 143 raw input

Table 2. Features Importance for the best in class model

Feature	E	A	C	S	O	Feature	E	A	C	S	O
Contacts mobile	7.4	2.3	2.7	3.3	2.4	Emails	1.3	0.7	1.4	1.4	1.4
Contacts	5.4	1.9	3.2	3.8	2.4	Contacts type work	1.1	0.7	0.7	0.7	1.4
Contact one month	4.3	4.0	2.8	2.5	2.2	Min install	1.1	1.1	1.9	2.1	2.1
Weekend ratio	4.1	2.7	2.7	2.9	2.1	Contacts photo	1.0	1.5	1.2	1.4	1.5
Battery level	3.6	3.6	3.3	3.5	2.4	Dtmf tone when dialing	1.0	0.2	0.4	0.4	1.8
Mean day	3.3	4.9	3.1	3.2	2.1	Ratio days per app	0.9	1.1	2.5	2.6	2.4
Boot count	3.2	3.9	2.8	3.1	1.8	Mobile net code	0.8	0.8	1.1	1.3	1.8
Contact six months	3.2	2.3	3.0	2.3	2.4	Density dpi	0.8	1.0	1.4	1.2	1.8
Max app install	3.2	4.4	3.6	3.5	2.3	Free size sd	0.8	0.8	2.9	2.9	2.3
Contact three months	3.1	2.8	2.6	2.7	2.3	Contacts type home	0.7	0.6	0.8	1.0	1.2
Screen brightness	3.1	3.5	3.5	3.1	2.0	Contacts ICE	0.5	0.3	0.2	0.3	0.7
User apps	3.1	3.7	3.0	3.1	2.0	UI mode	0.5	0.7	0.5	0.5	1.2
Work period ratio	3.0	5.3	3.3	3.1	2.2	Mute streams affected	0.4	0.3	0.6	0.6	1.7
App diff days	3.0	3.0	2.8	2.6	2.1	Data roaming	0.4	0.2	0.4	0.6	1.7
System apps	2.9	2.6	2.8	2.7	2.1	Font scale	0.4	0.8	1.7	0.9	1.6
Contact one week	2.9	3.4	3.5	3.5	2.2	Mobile data enabled	0.4	0.6	0.4	0.4	1.7
Available size sd	2.6	2.9	2.9	2.7	1.9	Color mode	0.4	0.4	0.5	0.6	1.4
Total size sd	2.5	2.9	2.5	2.8	2.0	Time mode 12 24	0.4	0.4	0.4	0.4	1.5
Contacts foreign	2.4	1.5	1.7	1.6	1.8	Contacts shared cost	0.3	0.4	0.6	0.5	1.5
All apps	2.4	3.1	2.8	2.7	2.1	Contacts type unidentified	0.2	0.2	0.2	0.2	0.9
Ratio apps per day	2.4	2.9	2.6	2.7	1.8	TTS default pitch	0.2	0.3	0.4	0.2	0.6
Duplicated contacts	2.3	2.1	1.9	1.8	1.7	Screen layout	0.2	0.5	0.2	0.3	0.8
Contact two weeks	2.2	2.5	2.6	2.4	2.2	Contacts address	0.2	0.4	0.5	0.6	1.1
Screen off timeout	2.2	1.4	1.5	1.4	2.0	TTS default rate	0.2	0.3	0.3	0.2	0.7
Contacts type mobile	1.8	2.6	3.0	3.6	2.6	TTS default synth	0.1	0.2	0.3	0.3	1.1
Contacts family	1.7	1.8	2.0	2.1	2.2	Battery saver mode	0.0	0.2	0.3	0.2	1.0
Contacts unknown	1.6	2.3	2.1	1.8	2.1	Contacts toll free	0.0	0.1	0.5	0.3	0.7
Contacts fixed	1.5	3.7	2.0	2.1	2.3	Contacts ussd	0.0	0.2	0.2	0.2	0.6
Contacts short number	1.5	1.2	1.5	1.3	2.1		100	100	100	100	100

data features, the presented model was built on 57. The remaining collected data (apps events, pictures, call logs) has not yet been used. The parameters of the best-performed models based on the three categories of android data are shown in Table 3. The final five models were evaluated on the test group (N = 100) against the assumed baseline. LGBM proved best for E and C, RF for A and O, ETC for S. Compared to the baseline, the best model is E and O. The model for A turned out to be the weakest - it was not possible yet to go beyond the assumed baseline. The specificity of the agreeableness dimension relating to the sphere of interpersonal contacts would require data related to such contacts. Perhaps the improvement will be brought by expanding the features with statistics from call logs and SMS.

Considering the importance of particular data types for creating the model itself, the system's information (e.g. battery level, screen brightness, and the amount of free memory) seems to be most significant Fig. 2. Information about installed applications comes second. Interestingly, there is little differentiation among the top 10 most important features. Lack of differentiation means there

Table 3. Comparison of best in class models performance with baseline (on hold out test sample N = 100)

The best in class:	LGBM	LGBM	RF	ETC	RF
Dimension:	E	C	A	S	O
Precision baseline	0.52	0.55	0.49	0.52	0.58
Precision model	**0.75**	**0.68**	**0.62**	**0.81**	**0.74**
F1 score baseline	0.61	0.63	0.58	0.60	0.66
F1 score model	**0.76**	**0.65**	**0.61**	**0.65**	**0.69**
Accuracy baseline	0.72	0.74	0.70	0.72	0.76
Accuracy model	**0.79**	**0.75**	0.70	**0.74**	**0.78**

are no unique traits to a given personality trait. Since these most critical traits are repeated for all dimensions of personality, these features are most related to user behaviour.

5.3 Method of Personalisation Based on UISP Model

The proposed concept of data-driven personalisation is investigating by implementing the user's personality data-driven model into the android application. The diagram (Fig. 3) shows the process flow. The user installs the application with implemented UISP model on the smartphone. The application counts the 57 statistics needed to calculate the profile. Furthermore, the services automatically adjust the appearance of the service, functionality, and communication to the user's personality profile. The profile remains secret and private because it is not shared with the application back-end and not send outside the device.

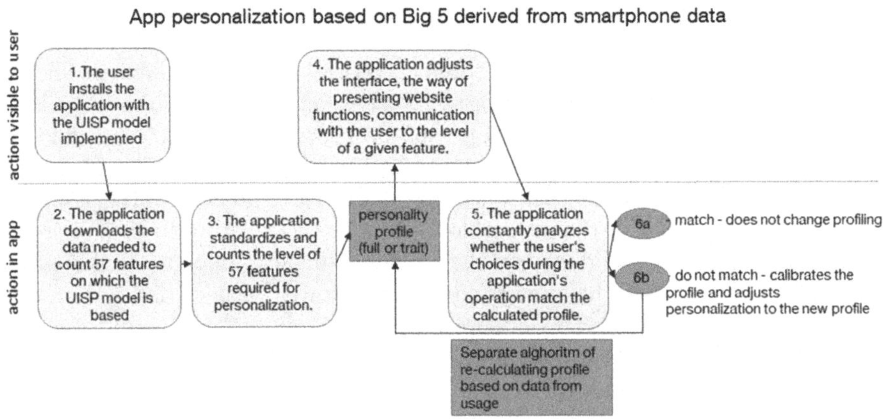

Fig. 3. General flowchart of personalization based on the personality profile (from UISP model) in the service

It is available only for this mobile application. The application has the same functionalities for everyone, but the service is delivered differently to the various users, thanks to the profile. Based on the Big 5 Theory, Introverts should receive an interface with fewer elements and subdued colours, while Extraverts receive an animated, more stimulating interface. The application adapts to the user's capabilities, e.g. to those with Low C, and it provides more messages reminding about the actions to be performed. The reinforcements (feedback) from the app are tailored to the needs of each user. For example, extroverts need more social-oriented communication. High C is related to the purpose and appreciation of the tasks. Adapting the service to different profiles requires additional research, which is developed for specific service functionalities.

The final UISP model (artefact 1 and personalisation logic (artefact 2) was implemented in the lab prototype application to allow evaluation. The user's personality profile, calculated automatically based on statistics from android data, is available immediately after installing the service (the calculation last c.a. 2 s). It was checked on twenty different handsets. This model implementation in the app is a sound basis for further research into determining personalisation preferred by different personality types.

The idea of personalising method (Fig. 3) is also assumed constant analyses of the user's choices during the application's. Based on the results of this analysis, the application will decide whether to continue profiling based on the initial profile or to run the profile re-calculation based on data from the service usage. Additionally, for personalisation purposes, it is enough to classify users into a low, medium, and high class of a given trait. Finally, since various personalised elements in the service are adjusted independently, potentially, the entire profile does not have to be included in the service. The five models can be created and used independently, e.g. adjusting interface graphics is significant high Extraversion and Openness. Thanks to that, no information about the user is transfer and stored outside the handset. All the data and the profile itself are stored locally.

6 Conclusion and Discussion

The article presents the work results aimed at creating a model for calculating the personality profile based on the data available during the installation of the service in a mobile phone. Furthermore, the concept of automatic personalisation of the service was also presented, and the mechanisms implemented in the smartphone app itself (front-end layer). The presented research resulted in the successful creation of both defined artefacts (1 and 2), although both require to be finally validated in UX tests. The new contribution of this study consists mainly in the fact that the possibility of calculating the personality profile from anonymous statistics available on the phone during the installation of the application has been shown and, it can be done with accuracy comparable to models based on large amounts of data despite distinguishing narrowly defined high and low class for personality traits. Therefore, this approach is more accurate for service-personalisation purpose. Furthermore, it also proposed using such an in-app model to personalise any service, which can be a base for new kind of user's personality-aware services.

The research was also aimed at confirming the suitability for profiling based on data from the moment of service installation, without the need to collect data logs from the services and test automatic analytic that can be implemented inside the service. Work on both the UISP model (extension with new data categories) and the first test application that uses the UISP model for personalisation is ongoing but is nearing completion.

Another UX research, in the experimental model, is planned to confirm the usefulness and value of the proposed solution for users. Currently, personalisation is determined based on theoretical descriptions of features that contain basic behavioural guidelines. Subsequent tests should be dedicated to the verification of the business use of such profiling.

References

1. Back, M.D., et al.: Facebook profiles reflect actual personality, not self-idealization. Psychol. Sci. **21**(3), 372–374 (2010)
2. Bin Tareaf, R., Alhosseini, S.A., Meinel, C.: Facial based personality prediction models for estimating individuals private traits. In: Proceedings - 2019 IEEE International Conference on Parallel and Distributed Processing with Applications, Big Data and Cloud Computing, Sustainable Computing and Communications, Social Computing and Networking, ISPA, BDCloud, SustainCom, SocialCom 2019 (2019)
3. De Raad, B.: Structural models of personality. In: The Cambridge Handbook of Personality Psychology (2012)
4. Eke, C.I., Norman, A.A., Shuib, L., Nweke, H.F.: A survey of user profiling: state-of-the-art, challenges, and solutions. IEEE Access **7**, 144907–144924 (2019)
5. Gjurković, M., Karan, M., Vukojević, I., Bošnjak, M., Šnajder, J.: PANDORA talks: personality and demographics on reddit (2020)
6. Hevner, A., Chateerjee, S.: Integrated Series in Information Systems, vol. 28 (2012)
7. Isabel Myers, P.M.: Gifts Differing: Understanding Personality Type (2010)
8. Judge, T.A., Ilies, R.: Relationship of personality to performance motivation: a meta-analytic review. J. Appl. Psychol. **87**(4), 797 (2002)
9. Kalimeri, K., Lepri, B., Pianesi, F.: Going beyond traits: multimodal classification of personality states in the wild. In: ICMI 2013 - Proceedings of the 2013 ACM International Conference on Multimodal Interaction, pp. 27–34 (2013)
10. Kern, M.L., et al.: From "sooo excited!!!" to "so proud": using language to study development. Dev. Psychol. **50**(1), 178–188 (2014)
11. Khan, A.S., Ahmad, H., Asghar, M.Z., Saddozai, F.K., Arif, A., Khalid, H.A.: Personality classification from online text using machine learning approach. Int. J. Adv. Comput. Sci. Appl. **11**(3), 460–476 (2020)
12. Krzeminska, I.: Data-based user's personality in personalizing smart services. In: Abramowicz, W., Corchuelo, R. (eds.) BIS 2019. LNBIP, vol. 373, pp. 686–696. Springer, Cham (2019). https://doi.org/10.1007/978-3-030-36691-9_57
13. Krzeminska, I.: Personalizing smart services based on data-driven personality of user. In: Debruyne, C., et al. (eds.) OTM 2019. LNCS, vol. 11878, pp. 199–203. Springer, Cham (2020). https://doi.org/10.1007/978-3-030-40907-4_21

14. Krzeminska, I., Rzeznik, J.: Personality-based lexical differences in services adaptation process. Technium: Rom. J. Appl. Sci. Technol. **3**(1), 61–73 (2021)
15. Levy, Y., J. Ellis, T.: Towards a framework of literature review process in support of information systems research. In: Proceedings of the 2006 InSITE Conference (2006)
16. Li, Y., Kazameini, A., Mehta, Y., Cambria, E.: Multitask learning for emotion and personality detection (2021)
17. Liu, L., Preoţiuc-Pietro, D., Samani, Z.R., Moghaddam, M.E., Ungar, L.: Analyzing personality through social media profile picture choice. In: Proceedings of the 10th International Conference on Web and Social Media, ICWSM 2016, pp. 211–220 (2016)
18. Liu, R., Hu, X.: A multimodal music recommendation system with listeners' personality and physiological signals. In: Proceedings of the ACM/IEEE Joint Conference on Digital Libraries in 2020 (2020)
19. Matthews, G.: Personality and information processing: a cognitive-adaptive theory. In: The SAGE Handbook of Personality Theory and Assessment: Volume 1 - Personality Theories and Models (2008)
20. Matz, S.C., Kosinski, M., Nave, G., Stillwell, D.J.: Psychological targeting as an effective approach to digital mass persuasion. Proc. Nat. Acad. Sci. U.S.A. **114**(48), 12714–12719 (11 2017)
21. de Montjoye, Y.-A., Quoidbach, J., Robic, F., Pentland, A.S.: Predicting personality using novel mobile phone-based metrics. In: Greenberg, A.M., Kennedy, W.G., Bos, N.D. (eds.) SBP 2013. LNCS, vol. 7812, pp. 48–55. Springer, Heidelberg (2013). https://doi.org/10.1007/978-3-642-37210-0_6
22. Ning, H., Dhelim, S., Aung, N.: PersoNet: friend recommendation system based on big-five personality traits and hybrid filtering. IEEE Trans. Comput. Soc. Syst. **6**(3), 394–402 (2019)
23. Sahu, Y., Ramani, Y., Parekh, V., Maru, N.: Personality prediction from social media images: a content driven approach. In: Proceedings of the 2019 6th International Conference on Computing for Sustainable Global Development, INDIACom 2019 (2019)
24. Stachl, C., et al.: Predicting personality from patterns of behavior collected with smartphones. In: Proceedings of the National Academy of Sciences of the United States of America (2020)
25. Stajner, S., Yenikent, S.: Why is MBTI personality detection from texts a difficult task? In: Proceedings of the 16th Conference of the European Chapter of the Association for Computational Linguistics: Main Volume, pp. 3580–3589. Association for Computational Linguistics, Online, April 2021
26. Xu, R., Frey, R.M., Ilic, A.: Individual differences and mobile service adoption: an empirical analysis. In: Proceedings - 2016 IEEE 2nd International Conference on Big Data Computing Service and Applications, BigDataService 2016, pp. 234–243. Institute of Electrical and Electronics Engineers Inc., May 2016
27. Zhang, Q., Esterwood, C., Yang, J., Robert, L.: An automated vehicle (AV) like me? The impact of personality similarities and differences between humans and AVs. SSRN Electron. J. (2019)

iCRM Workshop

ICRM 2021 Workshop Chairs' Message

Over the past years and in particular since the onset of the COVID-19 pandemic, the importance of digital media has increased. This not only applies to video platforms, which have grown strongly and contributed to managing the restrictions in personal contact, but also the broad mix of electronic channels, which are well known in the field of integrated customer relationship management (ICRM). Among the channels that support the key business processes for CRM, i.e. marketing, sales, and service, are online services and platforms, mobile devices, and social media as well as games and voice assistants with in-app purchases. According to eMarketer, electronic commerce sales—being the part of CRM that comprises electronic transactions—jumped from 3,354 billion USD to 4,280 billion USD in 2020, with a forecast to reach 4,891 billion USD in 2021 [1]. This represents an increase of 45.82%.

The use of social media for CRM is known as Social CRM and aims to improve the performance of a business by using this channel for building and maintaining relationships between the business and its customers. From an organizational perspective, Social CRM links the external world of social media with internal business processes in marketing, sales, and customer services [3]. However, social media is not limited to being a new communication and marketing channel. Rather, it is a lever for the digital transformation of existing CRM strategies, processes, and systems [4]. Integrated Social CRM may be conceived as an approach that aims to align these aspects in integrated and digitally supported use cases. It builds on a realm of information technology that helps to turn user-generated content (UGC) into actionable information for daily tasks, as well as strategic decisions [5, 6].

For the sixth time, the iCRM workshop aimed to shed light on current research and practical challenges regarding integrated Social CRM. The workshop attracted participants from different fields, such as marketing and relationship management, information systems design, and computational intelligence. While in the past, the notion of iCRM referred to intelligent CRM, the current workshop was organized with the emphasis on integrated (Social) CRM. Similar to previous years, all submissions underwent double-blind peer review and the authors were able to revise their manuscripts based on the feedback from the workshop, which was conducted virtually this year.

In total, three papers were accepted, which covered different perspectives. The first paper, "Social CRM as a Business Strategy: Developing the Dynamic Capabilities of Micro and Small Businesses", provides insights into the role and potential of Social CRM for small businesses based on a survey and four in-depth case studies with small Brazilian companies. The paper shows that those organizations see benefits in using social media for their businesses, but they often focus on marketing and lead generation, ignoring the potential for service and customer retention.

The second paper, "Understanding Customer Orchestration of Services: A Review of Drivers and Concepts", investigates concepts related to customer orientation, decentralization, and smart services in the platform economy and their implications on future CRM. The paper shows how decentralized platforms may shift the way companies establish relationships with their customers. The third paper, "Gaining Insights on Student Satisfaction by Applying Social CRM Techniques for Higher Education Institutions", examines the role of Social CRM for managing relationships between education institutions and students. It illustrates how these institutions may obtain insights into student satisfaction through an extensive analysis of social media activities and feedback by analyzing the data from two Brazilian universities.

The three papers highlight the potential of Social CRM in face of the increasing use of digital communication channels, providing insights in very different domains. While Social CRM is often examined in the context of medium and large enterprises, the potential and application areas for smaller enterprises and non-commercial organization are less understood. The workshop participants agreed that these organizations should use Social CRM in similar ways and that academia should collaborate with industry to leverage existing knowledge to adapt it to domain-specific application contexts in the various industries. Combining the technological and the managerial perspectives was recognized during the workshop as an important prerequisite for achieving integrated solutions in practice that help to build and foster relationships in a digital environment. This combination also helps to respond to the ever-changing customer behavior, especially now during the COVID-19 pandemic. Again, the sixth iCRM workshop was a community effort in extraordinary times and the valuable contribution of all authors and the Program Committee members in enabling this virtual event was highly appreciated.

Rainer Alt
Olaf Reinhold
Fabio Lobato

References

1. eMarketer. Retail Ecommerce Sales Worldwide, 2019–2024 (Dec 2020). https://www.emarketer.com/chart/242908/retail-ecommerce-sales-worldwide-2019-2024-trillions-change-of-total-retail-sales
2. Kemp, S.: Digital 2021 October Global Statshot Report (Oct 2021). https://datareportal.com/reports/digital-2021-october-global-statshot
3. Alt, R., Reinhold, O.: Social customer relationship management (Social CRM) - application and technology. Bus. Inf. Syst. Eng. **4**(5), 287–291 (2012)
4. Alt, R., Reinhold, O.: Social Customer Relationship Management - Fundamentals, Applications, Technologies. Berlin, Springer (2020). https://doi.org/10.1007/978-3-030-23343-3
5. Choudhury, M., Harrigan, P.: CRM to social CRM: the integration of new technologies into customer relationship management. J. Strateg. Mark. **22**(2), 149–176 (2014)
6. Trainor, K. J., Andzulis J., Rapp, A., Agnihotri, R.: Social media technology usage and customer relationship performance: a capability-based examination of social CRM. J. Bus. Res. **67**(6), 1201–1208 (2014)

Organization

Chairs

Olaf Reinhold	Leipzig University/Social CRM Research Center, Germany
Rainer Alt	Leipzig University/Social CRM Research Center, Germany
Fabio Lobato	Federal University of Western Pará, Brazil

Program Committee

Alireza Ansari	Leipzig University, Germany/IORA Regional Center for Science and Technology Transfer, Iran
Antonio Jacob Jr.	Instituto Federal do Maranhao, Brazil
Cristiana Fernandes De Muylder	FUMEC University, Brazil
Douglas Cirqueira	Dublin City University, Ireland
Emilio Arruda	FUMEC University/University of Amazon, Brazil
Fabio Lobato	Federal University of Western Para, Brazil
Flavius Frasincar	Erasmus University Rotterdam, The Netherlands
Gültekin Cakir	Maynooth University, Ireland
Jose Marcos de Carvalho Mesquita	University of Connecticut, USA
Julio Viana	Social CRM Research Center, Germany
Kobby Mesah	University of Ghana Business School, Ghana
Kwabena Obiri Yeboah	Catholic University College of Ghana, Ghana
Kwame Adom	University of Ghana Business School, Ghana
Luis Madureira	Universidade Nova de Lisboa, Portugal
Mattis Hartwig	University of Lübeck, Germany
Nino Carvalho	IPOG, Brazil/IPAM, Portugal
Omar Andres Carmona Cortes	Instituto Federal do Maranhao, Brazil
Paulo Andre Lima de Castro	Aeronautics Institute of Technology, Brazil
Rabi Sidi Ali	Takoradi Technical University, Ghana
Rafael Geraldeli Rossi	Universidade Federal do Mato Grosso do Sul, Brazil
Regine Vanheems	Université Jean Moulin Lyon 3/Orange, France
Renato Fileto	Federal University of Santa Catarina, Brazil
Sandra Turchi	Digitalents, Brazil
Thiago Henrique Silva	Federal University of Technology – Paraná, Brazil
Vicente Guerola Navarro	Universitat Politècnica de València, Spain
Winnie Ng Picoto	Technical University of Lisbon, Portugal

Social CRM as a Business Strategy: Developing the Dynamic Capabilities of Micro and Small Businesses

Isabelle da Silva Guimarães[1], Gustavo Nogueira de Sousa[2], Antonio Jacob Junior[2](✉), and Fábio Manoel França Lobato[1,2]

[1] Federal University of Western Pará, Santarém, Brazil
fabio.lobato@ufopa.edu.br
[2] State University of Maranhão, São Luís, Brazil
antoniojunior@professor.uema.br

Abstract. The global pandemic, caused by the spread of COVID-19, has altered the way people go shopping. In light of this, Social Media channels are an important means of sharing information about goods and services, and different kinds of brands. Since these channels are of considerable market significance, the authors of this paper decided to describe the results of a survey on how to use Social Media to improve customer relationship management processes in 31 companies. The focus was on digital marketing for micro and small businesses. In addition, an in-depth analysis was conducted of four companies, to determine the challenges and strategies in social customer relationship management adopted by micro and small businesses. The results show that this is still a new policy for micro and small companies, but has a great potential to boost sales, enhance customer loyalty and increase brand awareness. The lessons learned can assist policymakers in taking more suitable measures for strengthening this market sector.

Keywords: Digital marketing · Micro and small businesses · Social CRM

1 Introduction

The global pandemic caused by the spread of COVID-19, has greatly altered the way people go shopping [1]. Consumers have now shifted their purchases to e-commerce platforms, which are one of the main procurement mechanisms for sustaining the global economy [2]. According to [3], trade in the post-coronavirus era will no longer be the same. It is thus necessary for companies to invest time and resources in understanding their customers in the best possible way [4]. Social Media (SM) platforms feature prominently in undertaking, this task. In January 2021, SM platforms reported that they had over 4.2 billion active users [5]. It is worth noting that this was the most significant growth rate in three years and this sharp rise in User-Generated Content (UGC) may prove to be of the utmost importance for the development of goods and new brands by small, medium, and large-sized companies [6]. On the basis of User-Generated Content in social media, companies can make predictions about new trends and remain competitive in their market niche [7]. However, it is not feasible to

W. Abramowicz et al. (Eds.): BIS 2021 Workshops, LNBIP 444, pp. 161–173, 2022.
https://doi.org/10.1007/978-3-031-04216-4_16

retain traditional customer management strategies [8] and it has thus been found necessary to deploy standard Customer Relationship Management (CRM) processes that correspond to marketing, innovation, sales, user experience, and after-sales within the SM environment [9].

According to [10, 11] Social CRM can be defined as the adaptation of interactions between companies and customers through social media. It can also be defined as a set of technological processes that collect and analyze data to find out their customers' needs [12]. Moreover, with the knowledge generated from these data, companies can understand the patterns of behavior of their customers and objectively determine what their needs are [13]. In addition, marketing campaigns can be more efficient in making direct contact with their target audience and convert leads into customers [14]. This procedure is applicable to Micro and Small Companies (MSCs) [15]. It should be underlined how representative MSCs are to the Brazilian economy. Currently, small businesses account for 99% of the 6.4 million Brazilian companies and are responsible for providing about 52% of formal jobs in the private sector [16].

Despite this economic importance, this sector is experiencing a shortage of qualified professionals to fulfill their requirements and thus lack the expertise necessary to adopt Social CRM strategies in their policymaking. Moreover, in the existing literature there are only a few studies that are concerned with studying the adoption of Social CRM strategies by MSCs. Clearly, the need to analyze the adoption of Social CRM strategies by Micro and Small Companies and to seek opportunities for intervention, depends on the economic importance of the country. In response to the gaps in the literature, the following research questions have been defined: RQ1: Which aspects of Social CRM are important for MSCs? RQ2: How should MSCs adopt Social CRM strategies in their businesses?

To answer these questions, we conducted a 2-step investigation on the role of Social Media in improving customer relationship management, by determining the challenges and perspectives for Social CRM adopted by micro and small businesses. The first step consisted of conducting a survey of 31 companies, with the aim of establishing some quantitative factors. In the second step, we collected and analyzed data from four case studies involving companies from different markets. We also drew attention to some lessons learned and laid down some guidelines that can enable policymakers to create more suitable measures for boosting the MSC sector. The remainder of this paper is structured as follows: Sect. 2 conducts a brief survey of related works, followed by Sect. 3, which deals with the description of the methodology. In Sect. 4, the results of the research are examined and, finally, the conclusions are summarized.

2 Related Works

[17] examines how firms can improve CRM capabilities such as marketing and sales through the use of Social Media. His results demonstrate that social media can broaden the positive results of customer engagement and, hence, the firm′s performance. [18] investigated the use of Facebook Commerce (F-commerce) to determine how/how much it influences the organizational performance of micro and small companies.

Similarly, [19] sought to find out whether the adoption of Facebook has an influence on the company's performance. Based on their findings, the authors agreed that Facebook platforms for business processes help companies' performance management inputs and outputs. In addition, external pressures such as competition and consumer influence are characterized as factors that help improving relationships with the customers.

[20] The role of Managing Social CRMs in Social Information Systems (SIS) is based on four case studies. The authors' conclusions show that the systems have applications that are not restricted to strengthening the company-customer relationship but can also enable a business to increase its expansion capabilities. In the same niche, [21] surveyed the marketing technologies that are adopted by small businesses. The results show that managers see them as an opportunity to improve their relationships with customers. The authors also design a model built based on the researcher's insights that consists of collecting information, building awareness, and measuring results.

Following this, [18] carried out a study based on quantitative variables, obtained through an online survey, and qualitative variables, obtained from an exploratory search in state-of-the-art. [20] provided the results concerning Social CRM based on case studies conducted through f interviews, and based on quantitative and qualitative data. However, the approach was restricted to Large Companies, and thus avoids the real-life circumstances on which this study is based.

3 Methodology

This study was conducted both quantitatively, through the application of a survey and qualitatively, through case studies. After carrying out a review of the literature and defining the key concepts, the scope of the work and the nature of the research questions were defined. The scope was limited to MSCs and Individual Micro-Entrepreneurs (IMEs) who use SM in their businesses. The research questions that are outlined in the Introduction were defined with a view to filling in the gaps in the current literature. The quantitative data were collected through a self-administered online survey divided into three subsections: i) general information about the company (e.g., size, the activity segmentation, social media management, *etc.*); ii) social media management (e.g., the media used, published content, difficulties encountered, *etc.*), and iii) customer relationship management (e.g., perception of the CRM concept, software used, *etc.*). All the businesses involved were volunteers in this phase. The response method was divided between assays and the Likert-scale (from 1 to 5). The response time could vary between 5 and 10 min.

In the second phase, qualitative data were collected through the case study methodology adapted from [20]. Four MSCs were selected, who took an active part in the events scheduled by our research group. All the four companies belonged to different market niches to ensure unbiased results, (as described in the subsection - Case Studies). The data collection was conducted through semi-structured interviews which lasted between 50 and 60 min. Table 1 displays the script, which was defined in conjunction with the practitioners.

Table 1. Script for conducting semi-structured interviews.

Section	Question
Management data	General information about the interviewee
General information about social media	Who manages Social Media? Do you have external consultancy services? If so, how did the management take place? How did the procedure take place when using the networks?
Reasons for using social media in business	What are the perceived benefits of adopting social media in business? Were there any noticeable external factors that influenced the reason for using social media?
Investment and internal evaluation	How do you define the success of a publication? Do you invest money in Social Media? Does investing in Social Media/Networks make a difference to the business?
Use of Social CRM	Do you adopt CRM and/or Social CRM strategies in the company? How can Social CRM be conceptualized? What are the reasons for its adoption? Who is responsible for the strategy? How is media management carried out? Which social networks and metrics are most often used? What are the difficulties experienced?
Pre-in-post COVID-19	What are the strategies adopted/measures taken when facing a crisis? What is the outlook for the post-pandemic Era?

It should be noted that all the interviews had to be conducted by phone to avoid social contact, since this research phase coincided with the period of coronavirus restrictions. The results were first arranged individually in a feedback format for the interviewees, so that the positive and negative aspects of the business in the areas of customer relations could be highlighted. After this, the information collected was cross-checked to assess the quality of the companies' processes and the differences between them. The results were validated in partnership with specialists in Social CRM and Digital Marketing. The research results were collated and culminated in this research study.

4 Results

This section examines the results from the survey and interviews, and then contextualizes and discusses some of the significant findings.

4.1 Characterization of the Companies

This survey was an essential precondition for our attendance at some training courses provided by our research group. We adopted this approach instead of broadcasting the study because it allowed us to co-validate the answers. Moreover, it enabled us to become engaged with the open questions. We obtained 31 responses from MSCs, IME, and informal businesses that actively use Social Media in their business. Table 2 gives the primary information of the companies.

Table 2. General information of the companies.

Size	Total in numbers	Percentage
MSCs	5	16.1%
IME	14	45.2%
Informal	12	38.7%
Operating segment	Total in numbers	Percentage
Food service	8	25.8%
Advisory communication	6	19.3%
Entertainment	5	16.1%
Engineering Information Systems	5	16.1%
Clothing and handcrafts	4	12.9%
Beauty and well-being	2	6.4%
Veterinary service	1	3.3%

Table 3 shows that entrepreneurs regard the Internet as an essential channel for running their businesses. In response to this question, the entrepreneurs gave their answers on a Likert scale ranging from 1 to 5. For a better understanding of the result, the groupings made between 1 and 2 were categorized as "low relevance"; 3 and 4 as "medium relevance"; and 5 was "high relevance."

Table 3. The Internet and social media management.

Importance of the Internet for revenue management	Total in numbers	Percentage
Low relevance	1	3.2%
Medium relevance	3	9.7%
High relevance	27	87.1%
Who manages social media	Total in numbers	Percentage
The Owner/Manager	30	96.8%
Employee	1	3.2%
Outsourced	0	0%

In addition, in Table 3, it could be confirmed that, in almost all cases, the owner or manager was responsible for managing the company's social networks. Regarding the use of Social Media, the most widespread of the 31 companies, was WhatsApp, with 27 positive responses, followed by Instagram with 24 and Facebook with 22. This result confirms the findings of [22], who states that these platforms are adopted because they are free of charge and easy to access. This distribution can be explained by the fact that these three social media are among the most widely used by Brazilians, only behind YouTube [23].

Table 4. Types of content published on social media

Types of content	Total in numbers	Percentage
Sharing of goods	26	83.9%
Discounts	18	58.1%
Advertising	17	54.8%
Key information for customers	15	484%
Feedback	11	35.5%
Institutional material of the company	10	32.3%
Achievement designs	7	226%

In Table 4, it can be seen that the main purpose of using social media is to disseminate goods and services that have already been established in companies, followed by discount sharing and the creation of advertisements for new products. However, one of the least used topics is gathering customer feedback, which corresponds to less than 36% of the companies. One of the purposes of using social media for businesses is to bring the customers and company into closer proximity; for this reason, encouraging shared shopping strategies and a closer relationship with the company is of considerable importance for strengthening customer loyalty.

With regard to the difficulties encountered by managers in the use of online social networks, Fig. 1 shows that the most significant factor is the lack of knowledge of what tools are required for managing these platforms. In addition, a recurring complaint among entrepreneurs is the lack of time available for creating content and evaluating results. This can be explained by the information displayed in Table 3, which shows that in 96.8% of cases, the owner or manager is responsible for these tasks. Small business owners can carry out several tasks within their company, which tends to make them overloaded with work. This means that they do not have time to learn how to operate new tools or to deal with the reports they produce, which has a direct impact on the management of social media in business.

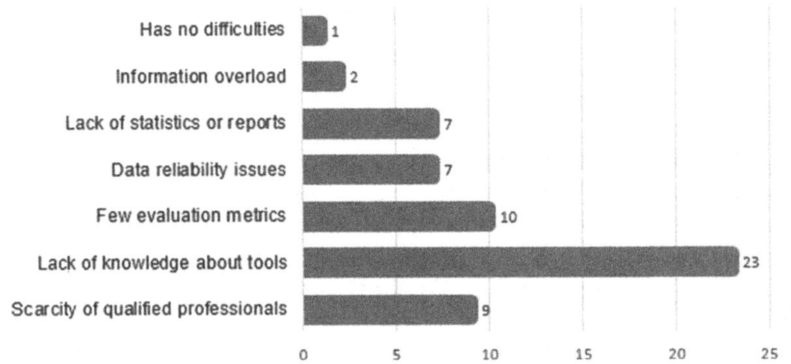

Fig. 1. Difficulties of managing results in Social Media.

In the case of Social CRM, Fig. 2 shows that 25 companies answered that they did not know how to conceptualize it. From the six that gave answers, only two stated that it is present in the entire process of forming a relationship with customers. The other four responses were only related to sales. Some examples can be read in Fig. 2.

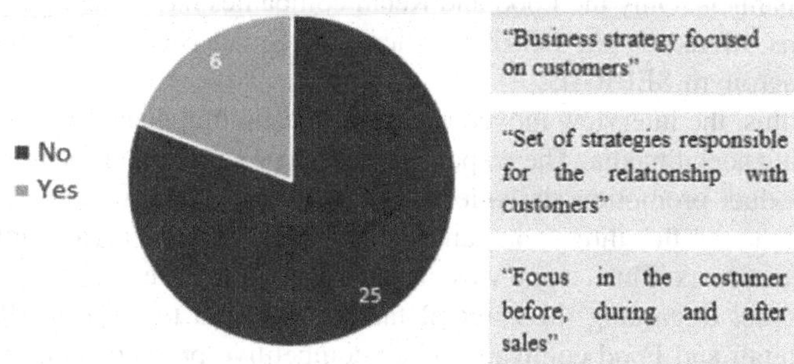

Fig. 2. Conceptualization of Social CRM by small businesses.

4.2 Case Studies

Four companies from different sectors were selected from the 31 companies that attended our courses to conduct the case studies. It is necessary to explain the omission of the names of the companies, as this was one of the criteria agreed upon between the researchers and the interviewees. Hence, they are simply designated as: **Retail Company** - a part of the retail footwear and clothing industry since 2009; **Food Company** - this operates in the sector that supplies prepared foods, beverages for sale, and the like. It has been in the market since 2018; **PET Company** - offering services for small animals, such as veterinary care and cosmetic procedures. It has been in operation since 2018; **Photography Company** - its activities are described as those of a filming service agency. It has been offering its services since 2018. This information was collected from the Federal Government Redesim free access platform, where it is possible to search in the National Register of Legal Entities by means of the companies' fancy names. All the participants (interviewees) were the owners of the companies. More information about the profile of each interviewee can be found in Table 5.

Table 5. General information about the interviewees.

	Retail	Food	PET	Photography
Degree	Administration	Psychology	Veterinary medicine	Veterinary medicine
Professional experience	Worked in a shoe store, in a real estate company, and seller of home appliances	No previous work experience	No previous work experience	Worked in his training area for two years
Favorite hobby	No -information	Enjoys reading and studying Gestalt Therapy	She loves going to the beach, but hasn't had have much time lately	Goes to the beach, rides a bike and goes out to bars

All of them have a higher education diploma, but only 'Retail' works in his area of training. In the case of the Food, Photography, and PET companies, the interviewees reported that they are responsible for managing the business's digital platforms. The Retail company has experienced two phases; initially, the owner was responsible for taking care of social networks, while currently, a third-party company manages their social media channels. Only the Food and Retail companies have taken part in external consulting firms offered by the Brazilian Micro and Small Business Support Service, known by its acronym SEBRAE.

Following this, the interview moved on to the questioning phase about the owners' reasons for using social media. The responses were very similar and focused on "brand presence," "product promotion strategies," and "appealing to customers." It was significant that none of the three companies cited social media as a communication channel with customers. Thus, a gap was detected here in the use of these platforms for business purposes. Regarding the external factors that lead to these platforms being adopted, the Retail and Food companies cited competitive pressure as having a direct influence on the concern with a digital presence. Photography reported that it uses social media to evaluate the work of competitors.

As a result, the most widely used metric by companies was the number of likes that a publication receives. Only Retail cited the number of people interested in a particular product posted on social media; this shows that social media is being used for trend identification, which guides the portfolio selection process. After this, it? the product? was entered in the section regarding Social CRM and its processes. First, respondents were asked whether they knew how to conceptualize Social CRM. None of the four could give an answer. This fact was evident when the Retail and PET companies stated that they have software that is responsible for cash flow, inventory control, providing a customer database, and sales control. The Food company also said it uses software to control orders. The Photography company reported that it started measuring customer satisfaction after taking part in a course related to this area. In other words, companies have internal CRM and Social CRM processes, but they still do not understand what they are and where they are located.

Regarding the difficulties experienced, the interviewee from the Food portfolio selection process spoke about the limitations of physical space, which is still too small to serve the public it serves. Another point was the administration of multiple virtual service channels. Orders arrive via WhatsApp, the phone, Instagram/Facebook messages, and application requests - in addition to on-site assistance. One of the solutions was to concentrate on orders and exclude calls and messages via Instagram and Facebook. Although some customers disliked this, there was an improvement in the speed of service and a reduction in any delays and errors in orders, which meant there was an increase in the level of satisfaction of the service users.

In the PET company, the most significant problems were related to the time management for the owner, who sometimes had to carry out several tasks and ended up being overloaded with work. An interesting dynamic in the relationship with the client was the use of WhatsApp as a tool for bringing together tutors and hosted pets – on a daily basis; the manager sends photos and videos of the pets (either hosted or in daycare) to each tutor. Some pictures were posted on Instagram, but this measure was canceled, because the tutors who did not have their pets included, complained to the manager. It should be noted that the manager stated that she did not feel comfortable using Facebook, which is why the company abandoned this platform. Photography cited seasonality as a problem faced by the company. In periods of low demand, one solution is product diversification - such as photography courses, an increase in promotional strategies, and sponsored posts to give greater exposure to these activities.

Finally, the owners were asked questions about the COVID-19 pandemic period. The PET company implemented a "Taxi Dog" service to reduce the effects of people being unable to circulate and restrictions imposed by social isolation. Retail is investing in online sales through Facebook, Instagram, and WhatsApp, and intensifying its campaigns carried out on this last platform. The Food company stated that it would continue with its delivery service. It will also (without moving in physical space) take time to renovate and expand its surroundings to accommodate customers through social distancing. The interviewee from Photography said it was necessary to postpone the work and, currently, they are betting on expanding the gastronomic photography market for restaurant delivery services.

After grouping the data and information obtained through the survey and case studies, individual reports were prepared for the four companies that took part in the interviews. Table 6 shows the information contained in the reports, which was also sent to the research partners for them to validate these suggestions.

Table 6. Social media assessment reports.

	Positive points	Points needing improvement
Retail	Investment in "Boosting posts," contracting an outsourced service for managing social media, using WhatsApp Business as a communication channel, and using software to manage internal processes	Need for automated responses in WhatsApp Business, interaction with the public in social media through stories, creation of a marketplace on social media, promotion of in-store discounts
Food	Use of delivery service platforms; interaction with the public in comments and reposts; and active presence on social media	Definition of digital presence and branding; studying ways to adopt WhatsApp Business as a communication channel; registration of company on Google My Business; adoption of software for publishing automation
PET	The excellent interval between posts, WhatsApp Business as a communication channel, request for customer feedback, continuous innovation in services	Automation of responses in WhatsApp Business, use of social media for market research, the problem of not feeding all the social media of the company, service standardization
Photography	Adaptation of the business in line with market trends, investments in "boosting posts," concern about optimizing the feedback process with customers, and market variability	They are studying ways to adopt WhatsApp Business as a communication channel with customers, building a database with customer information, and adopting anonymous feedback strategies

5 Lessons Learned and Future Perspectives

On the basis of the results, it can be seen that companies still use only a few of the functionalities made available within social media, for the purposes of Customer Relationship Management. For example, the metrics used by entrepreneurs are largely determined by the number of likes, which is just one of the data offered by the Facebook and Instagram platforms to assess the success of a publication. If they also had access to the publication's user engagement data, they could more broadly confirm what kind of content the public is receiving. As a result, they would be able to conduct more targeted and assertive campaigns and, hence, save time, money and effort, which are among the most painstaking tasks reported by managers.

A point that the four companies studied share in common is the use of WhatsApp Business as a business tool. This platform can assist them in several ways, such as automating frequent responses, which allows the manager to save time in the first contact with potential customers. In addition, WhatsApp Business can serve as a direct communication channel with customers to disseminate news and give information about discounts, as well as acting as a catalog of goods and services to optimize the work of the social media manager.

Another factor that was observed has a direct relationship with the branding of companies. According to [24, 25], branding can be defined as a set of strategies that define the product. It is everything that is involved, ranging from the colors and shapes needed to represent the value of goods and services, to sales practices. Moreover, in the case of small companies, this can make a huge difference since it serves to position the brand in the market and help to give it a distinct identity. This process can be initiated by defining the values inherent in the products, logo creation, and colors and visual elements which will serve to give the company its own identity. After these stages, the process of creating the visual identity can begin, and the way the company will communicate with its public will be defined. Then, we move on to the stage of forming a digital presence, where the company will set up its social media channels.

These channels are defined in accordance with the company's niche. For example, the food company cited iFood as one of its sales channels. In addition, in the case of the Retail company, the report sent to the manager indicated the shape of the store's marketplace, where he will be able to count on another online sales. This modality is known for bringing together in a single place several offers of goods and services from different companies, such as an online catalog (e.g., Amazon, Mercado Livre, OLX, etc.). Facebook currently has its own marketplace tool. The correct definition of social media can make a big difference to the volume of sales achieved by a company. As a means of assisting managers in making these adjustments, it is necessary to offer individualized courses and consultancies to guide them. It is also essential to assist them in understanding the metrics employed for measuring results and how to optimize their internal processes. As a result, it is expected to have a positive impact on its business in terms of the growth of brand exposure and an increase in sales.

6 Conclusion

Through a survey and with the aid of case studies, this paper conducted an analysis of both the quantitative and qualitative factors a that are involved in implementing Social CRM in Micro and Small Companies, a highly significant market sector for Brazilian trade. The purpose of these analyses was to determine which aspects of Social CRM are relevant to MSCs (RQ1) and how they are implemented in their business (RQ2).

With regard to RQ1, the results show that entrepreneurs view the Internet as a highly appropriate means of communication for their businesses, with Facebook, Instagram, and WhatsApp being the leading platforms. They use these media to publicize their products and make special discount offers, as well as disseminating news and holding sweepstakes. However, only a limited number of managers use them as a channel for obtaining feedback from their customers, which means that they end up neglecting the importance of fostering customer loyalty. The results also show that entrepreneurs are unaware of Social CRM concepts and applications. Additionally, the lack of this knowledge means that they are unable to achieve better results since they themselves are responsible for managing social media. As a solution to this problem, managers should be offered short and asynchronous training.

With regard to RQ2, the results obtained suggest that the main reason for using social media is directly linked to the presence of a particular brand. However, as in the previous phase, it became evident that managers still do not see social networks as a two-way communication channel and assume it should only be used to disseminate information about goods and services. One fact that attracted our attention was that entrepreneurs tend to implement CRM processes in their companies, such as by using software for cash and inventory management. At the same time, they stated that they do not know the Social CRM concept, which also leads to a neglect of related features. Despite this, the pandemic has forced micro and small companies to adapt to the new reality and demonstrate this by taking measures that include launching new service lines and offering discounts, in addition to offering new online sales and delivery services.

Based on our results, it is expected that this work has led to an awareness of the main gaps in the Customer Relationship Management sector in social media, which is shared by small companies. We concluded that Social CRM is still a new environment for micro and small companies, but has a great potential to boost sales, enhance customer loyalty and increase brand exposure. The lessons learned can guide policy-makers to take more suitable measures for strengthening this market sector. In future work, our aim is to plan a set of actions with an interventionist view to help entrepreneurs explore the functionalities of social media, and thus lead to the optimization of their internal and external processes.

Acknowledgments. This work was partly funded by the National Council for Scientific and Technological Development (CNPq) - DT-308334/2020-5; FAPESPA - Fundação Amazônia de Amparo a Estudos e Pesquisas (Grant PRONEM nº 045/2021); and FAPEMA - Fundação de Amparo à Pesquisa e ao Desenvolvimento Científico e Tecnológico do Maranhão. We would like to express our gratitude for this assistance.

References

1. Bartik, A.W., Bertrand, M., Cullen, Z., Glaeser, E.L., Luca, M., Stanton, C.: The impact of COVID-19 on small business outcomes and expectations. Proc. Natl. Acad. Sci. U.S.A **117**, 17656–17666 (2020)
2. Elrhim, M.A., Elsayed, A.: The effect of COVID-19 spread on the e-commerce market: the case of the 5 largest e-commerce companies in the world. SSRN Electron. J., 1–14 (2020). https://papers.ssrn.com/sol3/papers.cfm?abstract_id=3621166
3. Kim, R.Y.: The impact of COVID-19 on consumers: preparing for digital sales. IEEE Eng. Manag. Rev. **48**, 212–218 (2020). https://doi.org/10.1109/EMR.2020.2990115
4. Nisar, T.M., Prabhakar, G., Strakova, L.: Social media information benefits, knowledge management and smart organizations. J. Bus. Res. **94**, 264–272 (2019)
5. We Are Social: Digital 2021 - We Are Social. https://wearesocial.com/digital-2021. Accessed 27 Jan 2021
6. Alalwan, A.A., Rana, N.P., Dwivedi, Y.K., Algharabat, R.: Social media in marketing: a review and analysis of the existing literature. Telemat. Inform. **34**, 1177–1190 (2017)
7. Fernandes, L.C., Silva, J., Jacob, A., Lobato, F.: An extensive analysis of online restaurant reviews: a case study of the Amazonian Culinary Tourism. In: Proceedings of the 2020 Federated Conference on Computer Science and Information Systems, FedCSIS 2020, pp. 81–84. Institute of Electrical and Electronics Engineers Inc. (2020)
8. Lobato, F., Pinheiro, M., Jacob, A., Reinhold, O., Santana, Á.: Social CRM: biggest challenges to make it work in the real world. In: Abramowicz, W., Alt, R., Franczyk, B. (eds.) BIS 2016. LNBIP, vol. 263, pp. 221–232. Springer, Cham (2017). https://doi.org/10.1007/978-3-319-52464-1_20
9. Marolt, M., Zimmermann, H.D., Žnidaršič, A., Pucihar, A.: Exploring social customer relationship management adoption in micro, small and medium-sized enterprises. J. Theor. Appl. Electron. Commer. Res. **15**, 38–58 (2020)
10. Alt, R., Reinhold, O.: Social customer relationship management. In: An Introduction to Social Media Marketing, pp. 72–75. Routledge (2014)
11. Lobato, F.M.F., Silva Junior, J.L.F., Jacob, A., Lisboa Cardoso, D.: Social CRM: a literature review based on keywords network analysis. In: Abramowicz, W., Klein, G. (eds.) BIS 2020. LNBIP, vol. 394, pp. 237–249. Springer, Cham (2020). https://doi.org/10.1007/978-3-030-61146-0_19
12. Reinhold, O., Alt, R.: How companies are implementing social customer relationship management: insights from two case studies. In: 26th Bled eConference, pp. 206–221 (2013)
13. Felix, R., Rauschnabel, P.A., Hinsch, C.: Elements of strategic social media marketing: a holistic framework. J. Bus. Res. **70**, 118–126 (2017)
14. Rosenberger, M., Lehmkuhl, T., Jung, R.: Conceptualising and exploring user activities in social media. In: Janssen, M., et al. (eds.) I3E 2015. LNCS, vol. 9373, pp. 107–118. Springer, Cham (2015). https://doi.org/10.1007/978-3-319-25013-7_9
15. de León-Sigg, M., Vázquez-Reyes, S., Villa-Cisneros, J.L.: Factores que Afectan la Adopción de Tecnologías de Información en Micro y Pequeñas empresas: Un Estudio Cualitativo. RISTI - Rev. Iber. Sist. e Tecnol. Inf. **22**, 20–36 (2017)
16. SEBRAE: Pequenos negócios em números | Sebrae. https://www.sebrae.com.br/sites/PortalSebrae/ufs/sp/sebraeaz/pequenos-negocios-em-numeros,12e8794363447510VgnVCM1000004c00210aRCRD. Accessed 4 Feb 2021
17. Wang, Z., Kim, H.G.: Can social media marketing improve customer relationship capabilities and firm performance? Dynamic capability perspective. J. Interact. Mark. **39**, 15–26 (2017)

18. Longaray, A.A., Anselmo, C.R., Maia, C., Lunardi, G., Munhoz, P.: Análise do emprego do F-commerce como impulsionador do desempenho organizacional em micro e pequenas empresas no Brasil. RISTI - Rev. Ibérica Sist. e Tecnol. Informação **27**, 67–85 (2018)
19. Ainin, S., Parveen, F., Moghavvemi, S., Jaafar, N.I., Shuib, N.L.M.: Factors influencing the use of social media by SMEs and its performance outcomes. Ind. Manag. Data Syst. **115**, 570–588 (2015)
20. Rodrigues Chagas, B.N., Nogueira Viana, J.A., Reinhold, O., Lobato, F., Jacob, A.F.L., Alt, R.: Current applications of machine learning techniques in CRM: a literature review and practical implications. In: Proceedings - 2018 IEEE/WIC/ACM International Conference on Web Intelligence, WI 2018, pp. 452–458 (2019)
21. Alford, P., Page, S.J.: Marketing technology for adoption by small business. Serv. Ind. J. **35**, 655–669 (2015)
22. Baah-Ofori, R., Amoako, G.K.: Electronic Customer Relationship Management (E-CRM) practices of micro, small, and medium scale enterprises in Ghana. In: Strategic Customer Relationship Management in the Age of Social Media, pp. 72–94 (2015)
23. Imme, A.: As 10 Redes Sociais mais usadas no Brasil em 2020. https://resultadosdigitais.com.br/blog/redes-sociais-mais-usadas-no-brasil/. Accessed 31 July 2020
24. Swaminathan, V., Sorescu, A., Steenkamp, J.B.E.M., O'Guinn, T.C.G., Schmitt, B.: Branding in a hyperconnected world: refocusing theories and rethinking boundaries. J. Mark. **84**, 24–46 (2020)
25. Key, B.A., Tool, M.: Branding: A Key Marketing Tool. Springer, Cham (1992). https://doi.org/10.1007/978-1-349-12628-6

Gaining Insights on Student Satisfaction by Applying Social CRM Techniques for Higher Education Institutions

Gustavo Nogueira de Sousa[1(✉)], Fabio Lobato[2], Julio Viana[3], and Olaf Reinhold[3]

[1] State University of Maranhão (UEMA), São Luís, Brazil
sougusta@gmail.com
[2] Engineering and Geoscience Institute, Federal University of Western Pará (UFOPA), Santarém, Brazil
fabio.lobato@ufopa.edu.br
[3] Social CRM Research Center (SCRC), Leipzig, Germany
{julio.viana,olaf.reinhold}@scrc-leipzig.de

Abstract. Social Media and Customer Relationship Management (CRM) are already widely used in business settings, but other non-commercial sectors started only recently to adopt them. Among them are Higher Education Institutions (HEIs). Even though research shows positive effects on the quality of services, student satisfaction, and attractiveness towards international students, the adoption is very low. This research in progress reviews the state of research about Social CRM in HEIs and gives an example of the potential of social media for CRM approaches of HEIs by applying Social CRM concepts and techniques for better understanding the negative service experiences of students. By applying analytical Social CRM techniques on large amounts of User-Generated-Content (UGC) in complaint platforms the paper gives insights into problem chains inaccessible with manual methods. Based on the scarce research about Social CRM as well as the demonstrated potential of social media for CRM strategies of HEIs, this paper concludes with a call for further research on Social CRM in HEIs.

Keywords: CRM · Social media · Student satisfaction · Complaint management · Text mining · Topic modeling

1 Introduction

Social media has been gaining importance in higher education as it becomes a tool for interaction between institutions, lecturers and students. Researchers investigate the use and potential of social media in specific areas for several years. For example, as support in lectures [13], support in hybrid learning environments [28], for marketing purposes [27], or examined how students use them for study purposes [23]. However, their potential for building relationships with students

W. Abramowicz et al. (Eds.): BIS 2021 Workshops, LNBIP 444, pp. 174–185, 2022.
https://doi.org/10.1007/978-3-031-04216-4_17

and managing the student life cycle was only sparsely examined. A reason may be the fact that many universities do not compete over the increasing number of students or that the relationship between students and professors is more focused on teaching and students are expected to actively manage their study program. But as an increasing amount of young people, and so students, use social media more frequently, they also use them during their student lifecycle and expect higher education institutions (HEIs) to do the same.

Applying concepts and techniques developed for the management of customer relationship management (CRM) with the help of social media could help to manage this transformation. As in the industry, wherein many cases until 2005 the enterprise still owned the customer experience [18], the student experience is still often owned by HEIs. Universities already pay attention to social media to maintain communication [8], but use them only for dedicated tasks. Greenberg pointed out in 2010 [19] that enterprises need first to figure out the business models, applications, processes, and social characteristics that are required to actually implement the social CRM before social media customer service begins to happen. This counts now for HEIs as they need to figure out the application areas of social media within their student life cycle and service offerings.

This paper aims to initially explore the relevance of social media by analyzing the use of such channels in critical steps within the student life cycle, namely the handling of complaints as part of the service phase. As visible in a famous example of Social CRM [25], students will use social media if they are not satisfied with the service experience, regardless of whether HEIs are active on social media or not. Following the concept of Social CRM [5,19], higher education could build up its presence on social media platforms, provide services by using social media as a channel in workflows, learn from the content in social media and use these channels to perform collaborative tasks with students. Actively using social media and providing a satisfying service experience may decrease the number of complaints. As research about the management of complaints by HEIs via social media is scarce, this paper aims to show that students use social media for complaints, that a link between the number of complaints in social media with the active provision and management of social media by universities exists and to identify the core topics of student complaints. The research questions are:

- (RQ1) Are students using external and public platforms to complain about education-related services along their student life cycle?
- (RQ2) How can we derive information about major service quality issues affecting student satisfaction?
- (RQ3) What types of insights on customer satisfaction can HEIs managers expect from an analysis of external complaints?

The remainder of the paper is structured as follows. First, research about the role of customer satisfaction and Social CRM is reviewed and key elements for assessing customer satisfaction with the help of Social CRM are identified. Second, an experiment demonstrates an approach for analyzing customer satisfaction in higher education. Third, the results from the experiment are discussed, answering the research questions.

2 CRM and Social CRM in Higher Education

In this section, we discuss the CRM and Social CRM applied in high education institutions to improve the services and the students' satisfaction. We first outline the effects of the CRM on the service quality and students' satisfaction in HEIs, and before we discuss the potential of Social CRM to understanding customer satisfaction and to managing the service quality.

2.1 CRM Affects Service Quality and Student Satisfaction in HEIs

The application of CRM concepts in higher education was examined from different perspectives already. Rigo et al. (2016) [38] show that the main principles of CRM can also be applied to the context of HEIs and that HEIs must consider more stakeholders than just students in their CRM approach, calling for also using social media for linking and interacting with numerous stakeholders. Nair et al. (2007) [31] point out that HEIs first need to understand the student lifecycle ($Suspect \rightarrow Prospect \rightarrow Applicant \rightarrow Admitted \rightarrow Enrollee \rightarrow Alumni$) before successfully making use of a CRM approach.

By analyzing social media content, HEIs can improve their understanding of the student life cycle and optimize their CRM. Hrnjic (2016) [24] shows that student satisfaction is a good indicator for the successful adoption of CRM for the creation of a student-oriented environment and constantly adapting its processes. Critical elements are the university organization and management of teaching processes, academic staff skills and competencies, management board activities and institutional development, and quality of study materials used in the classes, and application of learning methods. Hrnijic (2016) [41] points out that universities that aim to achieve a leadership position in the higher education sector need to show an additional focus on reproducing highly skilled faculty staff that will have a capacity to improve teaching with regard to technology changes and market requirements. It is also important that the university board and leading people such as deans of HEIs understand the strategic dimension of CRM orientation at universities. Both call for HEIs to adopt social media as new technology and to develop an integrated management approach for social media and CRM. A study from Seemann et al. 2006 [41] shows the handling of students as customers provides a competitive advantage for higher education and enhances a college's ability to attract, retain and serve its customers. The benefits of implementing CRM in a college setting include a student-centric focus, improved customer data and process management, and increased student loyalty, retention, and satisfaction with the college's programs and services. As colleges increasingly embrace distance learning and e-business, CRM will become more pervasive. The COVID pandemic further increased this need. Badwan et al. (2017) [6] confirm this observation by showing that implementing electronic CRM can cause customer satisfaction, loyalty, retention, and high service quality as students pointed to be a customer.

CRM supports the understanding of customer expectations and thus provides a basis for service customization, which in turn can positively affect service quality. Wali points out that customer satisfaction and advocacy in HEIs depend on it. Wali et al. (2016) [43] show that an effective CRM program to improve service quality affects customer satisfaction and even has the ability to induce positive advocacy behavior from its international students. A key element is gaining and understanding customer's experience feedback.

2.2 New Potentials for Understanding Customer Satisfaction and Managing the Service Quality Arise from Social CRM

A key element of Social CRM is the interaction with stakeholders that influence the service system of a business and the knowledge derived from this interaction. As Greenberg (2009) [18] points out, the customer becomes the focal point of the ecosystem, and service providers can make use of social media to understand their role in the customer ecosystem. Social CRM provides the means for that, but as Meyliana et al. (2017) [40] a Social CRM model for HEI's is virtually non-existent. However, the first research points out, that applications, data, and information, adapted business processes, social media presences are among the critical success factors for social CRM in HEIs.

Following Meyliana et al. (2015) [30] many universities started with web 2.0 and social media adoption but focus mainly on real-time events webcast, widgets, and social networks for the users of the university website. The study of Oliveira et al. (2015) [32] shows that Social CRM propels HEI to engage in dialogical conversations and collaborative relationships. The use of social media platforms, allowing to reshape the HEI-student formal relationship, strengthening educational bonds through the development of dialogs, provides mutually beneficial value and, ultimately, allows for the growth of social and educational communities.

However, only a few examples have further investigated the potential of analytical Social CRM for HEIs, especially for assessing student satisfaction with experience service quality. Budiardjo et al. (2017) [9] show in general by mapping the student lifecycle with CRM processes and features of Social CRM that the latter can support core CRM processes of HEIs, and the application of Social CRM software can support operational, analytical, and collaboration tasks. Karna et al. (2015) [27] show how data analysis can provide further insights on candidates and help to individualize marketing activities. Ciqueira et al. (2017) [11] demonstrates how an analysis of an universities social network presence can help in understanding strength and weaknesses from the students perspective. But unlike Social CRM in a business context, the potential of UGC monitoring and mining for HEIs has not been extensively studied yet.

3 Improving the Understanding of Negative Service Experiences in HEIs with Analytical Social CRM Techniques

This section discusses the use of analytical Social CRM to understand the students' negative experiences in HEIs on social media platforms. In the following subsections, we present an overview of the process to analyze from social media platforms, the methodology to performs the analysis, the potential data source, and the potential analysis methods.

3.1 Complaint and Satisfaction Analysis in External Social Media

Following the concept of analytical Social CRM [37], building up a customer feedback and satisfaction analysis requires accessing the data of social media platforms where students share feedback and opinions. In a second step, this data needs to be turned into knowledge about customer satisfaction by either manual evaluation, observation of keywords, or applying methods for understanding at the semantic level. While the manual evaluation can provide rich insights, the potential amount of data makes it challenging for HEIs. Applying basic analyses such as looking for trending topics and sentiments of students might support HEIs to understand the behavior of their students. Although these analyses are supported by many social media monitoring tools, they have limits in uncovering larger patterns, such as the relationship between raised issues. Understanding previously unknown factors that affect service quality requires the application of data mining methods.

3.2 Process Design

In this scenario, it is critical to map relevant data sources and evaluate all pertinent analyses that improve customer satisfaction. One well-known data analysis methodology for conducting real-world projects is the Cross-Industry Standard Process for Data Mining (CRISP-DM). This methodology proposes a comprehensive process model for carrying out data mining projects. The process is divided into phases of *Business Understanding*, *Data Understanding*, *Data Preparation*, *Modeling*, *Evaluation*, and *Deployment* and is independent of the industry sector and the technology used [44].

In this research, CRISP-DM will be applied for the analysis of student complaints in Brazil. HEIs in Brazil provides a good example, because of the availability of well-used independent social media platforms for publicly raising and discussing complaints independently from specific industry sectors. Thus, insights on service quality and customer satisfaction outside of a HEIs direct control are accessible.

3.3 Potential Data Sources

The second phase of CRISP-DM is called "*Data Understanding*", in which it is possible to identify issues on data and provide initial insights into available data. However, there is plenty of platforms with a large volume of User-Generated Content (UGC) available with some challenges such as data diversity, unstructured data, missing data, *etc.* [29]. Despite these challenges, such platforms represent an interesting data source for providing business *insights* with reduced costs when compared with customer surveys and other market research strategies [7,42]. In the context of Brazilian HEIs relevant data sources are for example:

- ***Consumidor.gov*** [1] - Contains data referring to complaints reports about universities. The data types include *strings* (raw text), *timestamp* (Date and Time), and numerical features. This platform does not provide an *Application Programming Interface* (API) for data acquisition. However, considering that it is a public platform with open data (considering Brazilian Legislation), the data can be requested using the national accountability platform;
- **Ministry of Justice and Public Security** (MJSP) [3] - The MJSP portal provides details about the complaints published on the Consumidor.gov platform. Large parts of this data are in textual format. This platform provides an interface to get the available data;
- **Data from Higher Education Census** [2] - It is data related to the 2019 higher education census, and contains information about Brazilian students, courses, and high educational institutions. Most of the data are numerical features, requiring a mapping procedure with a data dictionary;
- **Data from Social Media platforms** - Contains information related to followers, publications, and Comments. However, it is important to verify the privacy rules and API restrictions of each platform.

3.4 Potential Methods for Analysis

"Data preparation" and "Modeling" phases of the CRISP-DM methodology represent the core for Computer Scientists [44]. The pre-processing methods aim to improve data quality, consequently, improving the reliability of the results [29]. Considering that most UGCs are encoded in text, there is a lot of effort on the development of *text mining* methods and pipelines [15]. The canonical pipeline includes the application of specific pre-processing methods for text data [4,17,20]. It can be followed by a data fusion/enrichment step, aiming to combine different data sources, which can reduce bias and uncertainties, increase reliability, and improve accuracy [35,45].

After the data preparation phase follows the analysis phase, which comprises common text mining tasks. For instance, **topic modeling** allows the identification of the most frequent topics and their terms, which would be difficult to discover through a manual process [34,39]. The discovered topics can be modeled as a *graph*, indicating the relationship between the topics, and allowing the identification of terms chains [14]. In addition to that, **sentiment analysis** can be used to extract sentiments through automatic polarity detection [26,36], and the text quality can be measured by the evaluation of the legibility and by the determination of the quality according to the expected in each phase of regular education [16,22,33].

4 Demonstration

This section presents an analysis of the textual content of students' complaints about two higher education institutions in Brazil, named *University A* and *University B*. These two universities have multiple internal channels of customer services available to students, which include the website online chats, email, telephone and social networks (see Table 1). On social networks, these universities have in general a high number of followers. However, *University A* has a low number of interactions on their content, opposite to *University B* with a very high number of interactions. Furthermore, their students are using external platforms for making complaints. As it is known that on average two-thirds of consumers check product, service, and brand evaluations before deciding on the purchase, [12,21] these external complaints can have an impact on reputation and in consequence the willingness of future students to join the university.

Table 1. Summary information about the universities

	University A	*University B*
Total of students [2]	302,841	393,578
Customer service channels	- Website - Phone - WhatApp - Social Networks: — Twitter with 8,804 followers — Instagram with 62,300 followers — Facebook with 617,702 followers - Reclame Aqui - Consumidor.gov	- Website - Phone - WhatApp - Social Networks: — Instagram with 283,000 followers — Facebook with 1,534,463 followers - Reclame Aqui - Consumidor.gov
Complaints on Consumidor.gov [a]	- Total of Complaints: 2,743 - Solution Rate: 63.7%	- Total of Complaints: 3,965 - Solution Rate: 57.2%
Complaints on ReclameAqui [b]	- Total of Complaints: 37,764 - Solution Rate: 58.6%	- Total of Complaints: 52229 - Solution Rate: 52%

[a] https://www.consumidor.gov.br/
[b] https://www.reclameaqui.com.br/

Following the methodology described before, the analysis comprised a preprocessing step and the analysis phase, using topic modeling and topic correlation methods. The data were collected on *Consumidor.gov* [1] from January to March of 2021, with a total of 652 complaints about *University A* and 803 about *University B*. Topic modeling was carried out using Negative Matrix Factorization (NMF) [10] and revealed the ten main topics in complaints about each university (see Fig. 1). Some topics are not exclusive, indicating that the two institutions facing the same type of problem, related to the "**Payments**", "**Attestation of degree**", and "**Classroom**".

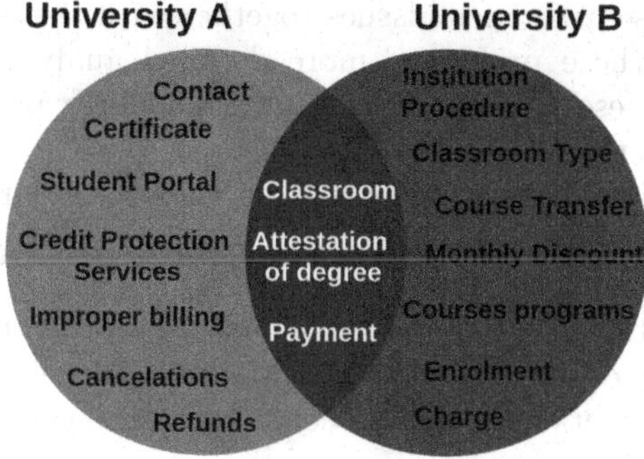

Fig. 1. Main Topics in complaints of the *University A* and *University B*

The correlation of the topics allows the identification of relationship degrees between the topics. Figure 2 was produced using open-source software called *Gephi*[1], and it shows the relationships between most frequent topics/aspects, and the strongest blue tone of the line indicates a stronger degree of correlation. In the topics of *University A* (see Fig. 2a) the formation of a chain of main problems refer to **Payments**, **Refunds**, and the **Contact**. Figure 2b shows the chain of main problems for *University B*, which are related to **Payments**, **Charges**, **Monthly Discount**, and **Enrollment**.

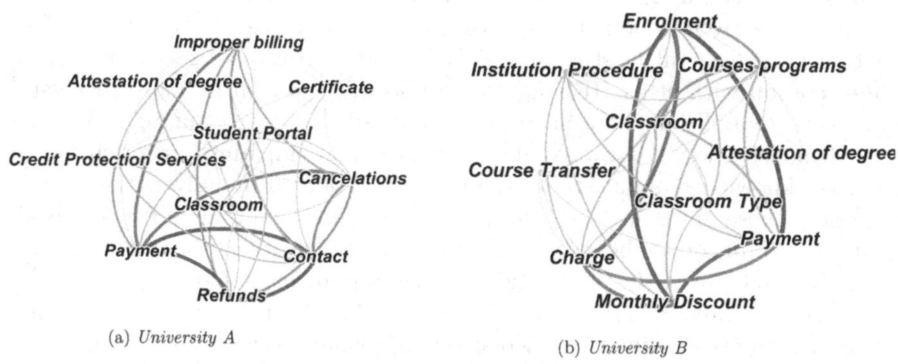

(a) *University A*

(b) *University B*

Fig. 2. Topic correlations of complaints

Modeling and correlation of topics give HEIs managers insights on potential service quality issues. But, they are also a starting point for planning actions to solve the problem chains that make up principal causes of dissatisfaction among students. For example, among the problems of *University A* are several types of issues related to finance and customer service. They indicate students

[1] https://gephi.org/.

are dissatisfied because of financial issues together with the university customer Service. For solving these problems, more detailed analysis can be made on complaint related to these areas to understand and identify products and services that are the source of the problems.

The demonstrated analysis provides insights based on a large number of complaints related to two universities. Such aggregated insights on major students' problems and their relations can support decision-making and is a basis for executing efficient actions for solving the chain of issues at the root of student complaints. Thus they can contribute to CRM efforts by helping and assisting in developing better products and services centered on the expectations and needs of students.

5 Conclusion and Implications

This paper investigates the potential of social media for improving service quality and student satisfaction as part of the ever-increasing adoption of CRM in HEIs. Building upon recent research about critical success factors of CRM for HEIs, the paper demonstrates how universities can improve their service system with insights from Social CRM analytics. Based on an examination of selected universities in Brazil, this research shows that students use external platforms to complain about their service experience, regardless if the university provides its own platforms (RQ1). With the application of data analysis and mining techniques employed in social CRM [15,37] in the enterprise context, major issues can be efficiently summarized (RQ2) from large data volumes. HEIs can learn from this data (RQ3) about the service quality experience of students and factors that negatively affect student satisfaction. It is also a basis for comparison and evaluation with other HEIs and an opportunity to identify more successful approaches. In addition, from the results produced by the text mining tasks used in the demonstration, it was possible to observe that complaints are a rich source to extract knowledge about students and services offered by universities.

Besides these direct insights, this research provides, on a more general level, a further example of the benefits of Social CRM for HEIs and the importance of further research about the application fields and implementation approaches for them. An integrated Social CRM allows to get data from internal and external sources and to use obtained knowledge for improving services. Many insights on the successful integration of social media with CRM in enterprises are available and HEIs can build on this extensive knowledge while adapting it to their specific environment.

This paper is an individual example of the potential of advanced semantic analysis in a limited case. Therefore the paper has limitations. The obtained results are not generalizable and demonstrate only how such insights that can be acquired by using the proposed approach. Also, the insights have not been evaluated with managers in HEIs in terms of novelty and meaningfulness compared to other sources. Therefore, this research is planned to be extended in the next step, by using the approach for a larger set of universities, expanding the insights by comparing the results with a higher number of institutions. Additionally, results will be validated with HEIs managers regarding the contribution from the applied method towards a better understanding of student satisfaction and potential improvements in service quality.

References

1. Brazilian Consumer complaints. https://www.consumidor.gov.br
2. Brazilian Higher Education Census. https://www.gov.br/inep/pt-br/areas-de-atuacao/pesquisas-estatisticas-e-indicadores/censo-da-educacao-superior/resultados
3. Data from Consumidor.gov.br - Datasets - Brazilian Open Data Portal. https://dados.gov.br/dataset/reclamacoes-do-consumidor-gov-br1
4. Allahyari, M., et al.: A Brief Survey of Text Mining: Classification, Clustering and Extraction Techniques (2017)
5. Alt, R., Reinhold, O.: Social CRM: challenges and perspectives. In: Social Customer Relationship Management. MP, pp. 81–102. Springer, Cham (2020). https://doi.org/10.1007/978-3-030-23343-3_4
6. Badwan, J.J., Al Shobaki, M.J., Abu-Naser, S.S., Abu Amuna, Y.M.: Adopting technology for customer relationship management in higher educational institutions (2017)
7. Bahtar, A.Z., Muda, M.: The impact of user - generated content (UGC) on product reviews towards online purchasing - a conceptual framework. Procedia Econ. Finance **37**, 337–342 (2016). http://www.sciencedirect.com/science/article/pii/S2212567116301344
8. Bonsón, E., Torres, L., Royo, S., Flores, F.: Local e-government 2.0: social media and corporate transparency in municipalities. Gov. Inf. Q. **29**(2), 123–132 (2012). https://doi.org/10.1016/j.giq.2011.10.001, https://www.sciencedirect.com/science/article/pii/S0740624X1200010X
9. Budiardjo, E.K., Hidayanto, A.N., Fitriani, W.R., Munajat, Q., et al.: Social CRM features identification for higher education. J. Eng. Appl. Sci. **12**(9), 2327–2333 (2017)
10. Chen, Y., Zhang, H., Liu, R., Ye, Z., Lin, J.: Experimental explorations on short text topic mining between LDA and NMF based Schemes. Knowl.-Based Syst. **163**, 1–13 (2019)
11. Cirqueira, D., et al.: Improving relationship management in universities with sentiment analysis and topic modeling of social media channels: learnings from UFPA. In: Proceedings of the International Conference on Web Intelligence. WI 2017, pp. 998–1005. Association for Computing Machinery, New York(2017). https://doi.org/10.1145/3106426.3117761
12. Constantinides, E., Holleschovsky, N.I.: Impact of online product reviews on purchasing decisions. In: Proceedings of the 12th International Conference on Web Information Systems and Technologies, pp. 271–278 (2016). https://doi.org/10.5220/0005861002710278, http://www.scitepress.org/DigitalLibrary/Link.aspx?doi=10.5220/0005861002710278
13. Dyson, B., Vickers, K., Turtle, J., Cowan, S., Tassone, A.: Evaluating the use of Facebook to increase student engagement and understanding in lecture-based classes. High. Educ. **69**(2), 303–313 (2015). https://doi.org/10.1007/s10734-014-9776-3, https://link.springer.com/article/10.1007/s10734-014-9776-3

14. Easley, D., Kleinberg, J.: Networks, Crowds, and Markets: Reasoning about a Highly Connected World (2010). http://www.cs.cornell.edu/home/kleinber/networks-book/
15. Fernandes, L.C., Silva, J., Jacob, A., Lobato, F.: An extensive analysis of online restaurant reviews: a case study of the Amazonian Culinary Tourism. In: Proceedings of the 2020 Federated Conference on Computer Science and Information Systems. FedCSIS 2020, vol. 21, pp. 81–84 (2020). https://doi.org/10.15439/2020F179
16. Flesch, R.F.: A new readability yardstick. J. Appl. Psychol. **32**(3), 221–33 (1948)
17. García, S., Luengo, J., Herrera, F., García, S., Luengo, J., Herrera, F.: Data Preprocessing in Data Mining, vol. 72 (2015)
18. Greenberg, P.: Social CRM comes of age. Sponsored by Oracle (2009)
19. Greenberg, P.: CRM at the Speed of Light: Social CRM Strategies, Tools, and Techniques. McGraw-Hill, New York (2010)
20. Han, H.J.S., Mankad, S., Gavirneni, N., Verma, R.: What guests really think of your hotel: text analytics of online customer reviews. Cornell Hospitality Rep. **16**(2), 3–17 (2016). http://scholarship.sha.cornell.edu/chrreports
21. He, L., Han, D., Zhou, X., Qu, Z.: The voice of drug consumers: online textual review analysis using structural topic model. Int. J. Environ. Res. Public Health **17**(10) (2020). https://doi.org/10.3390/ijerph17103648
22. Hirsch, M., Aggarwal, S., Barker, C., Davis, C.J., Duffy, J.M.: Googling endometriosis: a systematic review of information available on the Internet. Am. J. Obstet. Gynecol. **216**(5), 451-458.e1 (2017)
23. Hrastinski, S., Aghaee, N.M.: How are campus students using social media to support their studies? An explorative interview study. Educ. Inf. Technol. **17**(4), 451–464 (2012)
24. Hrnjic, A.: The transformation of higher education: evaluation of CRM concept application and its impact on student satisfaction. Eurasian Bus. Rev. **6**(1), 53–77 (2016)
25. Jarvis, J.: My Dell hell. The Guardian (2005). https://www.theguardian.com/technology/2005/aug/29/mondaymediasection.blogging
26. Jo, J.M., Ferreira, M.: An evaluation of sentiment analysis for mobile devices, October 2017
27. Karna, N., Supriana, I., Maulidevi, N.: Social CRM using web mining for Indonesian academic institution. In: 2015 International Conference on Information Technology Systems and Innovation (ICITSI), pp. 1–6. IEEE (2015)
28. Li, R.: Traditional to hybrid: social media's role in reshaping instruction in higher education. In: Digital Arts and Entertainment: Concepts, Methodologies, Tools, and Applications, pp. 387–411. IGI Global (2014)
29. Lobato, F., Pinheiro, M., Jacob, A., Reinhold, O., Santana, Á.: Social CRM: biggest challenges to make it work in the real world. In: Abramowicz, W., Alt, R., Franczyk, B. (eds.) BIS 2016. LNBIP, vol. 263, pp. 221–232. Springer, Cham (2017). https://doi.org/10.1007/978-3-319-52464-1_20
30. Meyliana, P., Hidayanto, A.N., Budiardjo, E.K.: Social media adoption for social CRM in higher education: an insight from Indonesian universities. Int. J. Synergy Res. **4**(2), 7 (2015)
31. Nair, C., Chan, S., Fang, X.: A case study of CRM adoption in higher education. In: Proceedings of the 2007 Information Resources Management Association International Conference. Citeseer (2007)
32. Oliveira, L.: Social student relationship management in higher education: extending educational and organisational communication into social media. In: 9th Annual International Technology, Education and Development Conference, IATED (2015)

33. Othman, I.W., et al.: Text readability and fraud detection. In: ISBEIA 2012 - IEEE Symposium on Business, Engineering and Industrial Applications, vol. 99, pp. 296–301 (2012)
34. Park, E.O., Chae, B.K., Kwon, J.: The structural topic model for online review analysis: comparison between green and non-green restaurants. J. Hospitality Tourism Technol. **11**(1), 1–17 (2018)
35. Qi, J., Yang, P., Newcombe, L., Peng, X., Yang, Y., Zhao, Z.: An overview of data fusion techniques for Internet of Things enabled physical activity recognition and measure. Inf. Fusion **55**, 269–280 (2020)
36. Ravi, K., Ravi, V.: A survey on opinion mining and sentiment analysis: tasks, approaches and applications. Knowl.-Based Syst. **89**, 14–46 (2015). https://www.sciencedirect.com/science/article/pii/S0950705115002336
37. Reinhold, O., Alt, R.: Analytical social CRM: concept and tool support. In: Bled eConference, p. 50 (2011)
38. Rigo, G.E., Pedron, C.D., Caldeira, M., Araújo, C.C.S.d.: CRM adoption in a higher education institution. JISTEM J. Inf. Syst. Technol. Manage. **13**(1), 45–60 (2016)
39. Roberts, M.E., et al.: Structural topic models for open-ended survey responses. Am. J. Polit. Sci. **58**(4), 1064–1082 (2014)
40. Sablan, B., Hidayanto, A.N., Budiardjo, E.K., et al.: The critical success factors (CSFS) of social CRM implementation in higher education. In: 2017 International Conference on Research and Innovation in Information Systems (ICRIIS), pp. 1–6. IEEE (2017)
41. Seeman, E.D., O'Hara, M.: Customer relationship management in higher education: using information systems to improve the student-school relationship. Campus-wide information systems (2006)
42. Vermeer, S.A., Araujo, T., Bernritter, S.F., van Noort, G.: Seeing the wood for the trees: how machine learning can help firms in identifying relevant electronic word-of-mouth in social media. Int. J. Res. Market. 1–17 (2019)
43. Wali, A.F., Wright, L.T.: Customer relationship management and service quality: influences in higher education. J. Customer Behav. **15**(1), 67–79 (2016)
44. Wirth, R.: CRISP-DM: towards a standard process model for data mining. In: Proceedings of the Fourth International Conference on the Practical Application of Knowledge Discovery and Data Mining, vol. 24959, pp. 29–39 (2000)
45. Zhang, J.: Multi-source remote sensing data fusion: status and trends. Int. J. Image Data Fusion **1**(1), 5–24 (2010)

Understanding Customer-Induced Orchestration of Services: A Review of Drivers and Concepts

Julio Viana[1(✉)], Rainer Alt[2], and Olaf Reinhold[1]

[1] Social CRM Research Center, Leipzig, Germany
julio.viana@scrc-leipzig.de
[2] Leipzig University, Leipzig, Germany

Abstract. Service Innovation plays an important role in research and practice and enabled the surge of new concepts that changes the focus from a product-oriented approach to a service-oriented approach. However, further developments place the customer in the center of company-client relationships. The recent advances in customer data analysis and the positioning of customers as company's co-creators led to the development of a new concept called Customer-induced Orchestration of Services. The novelty of the topic requires further studies and a deeper understanding of the interdisciplinary concepts around it. This paper identifies the main drivers and concepts, allowing a more holistic view on the topic. The results support further research, as well as the development of a framework or method for the application of Customer-induced Orchestration of Services, which enables more transparency and control for customers.

Keywords: Customer-induced orchestration of services · Service innovation · Customer orientation · Platform economy · Decentralization

1 Introduction

Scholars have been discussing the role of internet intermediaries [33], as added value to innovation [48] and strategy as an interdisciplinary agenda to define policies and decisions regarding the disclosure, conceal, bias and distortion of market information [17, 22], improving transparency. In this sense, traditional company-centric value creation (provider-dominant logic) gives room for personalized co-creation experiences and allow more transparency between companies and customers [34]. Transparency, by visually revealing operating processes to consumers and beneficiaries to producers, generates a positive feedback loop, since value is created from both sides [12]. In addition to improving transparency, the process of eliminating or reducing the intermediaries (disintermediation) contributes to lower costs for the customer and to increase the power from customers [39].

Despite the assumption that electronic marketplaces bring more transparency and, consequently, lower product prices [45], firms are sensitive to data access and disclosure rules required by the online exchange [55]. In this line, concerns regarding data privacy and protection for companies and consumers have been increasingly discussed,

W. Abramowicz et al. (Eds.): BIS 2021 Workshops, LNBIP 444, pp. 186–197, 2022.
https://doi.org/10.1007/978-3-031-04216-4_18

as new regulations take place worldwide. Therefore, society could benefit from an improvement in transparency and control when choosing and composing their portfolio of services.

Customer Orientation supports the rise of electronic markets (platform economy), which provide customized offers by integrating and analyzing customer data. The platform economy has changed the way people consume services related to hospitality, retail, transportation, music and video industries [38].

Despite this challenge, analyzing customer data, especially on online environments and social media, can support businesses to understand the behavior of their target and identify business potentials [5]. Hence, the analysis of customer behavior denotes a great potential for increasing customer orientation, as well as to develop innovative paths for decentralization, supporting users to control which company has their data, as well as the type of data they are sharing.

Following these discussions, recent developments in research regarding customer orientation and service innovation brought attention to the concept of Customer-induced Service Orchestration (COS) [7]. COS combines customer orientation with recent advances in Artificial Intelligence (AI) to provide service-oriented architectures for software development [29, 32]. Although the novelty aspect of the concept in research, patent applications show that large companies have been using AI methods for interoperating systems [14] based on Peer-to-Peer (P2P) service orchestration [10].

Despite the recent advances in service integration, the decentralization process comes with great challenges. Hence, this paper seeks to identify drivers and concepts related to the integration of systems for customer-induced orchestration of services, combining these interdisciplinary concepts.

As seen, provider-oriented platforms are dominant in the market. In addition, privacy issues and the lack of transparency and control from customers regarding their data is a growing challenges that platform companies should address. The novel COS concept indicates possible solutions by proposing decentralized platforms that integrate different systems and provides more control to customers when choosing their service providers. However, there is a lack in the literature of an approach that provides and understanding of all relevant drivers and concepts related to COS, as well as a method or design indications for building such a decentralized system.

Hence, this paper provides preliminary results of an ongoing study that seeks to develop design indications for system integration. It presents the current status of a research in progress by introducing drivers and concepts related to the topic of COS. Figure 1 depicts the research steps and the focus of this paper, while the next section introduces the identified drivers, related concepts and their integration for a holistic view on the topic.

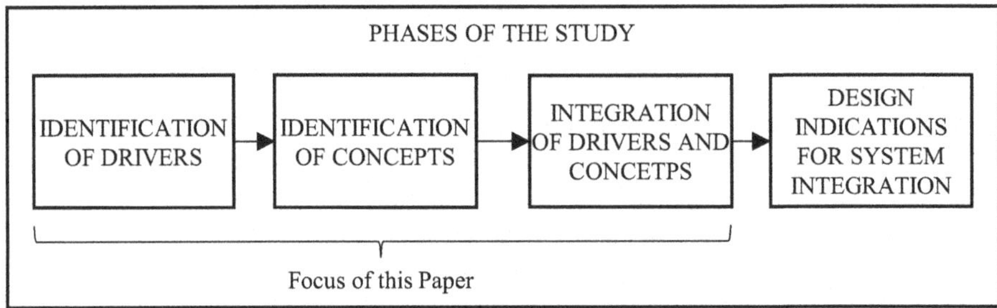

Fig. 1. Research steps

2 Methodology

This conceptual paper used literature review and desk research to identify the drivers around the concept of COS and its potential. As COS has not been fully discussed in the literature, the main article about the specificities of COS [7] was analyzed. The analysis indicated potential drivers related to COS. A backward research was then carried out to analyze the cited literature and to understand these drivers and their related concepts that support the development of decentralized systems. The drivers are aligned with the next convergence of the platform economy indicated by Alt (2021) [2], which points out that technologies that have been developed separately are now converging to an inter-organizational infrastructure. Besides the backward search, further articles and market information were added through desk research. The next section introduces the drivers and its main-related literature.

3 Drivers and Concepts of Customer-Induced Orchestration of Services

3.1 Customer Orientation

Customer orientation has been discussed in the business and marketing literature for the past decades, including measurements to evaluate its degree by salespeople [23, 41], in corporate culture [16] and effects from this orientation on customer perception and behavior [11]. Since the 1930s, intangible assets were considered relevant for innovation [42]. Following this lead, service innovation has integrated the research agenda for many years and its discussions include success drivers for market-creating innovations, such as scalable business model, customer experience management, employee performance, brand differentiation, etc. [8]. The term also evolved with the inclusion of approaches from different fields of knowledge and research, such as Economics, Marketing and Information Systems and the rise of new concepts, such as Service-Dominant Logic (SDL) [27] and Customer-dominant Logic (CDL) [20].

SDL defines service as the fundamental basis of exchange in which goods derive their value through use and, therefore, they provide a service based on operant resources - knowledge and skills [49]. It indicates that inside-out and outside-in "openness" plays a key role to develop continuously professional and technical services [24]. New technology companies are the ones profiting the most from this service integration, while ecosystems from incumbent companies are considered closed, slow and inflexible [53], requiring these companies to change their culture and move from product to service logic. Following SDL, Service-dominant Architecture (SDA) arose to define key building blocks of platforms, which include interconnectivity, scalability, modularity and interoperability. It includes discussions on the integration of service systems to the concept of SDL. A service system is considered a configuration of people, technology and other resources that provide each other value in the way of interaction with other service systems [46]. SDA aims at reusing and integrating technical functions [53] and comprises of three service systems and a data lake: system of interaction, which seeks real-time customer interaction; system of participation, which integrates external resources; system of operant resources, which refers to the capability of a business to integrate resources and the data lake referring to data exchange [28]. These systems require a high level of interaction and participation from customers, who become co-creators through their experience [51].

Although SDL and SDA place the customer as a co-creator, recent marketing research suggests that companies should move from a provider logic to a customer logic (CDL). CDL advises companies to consider the customer context in order to create value and points out the new challenges to service marketers, such as company involvement, company control in co-creation, visibility of value creation, scope of customer experience and character of customer experience [21, 50]. Following CDL concepts, overt and covert customer activities could provide relevant insights for service development. Additionally, CDL is in line with the process of service design that is used to understand how customers behave across a journey, their feelings and motivations [56]. In this sense, another interdisciplinary approach called Social Customer Relationship Management (Social CRM) encourages companies to look into customer data throughout the customer journey and combine it with external sources of data, especially from social information systems (e.g. social media) to improve the relationship with their clients [4, 57]. Nevertheless, computer science skills regarding artificial intelligence (AI) methods and machine learning techniques are important to collect and analyze this data [13, 37].

Despite the advances in the discussions around SDL, the concept might still focus on the provider perspective. While CDL suggests that consumers are in the center, managerial guidelines on how to practically implement the concept are still missing [20].

Following these discussions, recent developments in research regarding customer orientation and service innovation brought attention to the COS concept, which combines customer orientation with recent advances in AI to provide architectures and methods for software development [29, 32].

COS proposes that decentralized systems could combine different business ecosystems into one platform, allowing customers to orchestrate their services. These ecosystems consist of an economic community supported by a foundation of interacting organizations and individuals [30]. The tourism and transportation industry serves as an example of how this orchestration could take place. A customer, willing to go from point A to point B could select all different services, such as bike rental, car rental, train, bus and flight services, as well as additional services necessary along the way, on a unique platform [7]. These increasingly complex value systems could open up the possibilities for intermediaries that configure, offer and monitor solutions, which comprise products from different providers, as well as market places [3]. Current intermediaries, however, offer only little transparency over the criteria involved in the service creation process [7]. The integration of AI techniques might contribute to paving the road for new intermediaries able to integrate these different services. Nevertheless, scholars suggest a lack of research, which provides managerial guidelines or architecture to build such interoperating and decentralized systems efficiently [7, 20].

3.2 Platform Economy and Decentralization

With the rise of service-oriented platforms based on digital technologies, such as electronic markets (EMs) or "platform" companies, information technology (IT) plays an important role in providing and integrating ecosystems able to interoperate among themselves, according to customers' needs [27]. From the technological side, EMs comprise the application of IT to support communication and allocation purposes in an environment with multiple actors in one or multiple value chain [3]. Understanding the continuous adaptations, monitoring and improvements required by EMs becomes even more important as interconnections and diffusion of electronic agents increase [3].

In this sense, SDL and its architecture (SDA) outline the process of conceptualizing services in the digital economy. However, the need to focus on the client proposed by CDL instead of orientation on the provider side supports the idea of Customer-induced Orchestration, which suggests that decentralized systems could combine different services into one platform [7].

Linked data technologies support this decentralization process through its basis for application development [19]. In this line, Blockchain technology is considered very attractive to solve problems in different industries and sectors [15]. Several start-up companies, as well as large companies, are seeking to develop decentralized solutions using Blockchain [15], which is able to support the rise of digital institutions and infrastructure, such as markets, judiciaries, and payment systems, contributing to the process of disintermediation found in the online world [54]. Social information systems are relevant to pave the way for data integration, while Blockchain supports system decentralization by enhancing collaboration and fostering customization [26]. This influences the way COS-based enterprises communicate with their customers and sell their products.

These new processes highlight the need to study the coevolution of platforms, which takes into consideration their architecture, governance, environmental dynamics, theoretical lenses and evolutionary dynamics [47]. Coevolution has the potential to serve as a unifying framework for research in strategy and organization studies, as well as for reinterpreting, reframing, and redirecting the selection adaptation discourse [25]. In this sense, decentralization plays an important role in the next generation of integrated systems. Moreover, supply chain integration is able to support this system integration, as coordination modes are required to synchronize interdependent activities [44]. The integration of IT infrastructure enables supply chain process integration, which sustains gains in firm performance [35].

3.3 Smart Services

System integration and Customer Orientation pave the way for the rise of smart service systems, which derive from enhancements applied to products and services [6]. The co-creation of smart services, enabled by smart products, is based on monitoring, optimization remote control and autonomous adaptation of products [9]. The understanding of customer orientation is of paramount importance for today's service transformation and service economy with smart services [6].

Following the trend for improving customer orientation, collective intelligence can provide insights regarding a group of people. The term is used by technologists to refer to the combination of behavior, preferences, or ideas of a group that, consequently, generates novel perceptions [43]. The combination of AI and statistical methods allows the analysis of collective intelligence on collected data [43]. However, smart services using collective intelligence should deeply understand the overall system dynamics to provide safe and reliable features to its users [40].

4 Overview of Drivers and Concepts

Existing research on COS unveils four central drivers towards the design of systems that allows service orchestration, namely Customer Orientation, Platform Economy, Decentralization and Smart Services. The current knowledge base already provides a number of concepts related to each driver. Figure 2 summarizes related concepts identified in the literature review.

Customer Orientation and Customer Relationship Management (CRM) technologies have a positive association with the development of durable customer relationships [36]. This relationship is also improved by the possibilities of co-creation offered to customers in a COS scenario. For that, customer data derived from social information systems supports data integration through the concept of Social CRM and enables co-creation with customers and other relevant stakeholders. The Platform Economy serves as an example of this integration and the use of customer data to improve and customize smart services. Nevertheless, decentralization is in the core of COS as independent systems are necessary to link different service providers into one platform, improving transparency and allowing customers to orchestrate services according to their needs.

However, a holistic and integrated perspective seems necessary for understanding and designing COS systems. The literature review points towards a lack of research following such an integrated approach.

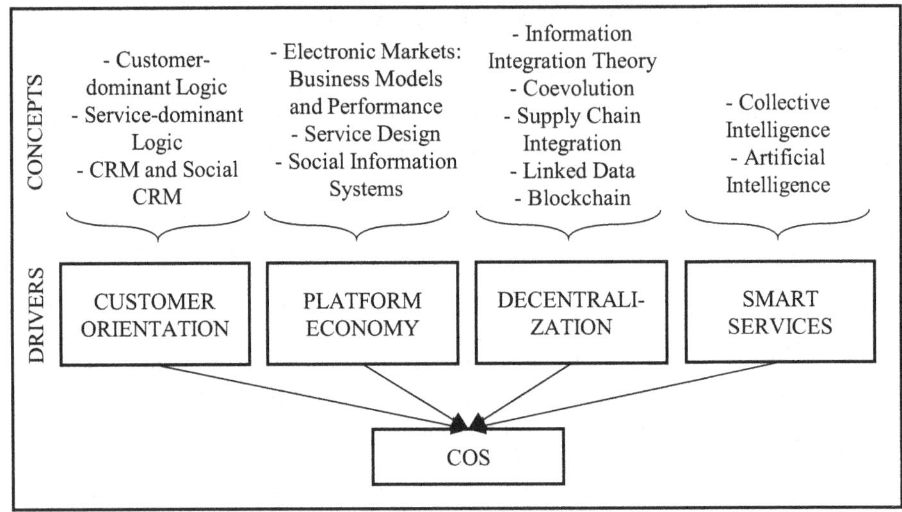

Fig. 2. Drivers and concepts for COS

5 Conclusions and Further Research

As seen, despite the recent advances in research, there is a lack of managerial guidelines and an architectural structure that supports companies to practically implement this integration. Integrated systems can increase transparency in processes and prices. Additionally, COS places consumers in the center of trading actions, enabling them to control the companies that will have access to their data and what type of private data are disclosed.

Different theories, methods and technologies support COS. At the same time, EMs are creating ecosystems that reduce costs and provide customers with new experiences [18]. These companies provide more than e-commerce, including cloud services, asset management, payment systems, logistics, media, etc. to their portfolio. The ability to integrate services and execute them based on a platform, allows these companies to create a system that is intelligent, open, connected and scalable [52].

Nevertheless, design indications with an interdisciplinary approach, combining different approaches from the fields of information systems, business, economics and computer science are necessary to guide new business models towards COS. Enabling customers to control and orchestrate services requires an understanding of their needs, which can be reached by collecting and analyzing customer data. Additionally, it is necessary to understand the demands for intermediation within the process of ecosystem integration. Scholars suggest a need for decentralization, but it requires an analysis of the integration process to identify how far this decentralization can reach. Concerning system integration, digital transformation has occurred in three different levels: (1) focus on software development in the 60s, (2) on business processes in the 80s and, (3) on values and the different business models that create innovative solutions since the 2000s [1]. Hence, the COS concept should consider all methods and techniques developed within these three levels of digital transformation, such as Agile development methods, business engineering and design thinking [1].

Regarding customer data, scholars identify the need for the design of services with transparency and data privacy in mind, providing customers with the appropriate value in exchange for data, educating them about its use, and allowing them to control it [31]. New decentralized systems could improve transparency through the integration of different service providers. Moreover, a user would be able to integrate different services into one system (Table 1).

Table 1. Main differences between COS and traditional approaches

Feature	COS	Traditional service development
Decentralization	Decentralized systems as intermediary does not offer directly the services	Centralized systems as one main service provider integrates additional services
Transparency	Data collection and processes are transparent	Some processes are transparent
Data Privacy	Customers in control of their data	Collection and analysis of customer data subject to regulations
Smart Services	Collective Intelligence and Social CRM as basis for the development of solutions	Focus on current Artificial Intelligence techniques to analyze collected data
Customer Orientation	Customer-dominant Logic	Provider or Service-dominant Logic

When compared to CRM and Social CRM concepts, COS reflects a new paradigm that will change the way service providers relate to consumers. If on one side Social CRM discusses strategies, processes and systems that support tasks related to marketing, sales and customer service, COS proposes a new level of customer orientation. In a COS scenario, customers would select and combine different service provider through a decentralized platform.

Nevertheless, the novelty aspect of COS points to a need for further research on the topic. This paper analyzed the existing literature and identified drivers and concepts of COS, taking into consideration the interdisciplinary aspect of the topic. The identified concepts were integrated, contributing to further research on COS. The increase on research on the topic is able to foster the development of an integration framework and design guidelines for COS-based systems. Institutions in different industries can benefit from knowledge on COS as a potential business trend.

The paper, however, has some limitations. A structured literature analysis could provide deeper insights on the concepts. Nevertheless, the novelty of the topic requires initial studies that connects related topics in a macro level as it is proposed and presented in this paper. Further studies could improve the relationship between the drivers and concepts and provide a detailed framework for the development of COS-related platforms.

Acknowledgement. The authors gratefully acknowledge the financial support of this research by the German Academic Exchange Service (DAAD) within the project 'Social CRM as Specialization Subject in Brazilian Universities' (57449332) and the Sächsische Aufbaubank and European Union within the ESF project SPE4CRM (100362354).

References

1. Alt, R.: Electronic Markets on digital transformation methodologies. Electron. Mark. **29**(3), 307–313 (2019). https://doi.org/10.1007/s12525-019-00370-x
2. Alt, R.: Electronic Markets on the next convergence. Electron. Mark. **31**(1), 1–9 (2021). https://doi.org/10.1007/s12525-021-00471-6
3. Alt, R., Klein, S.: Twenty years of electronic markets research—looking backwards towards the future. Electron. Mark. **21**(1), 41–51 (2011). https://doi.org/10.1007/s12525-011-0057-z
4. Alt, R., Reinhold, O.: Social-Customer-Relationship-Management (Social-CRM). Wirtschaftsinformatik **54**(5), 281–286 (2012). https://doi.org/10.1007/s11576-012-0330-6
5. Alt, R., Reinhold, O.: Social Customer Relationship Management. Fundamentals, Applications, Technologies, 1st edn. Management for Professionals. Springer, Cham (2020). https://doi.org/10.1007/978-3-030-23343-3
6. Alt, R., Demirkan, H., Ehmke, J.F., Moen, A., Winter, A.: Smart services: the move to customer orientation. Electron. Mark. **29**(1), 1–6 (2019). https://doi.org/10.1007/s12525-019-00338-x
7. Alt, R., et al.: Towards customer-induced service orchestration - requirements for the next step of customer orientation. Electron. Mark. **29**(1), 79–91 (2019). https://doi.org/10.1007/s12525-019-00340-3
8. Beverungen, D., Müller, O., Matzner, M., Mendling, J., vom Brocke, J.: Conceptualizing smart service systems. Electron. Mark. **29**(1), 7–18 (2017). https://doi.org/10.1007/s12525-017-0270-5
9. Bradley, W., Maher, D., Boccon-Gibod, G.: Interoperable systems and methods for peer-to-peer service orchestration. Patent US8234387B2 (2008)
10. Brady, M.K., Cronin, J.J.: Customer orientation. J. Serv. Res. **3**(3), 241–251 (2001). https://doi.org/10.1177/109467050133005
11. Buell, R.W., Kim, T., Tsay, C.-J.: Creating reciprocal value through operational transparency. Manage. Sci. **63**(6), 1673–1695 (2017). https://doi.org/10.1287/mnsc.2015.2411
12. Chagas, B.N., Viana, J., Reinhold, O., Lobato, F.M., Jacob, A.F., Alt, R.: A literature review of the current applications of machine learning and their practical implications. In: WEB, vol. 18, no. 1, pp. 69–83 (2020). https://doi.org/10.3233/WEB-200429
13. Cheyer, A.J., Guzzoni, D.R., Gruber, T.R., Brigham, C.D.: Service orchestration for intelligent automated assistant. Patent US8892446B2 (2014)
14. Deshpandé, R., Farley, J.U., Webster, F.E.: Corporate culture, customer orientation, and innovativeness in Japanese firms: a quadrad analysis. J. Mark. **57**(1), 23–37 (1993). https://doi.org/10.1177/002224299305700102
15. Granados, N., Gupta, A., Kauffman, R.J.: Research commentary—information transparency in business-to-consumer markets: concepts, framework, and research agenda. Inf. Syst. Res. **21**(2), 207–226 (2010). https://doi.org/10.1287/isre.1090.0249
16. Hardin, T.: Digital Platforms Trends 2018 (2018). https://blog.g2crowd.com/blog/trends/digital-platforms/2018-dp/. Accessed 15 Mar 2019

17. Heath, T., Bizer, C.: Linked data: evolving the web into a global data space. Synthesis Lectures on the Semantic Web: Theory and Technology, vol. 1, no. 1, pp. 1–136 (2011). https://doi.org/10.2200/S00334ED1V01Y201102WBE001
18. Heinonen, K., Strandvik, T.: Customer-dominant logic: foundations and implications. J. Serv. Mark. **29**(6/7), 472–484 (2015). https://doi.org/10.1108/JSM-02-2015-0096
19. Heinonen, K., Edvardsson, B., Mickelsson, K.-J., Strandvik, T., Sundström, E., Andersson, P.: A customer-dominant logic of service. J. Serv. Manag. **21**(4), 531–548 (2010). https://doi.org/10.1108/09564231011066088
20. Ibrahim Eldomiaty, T., Ju Choi, C.: Corporate governance and strategic transparency: East Asia in the international business systems. Corp. Gov. **6**(3), 281–295 (2006). https://doi.org/10.1108/14720700610671882
21. Kelley, S.W.: Developing customer orientation among service employees. J. Acad. Mark. Sci. **20**(1), 27–36 (1992). https://doi.org/10.1177/009207039202000103
22. Kowalkowski, C.: Dynamics of value propositions: insights from service-dominant logic. Eur. J. Mark. **45**(1/2), 277–294 (2011). https://doi.org/10.1108/03090561111095702
23. Lewin, A.Y., Volberda, H.W.: Prolegomena on coevolution: a framework for research on strategy and new organizational forms. Organ. Sci. **10**(5), 519–534 (1999). https://doi.org/10.1287/orsc.10.5.519
24. Liu, A., Zhang, D., Wang, X., Xu, X.: Blockchain-based customization towards decentralized consensus on product requirement, quality, and price. Manuf. Lett. **27**, 18–25 (2021). https://doi.org/10.1016/j.mfglet.2020.12.001
25. Lusch, R.F., Nambisan, S.: Service innovation: a service-dominant logic perspective. MIS Q. **39**(1), 155–175 (2015)
26. Lusch, R.F., Vargo, S.L., O'Brien, M.: Competing through service: insights from service-dominant logic. J. Retail. **83**(1), 5–18 (2007). https://doi.org/10.1016/j.jretai.2006.10.002
27. Mayer, P., Schroeder, A., Koch, N.: MDD4SOA: model-driven service orchestration. In: EDOC 2008. Proceedings: 12th IEEE International Enterprise Distributed Object Computing [Conference], Munich, Germany, 15–19 September 2008, pp. 203–212. IEEE Computer Society, Los Alamitos (2008). https://doi.org/10.1109/EDOC.2008.55
28. Moore, J.F.: Predators and prey: a new ecology of competition. Harvard Bus. Rev. **71**(3), 75–86 (1993)
29. Morey, T., Forbath, T., Schoop, A.: Customer data: designing for transparency and trust. Harvard Bus. Rev. **93**(5), 96–105 (2015)
30. Papazoglou, M.P., Traverso, P., Dustdar, S., Leymann, F.: Service-oriented computing: state of the art and research challenges. Computer **40**(11), 38–45 (2007). https://doi.org/10.1109/MC.2007.400
31. Pasquale, F.: Beyond innovation and competition: the need for qualified transparency in internet intermediaries. Nw. U. L. Rev. **104**, 105 (2010)
32. Prahalad, C.K., Ramaswamy, V.: Co-creating unique value with customers. Strategy Leadersh. **32**(3), 4–9 (2004). https://doi.org/10.1108/10878570410699249
33. Rai, P.: Seth: firm performance impacts of digitally enabled supply chain integration capabilities. MIS Q. **30**(2), 225 (2006). https://doi.org/10.2307/25148729
34. Rapp, A., Trainor, K.J., Agnihotri, R.: Performance implications of customer-linking capabilities: examining the complementary role of customer orientation and CRM technology. J. Bus. Res. **63**(11), 1229–1236 (2010). https://doi.org/10.1016/j.jbusres.2009.11.002
35. Rodrigues Chagas, B.N., Viana, J., Reinhold, O., Lobato, F., Jacob, A.F.L., Alt, R.: Current applications of machine learning techniques in CRM: a literature review and practical implications. In: IEEE/WIC/ACM International Conference on Web Intelligence (WI), Santiago, 12 Mar 2018–12 June 2018, pp. 452–458. IEEE (2018). https://doi.org/10.1109/WI.2018.00-53

36. Rosner, E.: The Dawn of the Platform Economy in Financial Services (2018). https://internationalbanker.com/technology/the-dawn-of-the-platform-economy-in-financial-services/. Accessed 14 Mar 2019
37. Sachse, S.: Customer-centric Service Management: Conceptualization and Evaluation of Consumer-induced Service Composition (2018)
38. Sassi, A., Zambonelli, F.: Coordination infrastructures for future smart social mobility services. IEEE Intell. Syst. **29**(5), 78–82 (2014). https://doi.org/10.1109/MIS.2014.81
39. Saxe, R., Weitz, B.A.: The SOCO scale: a measure of the customer orientation of salespeople. J. Mark. Res. **19**(3), 343–351 (1982). https://doi.org/10.1177/002224378201900307
40. Schumpeter, J.A.: Capitalism, Socialism and Democracy, 4th edn. Harperperennial, New York (1934)
41. Segaran, T.: Programming Collective Intelligence. Building Smart Web 2.0 Applications/Toby Segaran. O'Reilly, Beijing, Farnham (2007)
42. Simatupang, T.M., Wright, A.C., Sridharan, R.: The knowledge of coordination for supply chain integration. Bus. Process Manage. J. **8**(3), 289–308 (2002). https://doi.org/10.1108/14637150210428989
43. Soh, M.: Goh: electronic marketplaces and price transparency: strategy, information technology, and success. MIS Q. **30**(3), 705 (2006). https://doi.org/10.2307/25148746
44. Spohrer, J., Vargo, S.L., Caswell, N., Maglio, P.P.: The service system is the basic abstraction of service science. In: Proceedings of the 41st Annual Hawaii International Conference on System Sciences (HICSS 2008), Waikoloa, HI, USA, 1 July 2008–1 October 2008, p. 104. IEEE (2008). https://doi.org/10.1109/HICSS.2008.451
45. Tiwana, A., Konsynski, B., Bush, A.A.: Research commentary—platform evolution: coevolution of platform architecture, governance, and environmental dynamics. Inf. Syst. Res. **21**(4), 675–687 (2010). https://doi.org/10.1287/isre.1100.0323
46. Tran, Y., Hsuan, J., Mahnke, V.: How do innovation intermediaries add value? Insight from new product development in fashion markets. R&D Manag. **41**(1), 80–91 (2011). https://doi.org/10.1111/j.1467-9310.2010.00628.x
47. Vargo, S.L., Lusch, R.F.: Service-dominant logic: continuing the evolution. J. Acad. Mark. Sci. **36**(1), 1 (2008). https://doi.org/10.1007/s11747-007-0069-6
48. Vargo, S.L., Maglio, P.P., Akaka, M.A.: On value and value co-creation: a service systems and service logic perspective. Eur. Manag. J. **26**(3), 145–152 (2008). https://doi.org/10.1016/j.emj.2008.04.003
49. Warg, M., Engel, R.: Service-dominant architecture (SDA): a Building Block of Digital Transformation (2016). https://www.researchgate.net/publication/323027767_Service_Dominant_Architecture_SDA_a_Building_Block_of_Digital_Transformation
50. Warg, M., Frosch, M., Weiss, P., Zolnowski, A.: Becoming a platform organization: how incumbent companies stay competitive. Cutter Bus. Technol. J. **31**, 38–45 (2019)
51. Weiß, P., Zolnowski, A., Warg, M., Schuster, T.: Service dominant architecture: conceptualizing the foundation for execution of digital strategies based on S-D logic. In: Bui, T. (ed.) Proceedings of the 51st Hawaii International Conference on System Sciences (2018). https://doi.org/10.24251/HICSS.2018.204
52. Wright, A., de Filippi, P.: Decentralized Blockchain Technology and the Rise of Lex Cryptographia. SSRN J. (2015). https://doi.org/10.2139/ssrn.2580664
53. Zhu, K.: Information transparency in electronic marketplaces: why data transparency may hinder the adoption of B2B exchanges. Electron. Mark. **12**(2), 92–99 (2002). https://doi.org/10.1080/10196780252844535
54. Zomerdijk, L.G., Voss, C.A.: Service design for experience-centric services. J. Serv. Res. **13**(1), 67–82 (2010). https://doi.org/10.1177/1094670509351960

55. Zwikstra, H., Hogenboom, F., Vandic, D., Frasincar, F.: Connecting customer relationship management systems to social networks. In: Uden, L., Herrera, F., Bajo Pérez, J., Corchado Rodríguez, J.M. (eds.) 7th International Conference on Knowledge Management in Organizations: Service and Cloud Computing, vol. 172. Advances in Intelligent Systems and Computing, vol. 172, pp. 389–400. Springer, Heidelberg (2013). https://doi.org/10.1007/978-3-642-30867-3_35
56. Zomerdijk, L.G., Voss, C.A.: Service design for experience-centric services. J. Serv. Res. **13** (1), 67–82 (2010). https://doi.org/10.1177/1094670509351960
57. Zwikstra, H., Hogenboom, F., Vandic, D., Frasincar, F.: Connecting customer relationship management systems to social networks. In: Uden, L., Herrera, F., Bajo Pérez, J., Corchado Rodríguez, J.M. (eds.) 7th International Conference on Knowledge Management in Organizations: Service and Cloud Computing. Advances in Intelligent Systems and Computing, vol. 172, pp. 389–400. Springer, Heidelberg (2013). https://doi.org/10.1007/978-3-642-30867-3_35

QOD Workshop

QOD 2021 Workshop Chairs' Message

The Fourth Workshop on Quality of Open Data (QOD 2021), organized in conjunction with the 24th International Conference on Business Information Systems (BIS 2021), took place at the Königlicher Pferdestall at Leibniz University Hannover, Germany. The specific focus was on bringing together different communities working on quality in Wikipedia, DBpedia, Wikidata, OpenStreetMap, Wikimapia, and other open knowledge bases and data sources.

There were 10 papers submitted for the workshop and the Program Committee decided to accept five papers (an acceptance rate of 50%). There were 13 members in the Program Committee, representing 12 institutions from seven countries.

The first paper, "Data quality assessment of comma separated values using linked data approach", proposed a framework that converts files in CSV format into linked data with certain data quality requirements. Converted data are transferred to a knowledge graph based on the proposed ontology. Then, triples that have violated the data quality constraints are identified. In addition, the data quality metrics and corresponding dependencies were described.

The second paper, "A high-resolution urban land surface dataset for the Hong Kong-Shenzhen area", aimed to solve the problem of the lack of high-spatial-resolution land surface datasets of assured quality. This work identified datasets that provide a complete set of land surface information required for high-resolution urban climate modeling in the Hong Kong-Shenzhen area. Such datasets included land cover, vegetation coverage, urban morphology, artificial impervious area, and anthropogenic heat data.

The third paper, "Spatio-temporal data sources integration with ontology for road accidents analysis", proposed an approach based on a microservice architecture, in which each data source is mapped to a microservice that presents an ontological data model of the source. The work described the application of such an approach to road accident data in St. Petersburg, Russia. As a result, clusters of accidents were obtained by combining data about accident, road types, and weather conditions in the accident area.

The fourth paper, "Review of literature on Open Data for scalability and operation efficiency of electric bus fleets", provided an overview of existing literature regarding the utilization of Open Data by public transport operators to analyze the scalability and operation efficiency of electric buses. Different open data sources were analyzed and classified. The paper also provided examples of Open Data sources and platforms that might be used by decision makers.

The last paper, "Challenges of Mining Twitter Data for Analyzing Service Performance: A Case Study of Transportation Service in Malaysia", focused on problems related to data extraction from Twitter to understand public opinion. The work provided original evidence proving the potential of using social media data to assess public transportation services performance, which may vary depending on the demographics of the social media users.

Maribel Acosta
Włodzimierz Lewoniewski
Krzysztof Węcel

Organization

Chairs

Maribel Acosta	Ruhr University Bochum, Germany
Włodzimierz Lewoniewski	Poznań University of Economics and Business, Poland
Krzysztof Węcel	Poznań University of Economics and Business, Poland

Program Committee

Riccardo Albertoni	CNR-IMATI, Italy
Ioannis Chrysakis	Foundation for Research and Technology – Hellas (FORTH), Greece
Vittoria Cozza	University of Padua, Italy
Suzanne Embury	University of Manchester, UK
Ralf Härting	Hochschule Aalen, Germany
Antoine Isaac	Europeana and Vrije Universiteit Amsterdam, The Netherlands
Dimitris Kontokostas	University of Leipzig, Germany
Jose Emilio Labra Gayo	Universidad de Oviedo, Spain
Maristella Matera	Politecnico di Milano, Italy
Finn Årup Nielsen	Technical University of Denmark, Denmark
Matteo Palmonari	University of Milano-Bicocca, Italy
Simon Razniewski	Max Planck Institute for Informatics, Germany
Blerina Spahiu	Università degli Studi di Milano Bicocca, Italy

A High-Resolution Urban Land Surface Dataset for the Hong Kong-Shenzhen Area

Zhiqiang Li[1(✉)] and Bingcheng Wan[2]

[1] Institute of Space and Earth Information Science, The Chinese University of Hong Kong, Hong Kong 999077, China
PaterLee@link.cuhk.edu.hk

[2] State Key Laboratory of Atmospheric Boundary Layer Physics and Atmospheric Chemistry, Institute of Atmospheric Physics, Chinese Academy of Sciences, Beijing 100864, China

Abstract. In recent years, with fast-developing computational capabilities, high-resolution techniques have been widely employed in atmospheric models. Thus, researchers can apply these high-resolution models to produce detailed meteorological scenarios, which empowers studies of urban-scale climatology relying on finer grid spacing. The WRF ARW/Noah LSM/UCM model is often used in urban climate research. However, the default input land surface data in urban areas is not precise enough, especially for the data describing in China. This study was pertinent to the increasing presence of ambiguous modeling practices in urban-scale climatology because of the out-of-date land surface data with insufficient fixes. Given the lack of quality-assured high-spatial-resolution land surface datasets, we produced a high-resolution urban land surface dataset including the land cover, vegetation coverage, urban morphology, artificial impervious area, and anthropogenic heat data for the Hong Kong-Shenzhen area - one of the world's largest metropolitan areas and a unique pair of twin cities. In short, the high-resolution urban land surface dataset provided a complete set of land surface information required for high-resolution urban climate modeling in the Hong Kong-Shenzhen area, which is exceedingly rare, especially in which the data on detailed urban morphology and anthropogenic heat fluxes.

Keywords: Urban · Land surface · Hong Kong-Shenzhen area · Climate modeling

1 Introduction

In recent researches [2, 10, 15], rapid urbanization has become a popular research topic. In the 1990s, many studies have highlighted the unneglectable influence of socio-economic developments on urban climate [11]. These anthropogenic activities have changed the natural surface into the impervious surfaces rapidly and massively, leading to changes in surface sensible and latent heat fluxes. Thus, massive artificial changes in land surface play a vital role in the generation and development of the urban climate. The modelers should be fully aware that inaccurate urban land surface data could lead to a remarkable unreal divergence in urban climate modeling results.

W. Abramowicz et al. (Eds.): BIS 2021 Workshops, LNBIP 444, pp. 203–213, 2022.
https://doi.org/10.1007/978-3-031-04216-4_19

The development and opening of this dataset were triggered by the worrisome existence of a growing number of careless practices in urban climate modeling using the original land surface dataset provided by mesoscale meteorological models with minimal fixes and quality assurance, which brings significant risk to the reliability of model simulation results, especially for simulations having a high spatial-temporal resolution. The reliability of model simulations relies heavily on the quality of incoming data, and the land surface dataset is the most significant input dataset besides meteorological boundary conditions.

The Hong Kong-Shenzhen area experienced intensive urbanization in the last 50 years, which is an excellent example of urbanization in China for studying the impact of urbanization on the urban climate. We developed a sophisticated method for producing the high-resolution land surface dataset and applied it to the Hong Kong-Shenzhen area. The developed dataset has a high spatial resolution of 100m and contains all four major components required for high-resolution urban climate modeling – urban land cover, vegetation coverage, urban morphology, and anthropogenic heat. Some information, for example, the detailed urban morphology data, has limited access. Data from different data sources usually come with different spatial and temporal settings requiring careful upscaling or downscaling operations. There is limited access to carefully audited land surface datasets since sophisticated methodologies for developing such datasets are still lacking. Meanwhile, multi-source information about the land surface, such as the terrain, road network, and building layouts, needs to be gathered and quality-assured to fuel the development of a high-resolution land surface data input for urban climate simulations. It is costly to acquire all datasets required.

In summary, the developed dataset provides rare access to high-resolution, carefully audited information about detailed land surface characters that can be used as quality inputs for high-resolution urban climate simulation models to generate accurate simulation results.

2 Data Description and Development

2.1 Dataset Details

Identifier: https://doi.org/10.5281/zenodo.3687362.

Creator: Zhiqiang Li and Yulun Zhou.

Dataset correspondence: PaterLee@link.cuhk.edu.hk.

Title: High-resolution (100 m) urban land surface dataset for the Hong Kong-Shenzhen area.

Publisher: Zenodo.

Publication year: 2020.

Resource type: Dataset.

Version: 1.0

2.2 Geospatial Coverage

The Hong Kong-Shenzhen area is of unique research value for studies regarding urbanization and its environmental impacts. It is among the world's largest urban areas composed of a unique pair of twin cities. Hong Kong and Shenzhen are neighboring cities sharing similar meteorological backgrounds, but they have experienced drastically different urbanization processes. While Hong Kong urbanized gradually into a city world-famous for its strict preservation of natural parks and the consequent high-density urban form, Shenzhen transformed from a fishing village to a mega-city sheltering its 20 million residents in a blink of 50 years. Shenzhen has been listed among the fastest urbanizing cities globally, making the city a compelling area for studying the complex interactions between urbanization, development, and the associated environmental impacts. The high-density urban form existing in both cities puts high requirements on the spatial resolution of urban land surface datasets.

2.3 Data Sources

The development of high-resolution urban land surface datasets utilizes various external data sources, including government-funded data and open-source data. Below we provide a constructed view of data we used for developing the urban land surface datasets.

- Boundary of Shenzhen
- Boundary of Hong Kong
- 2005 PRD (Pearl river delta) land cover (CAS2005) [3]
- 2010 Shenzhen building morphology
- 2010 Hong Kong building morphology
- 2010 Shenzhen road network
- 2010 Hong Kong road network
- 2010 MODIS NDVI Shenzhen dataset [17]

The building morphology data in Hong Kong and Shenzhen are products of government-funded surveying. The data went through systematic quality control. Despite their decent quality, both building morphology data are limited for sharing due to privacy concerns. We utilized the high-quality building morphology data to derive high-resolution products of impervious area and urban land cover maps, along with the open-source road network data and the PRD land cover data (CAS2005) [3]. To refine the vegetation coverage data required by the climate models, we retrieved remotely sensed Normalized Difference Vegetation Index (NDVI) maps in 2010 capturing the Hong Kong-Shenzhen area provided by the Moderate Resolution Imaging Spectroradiometer (MODIS) [17] launched by NASA. We also utilized open-source geographical boundary data of Hong Kong and Shenzhen. We used the grid setting of the PRD Land Cover Data in all the output raster maps of the developed dataset.

2.4 Data Components and Development Methods

The high-resolution land surface dataset consists of six main categories of input data required by high-resolution climate simulation models (Table 1), including the basic data, land cover data, vegetation coverage data, urban morphology data, artificial impervious area data, and anthropogenic heat data. Each category is a composite of spatial features that collectively describes one aspect of the urban land surface characteristics of the Hong Kong-Shenzhen area in 2010. We provide three different spatial coverage settings, covering Shenzhen, Hong Kong, and both cities, respectively, for most spatial features. Moreover, we developed the data step by step (Fig. 1).

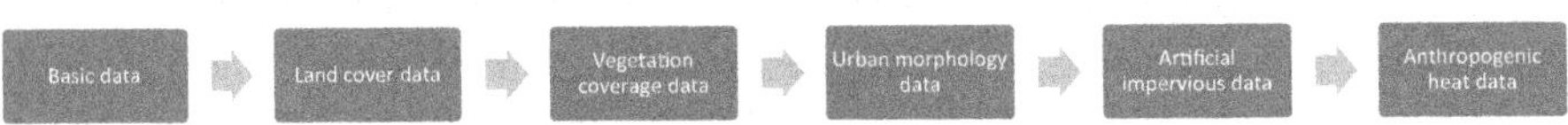

Fig. 1. The development process.

Table 1. Table captions should be placed above the tables.

Data categories	Spatial feature	Spatial coverage		
		Shenzhen	Hong Kong	PRD
Basic data	Geographical boundary	√	√	
Land cover data	Land cover			√
Vegetation coverage data	Monthly vegetation coverage	√		√
Urban morphology data	Building plan area	√	√	√
	Building height number mean	√	√	√
	Mean building height area weighted	√	√	√
	Building buffer area	√	√	√
	Building surface	√	√	√
Artificial impervious area data	Road area	√	√	√
	Building impervious area	√	√	√
	Artificial impervious area	√	√	√
Anthropogenic heat data	Annual mean	√	√	√

Geographical Boundary and Land Cover Data.
The basic data includes vector shapefiles representing the geographic boundaries of China, Shenzhen, and Hong Kong. Zhu et al. [24] also evaluated the land cover data (CAS2005) [3] provided by the Earth System Scientific Data Sharing Network (ESSDSN) as a decent quality product. Therefore, we directly used the 2005 PRD land cover (a component of CAS2005) as the PRD land cover data in the urban land surface dataset.

Monthly Vegetation Coverage.
First, we retrieved the monthly Shenzhen vegetation coverage maps from 23 frames of a subset of the 2010 Vegetation Indices 16-Day L3 Global 250m datasets [17] by using a typical algorithm for estimating the NDVI value of a grid. Then, we produced the monthly vegetation coverage map in the PRD area. Grids in Shenzhen were filled with the values of the corresponding grids in the monthly Shenzhen vegetation coverage maps, and grids in the outside area of Shenzhen were filled by the zonal mean values of the corresponding monthly Shenzhen vegetation coverage maps' grids with the same land cover category.

Urban Morphology.
First, we adapted the formulas provided by Burian et al. [1] and used them to produce the sets of Shenzhen and Hong Kong building parameter maps based on the 2010 Shenzhen building morphology and the 2010 Hong Kong building morphology, respectively. Then, we produced the PRD building parameter maps. Grids in Shenzhen and Hong Kong areas were filled with the values of the corresponding grids in Shenzhen and Hong Kong parameter maps, and the grids of the remaining area were filled with the zonal mean value of the corresponding Shenzhen building parameter maps' grids with the same land cover categories of urban construction land and rural settlement.

Impervious Area.
First, we produced the Shenzhen and Hong Kong building buffer area maps by calculating the buffer area of buildings in the grids of Shenzhen and Hong Kong areas according to the Hong Kong Building (Planning) Regulations [7] and Shenzhen Urban Planning Standard and Regulation [21]. We also produced the Shenzhen and Hong Kong road area maps based on the 2010 Shenzhen road network map and the 2010 Hong Kong road network map. Then, we produced Shenzhen and Hong Kong artificial impervious area maps by the grid-aggregated operation, which sums up the building buffer area, building plan area, and road area of the corresponding grids in the building buffer area plan area, and road area maps, respectively. Finally, we produced the PRD impervious area map. Grids in Shenzhen and Hong Kong areas were filled with the values of the corresponding grids in Shenzhen and Hong Kong artificial impervious area maps, and the grids of the remaining area were filled with the zone mean values of the Shenzhen artificial impervious area map's grids with the same land cover categories.

Anthropogenic Heat.
We summarized the methods in the previous studies [4, 5, 9, 16, 19, 22] to develop a method to estimate the anthropogenic based on the census and statistic data (Hong Kong [8] and Shenzhen [18, 20]) We first developed the Shenzhen and Hong Kong population maps. Second, we produced the Shenzhen and Hong Kong annual mean anthropogenic heat maps based on the Shenzhen and Hong Kong population maps. Third, we produced the PRD annual mean anthropogenic heat map by filling the Shenzhen and Hong Kong areas with the annual mean anthropogenic heat maps in Shenzhen and Hong Kong, respectively, and the grids outside Shenzhen and Hong Kong area with the corresponding mean value of urban construction land or rural settlement in the Shenzhen annual mean anthropogenic heat map. Last, the PRD annual maximum anthropogenic sensible and latent heat maps were produced based on the PRD annual mean anthropogenic heat map.

2.5 Data Quality Control

A sophisticated data quality control procedure was implemented throughout the data development of the high-resolution land surface dataset. For more details, please refer to the companion paper – "Incoming data quality control in high-resolution urban climate simulation: Hong Kong-Shenzhen area urban climate simulation as a case study using WRF/Noah LSM/SLUCM model (Version 3.7.1)" [12].

3 Data Access

We provided all data in GeoTIFF format for direct compatibility with a rich family of mapping and spatial data editing software such as the recommended ArcGIS and Matlab. Li et al. [12] also developed and opened sourced a format conversion tool that converts data in GeoTIFF format into the NetCDF format required by climate simulation models such as the Weather Research and Forecast (WRF) model, which includes two processing packages (geo_data_refinement processing package and wrf_input_refinement processing package). These two passages can be used for the high-resolution urban land surface dataset to refine the inner-most-domain's primary files (interpolated geo-data and the WRF input data files). The geo_data_refinement processing package is responsible for the interpolated geo-data refinement by revising the related fields of the geo_em.d01.nc, geo_em.d02.nc, geo_em.d03.nc, and geo_em.d04.nc. The wrf_input_refinement processing package was used for employing the spatial refined anthropogenic heat data and the NUDAPT [6] format urban morphology data. Specifically, it extracted the 2D refined data (urban morphology and anthropogenic heat data) from the urban land surface dataset and interpolated the data into the inner-most domain's grid by filling the corresponding 2D array variables in the file of wrfinput_d04. For downloading these two packages, please refer to https://doi.org/10.5281/zenodo.3996876.

While climate modelers can use the open-source tools to generate inputs for climate models, other researchers can easily use the data in GeoTIFF format for empirical studies regarding high-resolution urban morphology in one of the world's most compelling high-density urban areas. All geospatial data use the following geospatial reference systems:

- Geographic Coordinate System (GCS): World Geodetic System 1984 (WGS 1984, EPSG:4326)
- Projected Coordinate System (PCS): Universal Transverse Mercator Zone 49N (UTM 49N, EPSG:32649)

To access the data, Fig. 2 illustrates the complete data directory of the high-resolution land surface dataset.

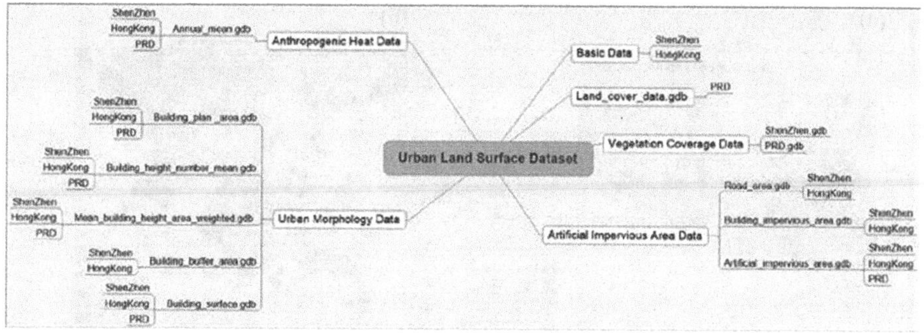

Fig. 2. The data directory of the high-resolution urban land surface dataset.

4 Potential Dataset Use and Reuse

This data provides rare access to detailed information on high-resolution land surface characteristics in the Hong Kong-Shenzhen area, which can be used as direct input for urban climate simulation models to account for the urbanization impacts. In the emergence of global climate change, the dataset can be used to study the environmental impacts of urbanization, understand the connecting mechanisms [13, 14], and derive policies that balance development and environmental impacts [23].

The dataset can also be used to study the necessity and impact of using high-resolution land surface characteristics as inputs for urban climate simulations by the WRF ARW model. Li et al. [13, 14] used the land cover and impervious area data (Fig. 3) to refine the WRF model default land cover data through two steps of the audition. First, Li et al. [13, 14] audited the extent of urban areas using the land cover map provided by CAS2005. Then, Li et al. [13, 14] further classified urban areas into three different urban levels – high-density, mid-density, and low-density urban areas based on the impervious area map. Li et al. [12] compared the simulation outcomes from a climate model (WRF v3.7) using the proposed high-resolution dataset instead of the default input data having coarse resolutions.

We compared the default land cover data with the refined one (Fig. 4). We found that the default land cover data had many false classifications in urban and nonurban. In the fast-urbanizing city of Shenzhen, we observed significant differences between the original land cover product provided by NCAR and the refined land cover product (Fig. 4a), which confirmed that the land cover map provided by NCAR is out-of-date. The original land cover data misclassified 208 units of urban lands to nonurban lands, and 164 units of nonurban lands were found misclassified to urban lands. These changes were not distributed equally in space. Most of the changes occurred in the fast-developing areas in Shenzhen, such as the west side, making local data distortion more severe than the average level. Moreover, the refined land cover product contains three different urban categories - low-intensity residential, high-intensity residential, and commercial/industrial/transportation, instead of the one single urban category in the original data product that ignores the significant differences in surface characters among urban lands. We also found that the default land cover data missed many details compared with the refined one.

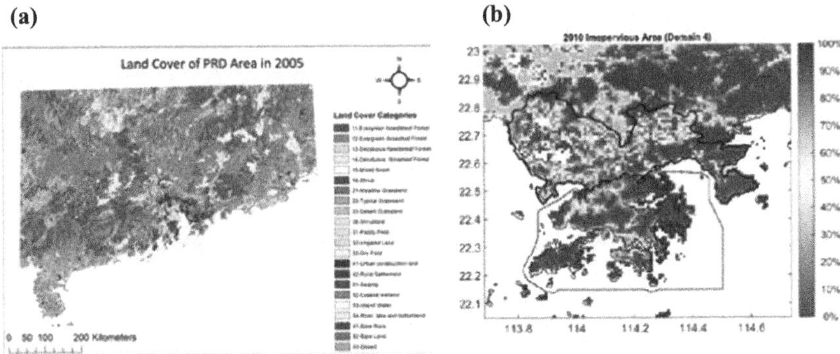

Fig. 3. The land cover (a) and impervious area (b) data.

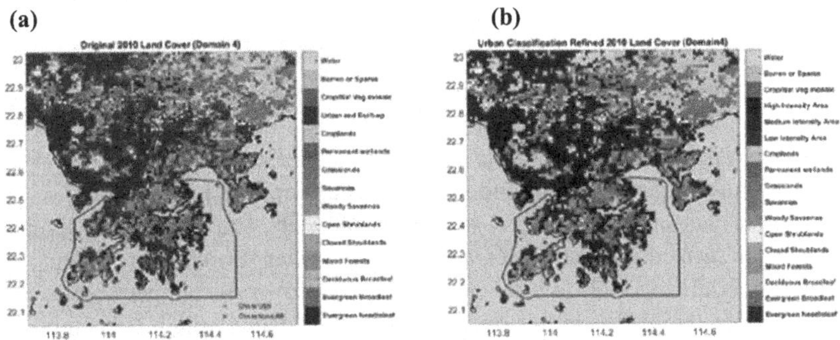

Fig. 4. The default land cover data of the WRF model (a), and the adjusted land covers based on the urban land surface dataset (b), data source: [12–14].

Besides, the developed dataset can be applied in other socio-economic researches, such as the retrieval of street-level green views from the vegetation coverage data, the estimation of sky view factor in the urban area, the relation between the housing price and building density, and the land cover change of urbanization. For inspiration of new research direction, we presented the land cover of Shenzhen in 1979 and 2010 (Fig. 5) and two major indicators of building morphology: the building plan area fraction and the building height number mean (Fig. 6). For other indicators of building morphology, please refer to Li et al. [12] paper.

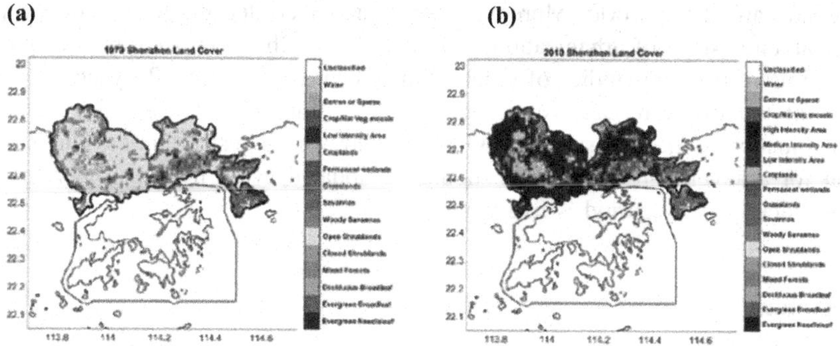

Fig. 5. The Shenzhen land cover in 1979 (a) and 2010 (b) data source: Li et al. [13].

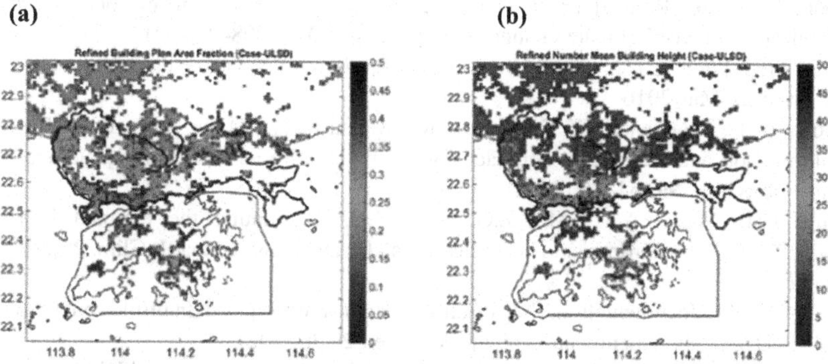

Fig. 6. The building plan area fraction (a) and the building height number mean (b) data source: Li et al. [12].

5 Conclusions

Throughout the development of the high-spatial-resolution data, we have experienced multiple challenges that need more attention in future studies. The primary challenge is gathering various information about urban land characteristics covering the same study period from numerous reliable sources for land surface data development, which is resource-heavy and time-costly. We collected data from the government field-surveyed data sources, openly accessible satellite imageries, and statistical yearbooks.

The Hong Kong-Shenzhen area has the unique research value to understand urbanization and its various impacts worldwide since the study area is one of the largest urban areas in the world (both in population and size). The two neighboring cities also have the exact geographic locations and meteorological conditions but the entirely different magnitude, speed, and mode of urbanization and dramatically different urban forms. Studying and comparing the differences in urbanization and its environmental impacts from various perspectives in the two cities could provide intriguing insights that benefit future sustainable urbanizations, especially in the Global South – the less-developed half of the world. Moreover, the dataset recorded the texture of the land surface at a milestone of urbanization, especially in the Shenzhen area. In our opinion, it can be used in future studies of urban climate change in at least 20 years.

As for the future research directions, we first suggested investigating the cost-benefit equation of the land surface dataset. Second, we suggested that the land surface dataset for a developing urban area should be updated every five years because the typical official survey period is five years.

References

1. Burian, S.J., Brown, M.J., Augustus, N.: Development and assessment of the second generation National Building Statistics database. In: Seventh Symposium on the Urban Environment, pp. 10–13 (2007)
2. Chen, B., et al.: Refined urban canopy parameters and their impacts on simulation of urbanization-induced climate change. Urban Climate **37**, 100847 (2021)
3. PRD Area Land Cover 2005 [Data file]. http://www.geodata.cn/data/dataresource.html. Accessed 22 Mar 2016
4. Chow, W.T., Salamanca, F., Georgescu, M., Mahalov, A., Milne, J.M., Ruddell, B.L.: A multi-method and multi-scale approach for estimating city-wide anthropogenic heat fluxes. Atmos. Environ. **99**, 64–76 (2014)
5. Feng, J.M., Wang, Y.L., Ma, Z.G., Liu, Y.H.: Simulating the regional impacts of urbanization and anthropogenic heat release on climate across China. J. Clim. **25**(20), 7187–7203 (2012)
6. NUDAPT 44 documentation. https://ral-dev.rap.ucar.edu/sites/default/files/public/product-tool/NUDAPT_44_Documentation.pdf. Accessed 29 Jul 2021
7. Hong Kong building (planning) regulations. http://www.legislation.gov.hk/blis_pdf.nsf/6799165D2FEE3FA94825755E0033E532/25C2868DA2669A12482575EE003F079B/$FILE/CAP_123F_e_b5.pdf. Accessed 22 Dec 2016
8. Hong Kong energy statistics 2010 annual report. http://www.statistics.gov.hk/pub/B11000022010AN10B0100.pdf. Accessed 22 Dec 2016
9. Hu, X., Li, A., Chen, H., Xin, D., Zhang, G., Zheng, B.: General principles for calculation of the comprehensive energy consumption (GB/T 2589–2008). In: National Standards of China, China Standards Press, Beijing (2008)
10. Kumar, M., Deka, J.P., Kumari, O.: Development of water resilience strategies in the context of climate change, and rapid urbanization: a discussion on vulnerability mitigation. Groundwater Sustain. Dev. **10**, 100308 (2020)
11. Lo, J. C. F., Yang, Z. L., Pielke sr, R.A.: Assessment of three dynamical climate downscaling methods using the Weather Research and Forecasting (WRF) model. J. Geophys. Res. Atmosph. **113**(D9) (2008)
12. Li, Z., Wan, B., Zhou, Y., Wong, H.: Incoming data quality control in high-resolution urban climate simulation: Hong Kong-Shenzhen area urban climate simulation as a case study using WRF/Noah LSM/SLUCM model (Version 3.7.1). Geosci. Model Dev. 1–13 (2020)
13. Li, Z., et al.: The impact of urbanization on air stagnation: Shenzhen as case study. Sci. Total Environ. **664**, 347–362 (2019)
14. Li, Z., Zhou, Y., Wan, B., Chung, H., Huang, B., Liu, B.: Model evaluation of high-resolution urban climate simulations: using the WRF/Noah LSM/SLUCM model (Version 3.7.1) as a case study. Geosci. Model Dev. **12**, 4571–4584 (2019)

15. Meng, L., Sun, Y., Zhao, S.: Comparing the spatial and temporal dynamics of urban expansion in Guangzhou and Shenzhen from 1975 to 2015: a case study of pioneer cities in China's rapid urbanization. Land Use Policy **97**, 104753 (2020)
16. Moriwaki, R., Kanda, M., Senoo, H., Hagishima, A., Kinouchi, T.: Anthropogenic water vapor emissions in Tokyo. Water Resour. Res. **44**(44), 150–176 (2008)
17. NASA EOSDIS Land Processes DAAC, USGS earth resources observation and science (EROS) Center.: Vegetation Indices 16-Day L3 Global 250m (MOD12Q1) [Data files] (2011). https://lpdaac.usgs.gov/dataset_discovery/modis/modis_products_table/mod13q1. Accessed 22 Mar 2016
18. China statistical yearbook (2011). http://www.stats.gov.cn/tjsj/ndsj/2011/indexeh.htm. Accessed 10 Mar 2015
19. Sailor, D.J., Lu, L.: A top–down methodology for developing diurnal and seasonal anthropogenic heating profiles for urban areas. Atmos. Environ. **38**(17), 2737–2748 (2004)
20. Shenzhen annual statistical year book (2011). http://www.sztj.gov.cn/xxgk/tjsj/tjnj/201202/t20120221_1806408.htm. Accessed 22 Dec 2016
21. Shenzhen urban planning standard and regulation (2014). http://www.szpl.gov.cn/xxgk/csgh/ghbz/201404/P020140416538509658678.pdf. Accessed 22 Dec 2016
22. Shi, G., Dai, T., Tan, S., Shen, Y., Wang, B., Zhao, J.: Preliminary estimate of the global average annual climate forcing resulted from anthropogenic heat release. Adv. Clim. Chang. Res. **6**(2), 119–122 (2010)
23. Zhou, Y., Huang, B., Wang, J., Chen, B., Kong, H., Norford, L.: Climate-conscious urban growth mitigates urban warming: evidence from Shenzhen China. Environ. Sci. Technol. **53**, 11960–11968 (2019)
24. Zhu, Y., et al.: Earth system scientific data sharing research and practice. J. Geo-Inf. Sci. **12** (1), 1–8 (2010)

Review of Literature on Open Data for Scalability and Operation Efficiency of Electric Bus Fleets

Tomasz Graczyk[1,2(✉)], Elżbieta Lewańska[2], Milena Stróżyna[2], and Dariusz Michalak[1]

[1] Solaris Bus and Coach, Bolechowo, Poland
tomasz.graczyk.sci@gmail.com
[2] Poznań University of Economics and Business, Poznań, Poland

Abstract. Open data is an integral part of Smart City projects carried out around the world. A public transport network is widely used when it is safe, well designed and reliable. The development and maintenance of the urban transport are key items in city budgets. Decisions regarding changes and the future organization of the public transport are supported mainly by Intelligent Transport Systems (ITS). There are many challenges, and one of them is the bus electro-mobility revolution. European leaders encourage a faster transformation to sustainable economy by introducing incentives and directives followed by EU funding. As a result cities replace aged fleets of diesel buses with the electric ones. A zero emission buses network is a new technology for public operators. It involves investments in chargers integrated to electric grids and the introduction of new maintenance processes. Each investment project that aims to introduce that innovative eco-friendly solution is preceded by feasibility studies. Total Cost of Ownership (TCO) is one of the crucial measure for making business decision. To calculate a proper configuration of chargers and fleet of buses, knowledge of specific operational conditions is necessary. That includes analysis of Open Data such as route characteristics, weather and dynamic traffic conditions. The paper review the existing literature with regard to utilization of Open Data by public transport operators to analyze scalability and operation efficiency of the electric buses. The Open Data used in the recent studies are characterized, classified and analyzed. Moreover, the examples of Open Data sources and platforms that might be used by decision makers are provided.

Keywords: Open Data · Public transport · Smart city · Battery Electric Bus · Electro-mobility · Large-scale information extraction · Geospatial data · Data enrichment

1 Introduction

Battery Electric Buses (BEB) adoption is accelerating all over the European cities (Table 1). Following the Paris climate agreement, the European Parliament & Council adopted the revised Clean Vehicle Directive [52] that promotes clean mobility solutions in public procurement tenders, encouraging European cities to deploy zero-emission

W. Abramowicz et al. (Eds.): BIS 2021 Workshops, LNBIP 444, pp. 214–226, 2022.
https://doi.org/10.1007/978-3-031-04216-4_20

vehicles. The most developed EU-23 countries set the national targets for clean buses to 45% in 2025 and 65% in 2030. One of the main pillar of the initiative is deployment of a platform where public authorities, public transport operators, manufacturers and financial organizations can come together with the aim to better exchange information and organize relevant actors [27].

The aim of the paper is to review the latest scientific publication in the field of electric bus public transport and extract which Open Data is used by researchers on different continents in their optimization models. The identified Open Data are then classified and characterized. Moreover, the paper aims at identifying which Open Data platforms and sources would be beneficial for decision makers.

The rest of the paper is structured as follows: in Sect. 2 review's methodology is presented, followed by a description of data used by public transport operators (Sect. 3), classification of Open Data used in the existing literature (Sect. 4) and a review of Open Data sources (Sect. 5). The paper ends with conclusions and indication of the future work.

Table 1. Total number of Battery Electric Buses (BEB) and Plug-in Hybrid Electric Buses (PHEB) in EU

Year	BEB	% BEB change	PHEB	% PHEB change	Total
2016	686	63%	304	110%	990
2017	888	29%	445	46%	1333
2018	1608	81%	486	9%	2094
2019	3636	126%	525	8%	4161
2020	5311	46%	550	5%	5861

Source: European Alternative Fuels Observatory (eafo.eu)

2 Literature Review Methodology

The concepts that were in the scope of the presented literature review cover electric bus and zero emission public transport. Google Scholar database has been used as a search platform. Keywords used included "electric bus", "electric buses", "BEB", "optimization of bus fleets", "Open Data electric bus" and "zero emission buses". The search results contained numerous items with the tendency to increase in a last few years. Because the electric bus technology is a modern field, we decided to narrow the review to publication from the last 4 years (2018–2021). At first we analyzed over 150 abstracts of found papers. Exclusion criteria included such topics as battery swapping, environment impact, cost comparison to diesel buses, charging technology, battery technology, electric grid impact, hybrid vehicles and charging scheduling. Initially we planned to focus on Open Street Maps but it appeared that weather data, traffic jams and number of travelling passengers are heavily used in research as well. We ended up with 25 articles where we followed forward and backward references. The final set of articles was further analyzed and used as a basis for the review presented in the paper.

3 Overview of Open Data Used by Public Transport Operators

By 2050 nearly all cars, vans, buses as well as new heavy-duty vehicles will be zero-emission according to EU Sustainable and Smart Mobility Strategy [31]. For the land transport an intermediate milestone has been set for 2030 and in practice this means that 100 European cities will be climate neutral. In terms of zero-emission vehicles it counts for 30 million zero-emission units on European roads. Additionally 3 million public chargers will be installed [31].

Our research focuses on buses, which are the most popular type of public passenger transport in Europe. Urban and sub-urban buses account for more than half of public passenger journeys (measured as passenger-kilometers) [32].

The term "smart" contained in the European transport strategy highlights the need of innovation and digitalization, foresees connected and automated multimodal mobility, and assumes new ways of usage of data and Artificial Intelligence (AI).

After the review of the literature, we know that in order to make optimal investments decisions, it is necessary to integrate and analyze data from various sources (Fig. 1).

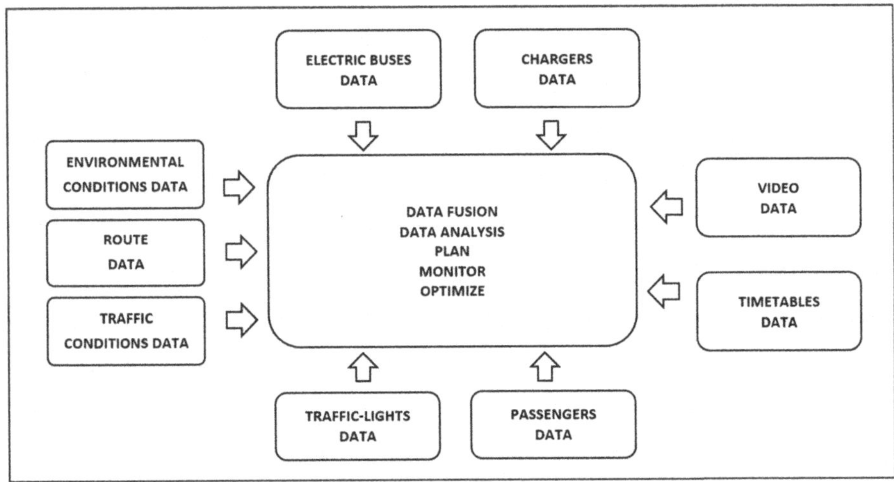

Fig. 1. Data used by bus operators. Source: Own work

For many years large cities have had Intelligent Transport Systems (ITS) [29] that ensure the connection between the public operator control center and cruising buses. These systems, depending on the city, manage passenger information on the bus stops and inside vehicle. Integration is also done usually with monitoring cameras and the GPS system. Selected buses have also passenger counting systems, which are not installed in every bus due to their high cost. Some cities also have Intelligent Traffic Control systems. In addition, in recent years in European cities, along with the increase in the number of electric city buses, the number of bus chargers has been dynamically growing, which must also be integrated on-line to maintain their reliability.

Operators also require from manufacturers that they install computer data loggers in their vehicles. It is an electronic device that records diagnostic data over time or in relation to location. This data can be extracted from the bus by connecting specialized tools used in the automotive sector. Transforming this data for business analysis purposes is not an easy process, since it requires hiring qualified experts and physical access to each bus. These tools were created for the purpose of technical diagnosis of specific faults, occurring sporadically, and not with the aim to constantly send them to the data center.

With the advent of new electric bus technology, manufacturers and operators have started to invest in the new fleet monitoring systems in order to be able to jointly optimize maintenance, detect potential faults and take corrective actions. The goal of these kind of systems is to deliver packets of information related to the state of the bus and its components (e.g., battery, engine, propulsion, air-conditioning, doors). The aggregation of this data is a key aspect for further projects and source of lesson learned. As a result, whenever the need arises, individual stakeholders of the bus operator, charger and support teams have access to the near-real time data. They are able to access historical information and gain advantage of what has changed over time.

China is the undisputed world leader in the operation of electric buses and we noticed a large number of publications from Asia. The city that stands out from other agglomerations is Shenzhen with more than 20 000 units [33]. The review showed that the European studies focus mostly on research up to a few dozen of electric buses, and Chinese studies in some cases cover the entire networks of public bus transport including several hundreds or even thousands of vehicles [14].

Most of the large European cities have a long-term view of the upcoming challenge to deploy several hundreds or even thousands of electric buses. We have mentioned already what are the drivers for the electro mobility revolution but there are also some barriers. Bloomberg's report (2018) mentions several factors that are holding back multibillion programs. The first one is a high upfront cost mainly due to the high prices of batteries and chargers. The second barrier is flexibility and operational experience due to the limited capacity of the battery and hence a range of the vehicle. As a result, it is difficult to incorporate them in the existing network of bus routes operated for 24 h. The third difficulty is charges and grid issues. Potential power outages and a space required for installing the plug-in chargers or high-power pantographs on bus stops are also listed [33]. Because of the high power of chargers, the new member joins this complex discussions - the energy distributor which asks the challenging question related to a long term plan of energy demand. They need answers related to charger characteristics to redesign their networks of energy transformers.

Despite all these difficulties, stakeholders strive to prepare a long-term plan for sustainable public transport. Such plans are based on milestones in which the measure of success is the number of vehicles and chargers deployed and operating smoothly. The first step is the planning and arrival of the first charger and several electric buses on a single route. The second milestone is the implementation of several dozen vehicles and several chargers in various locations. High maturity is achieved when several hundred electric vehicles operate in the city.

Public operators need reliable partners to achieve these milestones, and a dialogue with them begins after the publication of a public tender. A key aspect in the bidding process is the technical feasibility study and adjustment of the electro-mobility solution to the city environment due to the limitations of the new technology mentioned earlier. For the solution to be reliable, scalable and financially optimal, it is necessary to select the charging option, battery technology and receive manufacturer's declaration of the expected energy consumption. To make a correct calculation, experts need to know how the buses will be operated. Operating conditions understood as routes, traffic jams, timetables and weather conditions. Data quality excellence is crucial for sophisticated analytical models and configurators which as a result recommend the optimum solution. The competitive advantage in these calculations determines who is the winner in the tender.

After the design and configuration is agreed, contract is signed, buses are manufactured and integrated to smart city network. From now on vehicles and chargers transfer the near-real time data to data centers where group of experts monitor their effectiveness. It is their daily job to understand how extreme weather, mass city events, route characteristics and traffic condition affect the operation of new technology. Optimization efforts begins.

4 Review of Open Data

As indicated in the previous section, understanding the operating conditions is a key aspect in this complexity management puzzle. Open Data platforms might be one of the key sources used by stakeholders to make informed decisions on investment, utilization and maintenance of an electric bus fleets. The conducted analysis of the recent studies shows various models, frameworks and methods that use diverse information to enhance risk management and planning process. Based on the literature review, we have distinguished 4 Open Data classes and 14 variables (Table 2). Environmental conditions class contains 4 variables, route characteristics category carries 6 variables, traffic condition and operation time have 2 variables. Different optimization criteria are used in the studies, that is route planning [9, 11, 20, 23], cost and TCO models [22, 24], heat load and passenger comfort [17, 18, 21, 22]. Most of the researchers focus on looking for influential factors on energy demand. On one hand, application of energy measure is crucial for network timetable effectiveness, scheduling of charging events and predicting impact on electric grid. On the other hand, it can be used to monitor battery safety, degradation and predict the end of battery life cycle.

In Table 2 we present a proposal of classification of all Open Data identified in the existing studies.

Table 2. Open data variables used in the existing studies

Category	Variable	How often used	References to publications
Route characteristics	Trip distance; Route length; Traveled distance	24	[1, 2, 4–25]
	Driving direction	5	[11, 13, 14, 18, 23]
	Bus stops; Location of bus stops; Number of bus stops	9	[2, 7, 10, 11, 13, 14, 17, 20, 23]
	Traffic lights; Location of traffic lights; Number of traffic lights	6	[1, 7, 10, 11, 15, 23]
	Intersections; Location of intersections; Stop signs; Stops per km	7	[4, 6–8, 11, 15, 19]
	Road grade; Elevation Route profile	14	[2, 7–17, 19, 20]
Operational environmental conditions	Ambient temperature;	16	[1, 4, 7, 9, 10, 12–14, 16–19, 21–23, 25]
	Humidity; Wet-dry conditions; Precipitation;	8	[1, 4, 9, 15, 17–19, 23]
	Atmospheric pressure; Air pressure; Air density	6	[1, 7, 11, 14, 15, 17, 18]
	Wind speed; Wind direction	7	[7, 8, 11, 15, 16, 18, 21]
Operating time, Timetable	Departure time; arrival time; Time on bus stop; daytime trip; Night-time trip;	9	[2, 10, 11, 13, 14, 18, 23–25]
	Bus operating days; Monday-Friday; Weekends	4	[11, 14, 18, 23]
Dynamic traffic conditions	Traffic speed; Average travel speed; Maximum speed; Speed of travel; Vehicle speed	20	[1, 2, 4–8, 10–16, 19, 21–25]
	Passengers load; Number of passengers	12	[1, 2, 4, 8, 10–12, 14, 18–21]

Source: Own work

4.1 Description of Variables and Their Importance

BEB Route Characteristic

Traveled Distance

The distance is present in the 24 papers that have been reviewed. The distance traveled by an electric car is currently the main barrier to a wider adoption. By increasing the range, the size of the battery and its weight increases as well, what has an impact on the structure of the vehicle, the price and the number of passengers that the bus can transport. As buses run on mapped routes with designated recharging points, analytical models use the distance to best match a bus to a route (or route to a bus). The research indicates various types of distance covered, such as the average distance per day [1], the distance of the bus line [4], the distance between stops [2, 13] and even the distance to the charger. [24].

Driving Direction

When the bus route is known, scientists indicate the driving direction characteristics as an additional variable to enrich the analytical model. This is because traffic flows towards the city center and back to the suburban areas following people commuting to and from work. [3, 23] The influence of solar radiation on the energy consumption of air-conditioning systems has been interestingly presented in studies in Turkey [18]. If one combines the driving direction of the vehicle, sunlight and the technology of solar panels mounted on some buses (e.g., Warsaw, Poland), another field for analysis appears. The influence of the slope of the road which is inherently related to the direction of travel is discussed as a separate parameter.

Bus Stops

There are 19,500 bus stops in London [22]. In the electro-mobility era, each bus stop is a candidate to have a battery charger installed, especially when the knowledge of the nearest energy transformer characteristic is known. In many publications the bus stop act as an intermediate measuring point for passenger load [20, 23], punctuality KPI according to a timetable [10] or altitude reference point [2, 13]. Chinese studies in energy saving recommend the optimal number of bus stops per trip length [14, 23]. With the exact knowledge of the bus stops network, city planners can prioritize electrification of lines.

Traffic Lights

A study that focuses on smart traffic lights and their impact on the electric buses is presented by Connor [15]. Research shows the potential of 27% cost reduction which can be achieved by implementation of predictive driving techniques that integrates data from connected buses and traffic lights. The cost reduction is so significant because in addition to energy costs, it also focuses on the degradation aspect of the battery. Additionally, the aspect of the number of street light cycles needed by a bus to cross an intersection appeared in Swiss research [10].

Stops, Intersections

Kivekas et al. [19] consider a stop as a pause maneuver where the vehicle is stationary for two seconds or more. Although the electric buses have regeneration braking systems, the effect of stops is still significant in the energy consumption. An estimation using an average speed could be only 30% of the total energy used by the bus [11]. It needs to be mentioned that the heavy-duty vehicles suspension systems are responsible for a significant energy load and they are utilized during the braking event [4]. A predictive driving is one of the technique used to minimize this effect [12, 19]. It can reduce the average energy consumption by 9,5% [19]. A known technique used by planners is the separation of bus lanes, what is especially important for the electric buses going uphill. An elevation of a route aggravates the process of a heavy vehicle stopping in terms of the power consumption and the battery degradation. Ensuring unimpeded passability of bus lanes is a recommendation of Zhang [6]. It appears that a good knowledge of stop events could enrich advanced analytical models.

Grade of the road has a high impact on the energy demand, e.g., as showed by [13] 25 m difference over 386 m of the elevation between consecutive stops brings 18–23% change in the energy demand. Studies made in Beijing, that are beneficial for city planners, issues also a recommendation that the transit routes should be planned on downhill road segments for energy savings. The energy recovery systems are highly efficient when going downhill [14].

BEB Operational Environmental Conditions

The Ambient Temperature

Many studies have emphasized the influence of temperature on the bus operation. They are of particular importance when the temperature reaches extremely high and low values. Cigarini [17] in his experimental study highlights that heating and air-conditioning systems may reduce the travel range of the electric bus of a factor of 2.0.

Other studies also confirm the large amount of energy needed to ensure thermal comfort [1, 4] and call for a proper calculation of the thermal load during the feasibility study and design phase [18]. Luo [21] proposes the introduction of a new control method that takes into consideration operational condition of electric bus auxiliary systems in which the energy consumption drops by 18,56%.

Another angle is the effect of temperature on the dynamic capacity fading and consequently drop of the vehicle range. For a large-scale public transportation network it would mean additional possibility to optimize the cost of battery replacement with the support of dynamic programming presented by Wang [9].

Precipitation and humidity affects negatively energy recovered from the regenerative braking [1], but it does not mean that overall energy consumption of the bus is higher. Li [23] in his model finds that in wet conditions overall consumption of the energy drops, due to the frictional resistance [23]. Overall the wet conditions are more risky for the safety of passengers and road users, they are part of studies related to driver behavior and risk management. Humidity is also important factor for intelligent control and settings of air-conditioning systems [17, 18].

Atmospheric pressure and wind, that impact on the energy consumption, can be also found in the studies. Buses have large frontal surface area which is important factor of aerodynamic formulas [7, 14, 19]. The pressure effect was widely described in Romanian studies [1]. The importance of this variable increases with the development of dedicated bus lines, where vehicles achieve higher speeds.

BEB Dynamic Traffic Condition

Timetable Departure Time, Departure Day
Nicolaides [22] in his techno-economic analysis shows that the BEB energy supply system is highly sensitive to operational factors such as route, charging technology and timetable. For mixed bus fleet Ma recommends usage of diesel buses during the non-peak hours and electric ones during the peak hours [14]. Sweden TCO studies highlight the cost variations related to "departure per hour" measure [24]. Typically, timetable lists times when a bus is scheduled to arrive at and depart from bus stop and depot. An electrified bus schedule has an additional complex dimension, and it is the charging event. When a bus operator chooses the overnight charging technology, buses are charged couple hours overnight so they cannot be used for service. In that case planning algorithms need to be adjusted and constraints applied for fleet size. Additionally an appropriate number of chargers needs to be planned to allow parallel charging of multiple buses. That must be also aligned with an energy distributor. There is another option, namely the opportunity charging. In that case planning models need to be adjusted to optimize multiple shorter power loads per day. The accurate in time charging events need to be scheduled for each day of workweek and weekend. In the long term taking into consideration hundreds of chargers, couple thousands of operating buses and energy grid safety.

BEB Dynamic Trip Characteristic

Traffic Conditions and Passenger Load
In most cases the public transport systems run along predetermined routes and schedules. The number of buses operating at a given hour in the city space depends on the rhythm of city life. Bus operators try to adapt their fleet's operation to these rules. They use data from traffic sensors and passenger counting systems to make improvements and ensure passenger comfort. Collected and retained traffic data is used to forecast expected traffic, allows to reveal patterns [3] and might be used by the predictive algorithms [8]. Basma [4] describes the behavior of auxiliary systems in traffic conditions, it is important because they consume vast amount of energy during the extreme weather. A significant difference in the electricity consumption caused by passengers commuting to work describes Li [23]. Integration of the passengers data into synthesized driving cycle and Monte Carlo simulation has been proposed by Liu [20]. Dynamic passenger load appears also in TCO calculation [22].

5 Open Data Platforms

In parallel to the analysis of the electric bus related research, we started an investigation of Open Data platforms to find the listed variables impacting electro-mobility solutions. The results of this search are presented in Table 3.

We naturally resorted to OpenStreetMaps (OSM) in our search for route characteristics. OSM led us to Overpass read only API (formerly known as OSM Sever Side Scripting) and OpenElevation. Weather information are available in multiple resources. We decided to focus on YR [39] served by Norvegian Meterological Institue, World Weather provided by World Meteorological Organization [40] and Open Weather Map [41]. For timetable and dynamic traffic condition we found only resources related to a given city but it should be noticed that The World Bank has been working on Open Traffic initiative since 2016 [50].

There are three principal constituents in any successful Open Data Ecosystem that might be distinguished [53]:

- Open Government Data (OGD) - data produced, collected or paid for by the public sector,
- Open Business Data (OBD) – data produced or collected by the private sector and published freely and openly,
- Open Citizen Data – the personal and non-personal data of individual citizens published into the open domain.

Table 3 presents the results of the analysis and the characteristics of selected open data platforms.

Table 3. Analysis of the selected open data platforms

Class	Source of data	Cost	License type	Open data type	Data format
Route characteristics	OverpassAPI [34]	Free	Affero GPL v3;	OCD	API
	OpenStreetMap [36]	Free	Creative Commons BY-SA 2.0;	OCD	API
	Open-Elevation [37]	Free	GNU LGPL	OCD	API
	Elevation_1.1.3 [38]	Free	Apache License Version 2.0;	OCD	API
Operational environmental conditions	YR[39]	Free	Terms of service	OGD	XML GRIB
	World Weather [40]	Free	Level AA Conformance to Web Content Accessibility Guidelines 2.0;	OBD	JSON
	Open Weather Map[41]	Free (limit)	Creative Commons BY-SA 2.0;	OBD	API CSV JSON
Operating time, Timetable	Timetables GTFS [47] (Poznan, Poland)	Free	Terms of service	OGD	GTFS
	Open Data [48] (Curitiba, Brazil)	Free	Creative Commons Attribution 4.0 International	OCD	API
	Kaggle Singapore [51]	Free	Creative Commons(CC0) Public Domain	OCD	CSV
Dynamic	Traffic Torino [44]	Free	Described as Open Data	OGD	XML

(*continued*)

Table 3. (*continued*)

Class	Source of data	Cost	License type	Open data type	Data format
traffic conditions	Open Data Ireland [42]	Free	Creative Commons Attribution 4.0 International (CC BY 4.0)	OGD	CSV JSON XML
	OpenRouteService [35]	Free (limit)	Creative Commons BY 4.0;	OBD	API
	Open Data Baidu [43]		No cost for research and personal uses	OBD	API
	Open Data Hamburg [45]	Free	Terms of service	OGD	XML JSON
	GoogleDataSet [46]	Free	Commercial use Non Commercial use	OBD	N/A

Source: Own work

6 Conclusions

In this article, we present Open Data from the perspective of the city public transport. We review scientific research from the last few years in the field of the electric buses and indicate the types of Open Data that are of interest to scientists, bus operators and vehicle manufacturers. We distinguish route characteristics, operational environmental conditions, timetable and dynamics traffic conditions as the most popular categories of Open Data. The impact of these factors on the operation of the electric buses is significant, which is confirmed by analytical models of the reviewed studies. We mention the aspect of European Union directives that incentive cities to take urgent actions in the field of clean transport. We also present the aspect of the diverse data and information that are necessary for stakeholders to make short- and long-term decisions. In addition, we analyze the selected Open Data platforms from which this data can be extracted. We also emphasize that utilization of Open Data will become more and more important along with the dynamically growing number of electric buses on the streets of European cities. Our future research will focus on obtaining data from open sources and transport network. This will allow us to create advanced analytical models supporting change and risk management processes.

References

1. Iclodean, C., Cordos, N., Todorut, A.: Analysis of the electric bus autonomy depending on the atmospheric conditions. Energies **12**(23) (2019)
2. Pamuła, T.: Application of a neural network for the estimation of energy consumption for a bus line in public urban transport. WUT J. Transp. Eng. **128**, 19–28 (2020)
3. Biuk-Aghai, R.P., Kou, W.T., Fong, S.: Big data analytics for transportation: Problems and prospects for its application in China. In: Proceedings - 2016 IEEE Region 10 Symposium TENSYMP 2016, no. February 2019, pp. 173–178 (2016)

4. Basma, H., Mansour, C., Haddad, M., Nemer, M., Stabat, P.: Comprehensive energy assessment of battery electric buses and diesel buses. In: ECOS 2019 - Proceedings of the 32nd International Conference on Efficiency, Cost, Optimization, Simulation and Environmental Impact of Energy Systems, pp. 275–287 (2019)
5. Lin, K.C., Lin, C.N., Ying, J.J.C.: Construction of analytical models for driving energy consumption of electric buses through machine learning. Appl. Sci. **10**(17) (2020)
6. Zhang, Y., Yuan, W., Fu, R., Wang, C.: Design of an energy-saving driving strategy for electric buses. IEEE Access **7**, 157693–157706 (2019)
7. Vepsäläinen, J., Kivekäs, K., Otto, K., Lajunen, A., Tammi, K.: Development and validation of energy demand uncertainty model for electric city buses. Transp. Res. Part D Transp. Environ. **63**, 347–361 (2018)
8. Lajunen, A., Kivekäs, K., Baldi, F., Vepsäläinen, J., Tammi, K.: Different approaches to improve energy consumption of battery electric buses. In: 2018 IEEE Vehicle Power and Propulsion Conference, VPPC 2018 – Proceedings (2019)
9. Wang, J., Kang, L., Liu, Y.: Effects of working temperature on route planning for electric bus fleets based on dynamic programming. Chem. Eng. Trans. **76**, 907–912 (2019)
10. Scarinci, R., Zanarini, A., Bierlaire, M.: Electrification of urban mobility: the case of catenary-free buses. Transp. Policy **80**, 39–48 (2019)
11. Basso, R., Kulcsár, B., Egardt, B., Lindroth, P., Sanchez-Diaz, I.: Energy consumption estimation integrated into the electric vehicle routing problem. Transp. Res. Part D Transp. Environ. **69**, 141–167 (2019)
12. Bunzel, A., Baker, B.: Energy consumption of electric city buses: determination as a part of a technological and economic evaluation of bus lines with regards to their electrifiability. In: 2018 IEEE International Conference on Electrical Systems for Aircraft, Railway, Ship Propulsion and Road Vehicles and International Transportation Electrification Conference, ESARS-ITEC 2018 (2019)
13. Pamuła, T., Pamuła, W.: Estimation of the energy consumption of battery electric buses for public transport networks using real-world data and deep learning. Energies **13**(9) (2020)
14. Ma, X., Miao, R., Wu, X., Liu, X.: Examining influential factors on the energy consumption of electric and diesel buses: a data-driven analysis of large-scale public transit network in Beijing. Energy **216** (2021)
15. Connor, W.D., Wang, Y., Malikopoulos, A., Advani, S.G., Prasad, A.K.: Impact of connectivity on energy consumption and battery life for electric vehicles. IEEE Trans. Intell. Veh. (2020)
16. Wang, J., Liu, K., Yamamoto, T.: Improving electricity consumption estimation for electric vehicles based on sparse GPS observations. Energies **10**(1) (2017)
17. Cigarini, F., Fay, T.A., Artemenko, N., Göhlich, D.: Modeling and experimental investigation of thermal comfort and energy consumption in a battery electric bus. World Electr. Veh. J. **12**(1), 1–22 (2021)
18. Nicem Tanyeri, M., Çağlar Başlamişli, S.: Prediction of the annual heat load of an articulated electric urban bus. Isi Bilim. Ve Tek. Dergisi/J. Therm. Sci. Technol. **40**(1), 27–36 (2020)
19. Kivekas, K., Lajunen, A., Baldi, F., Vepsalainen, J., Tammi, K.: Reducing the energy consumption of electric buses with design choices and predictive driving. IEEE Trans. Veh. Technol. **68**(12), 11409–11419 (2019)
20. Liu, X., Ma, J., Zhao, X., Du, J., Xiong, Y.: Study on driving cycle synthesis method for city buses considering random passenger load. J. Adv. Transp. 2020 (2020)
21. Luo, Y., Tan, Y.P., Li, L.F.: Study on saving energy for electric auxiliary systems of electric bus. Energy Sour. Part A Recover. Util. Environ. Eff. (2020)

22. Nicolaides, D., Madhusudhanan, A.K., Na, X., Miles, J., Cebon, D.: Technoeconomic analysis of charging and heating options for an electric bus service in London. IEEE Trans. Transp. Electrif. **5**(3), 769–781 (2019)
23. Li, P., Zhang, Y., Zhang, K., Jiang, M.: The effects of dynamic traffic conditions, route characteristics and environmental conditions on trip-based electricity consumption prediction of electric bus. Energy **218** (2021)
24. Grauers, A., Borén, S., Enerbäck, O.: Total cost of ownership model and significant cost parameters for the design of electric bus systems. Energies **13**(12) (2020)
25. Wang, S., Lu, C., Liu, C., Zhou, Y., Bi, J., Zhao, X.: Understanding the energy consumption of battery electric buses in urban public transport systems. Sustain. **12**(23), 1–12 (2020)
26. European Alternative Fuels Observatory, AF fleet Electricity (2020) (eafo.eu)
27. European clean bus deployment initiative. https://ec.europa.eu/transport/themes/urban/clean-vehicles-directive_en
28. European automobile manufacturers association. https://www.acea.be/automobile-industry/buses [visited 2020]
29. Seredynski, M., Viti, F.: A survey of cooperative ITS for next generation public transport systems. In: IEEE Conference on Intelligent Transportation Systems, Proceedings, ITSC (2016). Accessed 15 Apr 2021
30. Acea. https://www.acea.be/automobile-industry/buses. Accessed 15 Apr 2021
31. Mobility Strategy | Mobility and Transport (europa.eu), April 2021
32. Energy, transport and environment statistics handbook, Eurostat (2020 edition)
33. Electric Buses in Cities Driving Towards Cleaner Air and Lower CO2, Bloomberg (2018)
34. Overpass. https://www.overpass-api.de/. Accessed 15 Apr 2021
35. OPS. https://openrouteservice.org/. Accessed 15 Apr 2021
36. OPM. https://www.openstreetmap.org/. Accessed 15 Apr 2021
37. Open elevation. https://open-elevation.com/. Accessed 15 Apr 2021
38. PyPi. https://pypi.org/project/elevation/. Accessed 15 Apr 2021
39. YR. https://www.yr.no/. Accessed 15 Apr 2021
40. World weather. https://worldweather.wmo.int/. Accessed 15 Apr 2021
41. Open weather. https://openweathermap.org/. Accessed 15 Apr 2021
42. Ireleand Gov. https://data.gov.ie/dataset?q=traffic. Accessed 15 Apr 2021
43. AI Baidu. https://ai.baidu.com/broad/introduction?dataset=traffic. Accessed 15 Apr 2021
44. Opendata Torino. http://opendata.5t.torino.it/get_fdt. Accessed 15 Apr 2021
45. GEO Hamburg. https://suche.transparenz.hamburg.de/dataset/geo-online-portal-hamburg. Accessed 15 Apr 2021
46. Google data set search. https://datasetsearch.research.google.com/search?query=traffic. Accessed 15 Apr 2021
47. ZTM. https://www.ztm.poznan.pl/pl/dla-deweloperow/gtfsFiles. Accessed 15 Apr 2021
48. Zenodo. https://zenodo.org/record/4654068#.YIWd5ZAzaUk. Accessed 15 Apr 2021
49. World bank. https://www.worldbank.org/. Accessed 15 Apr 2021
50. Open traffic. https://opentraffic.io/. Accessed 15 Apr 2021
51. Kaggle. https://www.kaggle.com/gowthamvarma/singapore-bus-data-land-transport-authority. Accessed 15 Apr 2021
52. Eur-Lex. https://eur-lex.europa.eu/eli/dir/2019/1161/oj. Accessed 15 Apr 2021
53. Delloite. https://www2.deloitte.com/content/dam/Deloitte/uk/Documents/deloitte-analytics/open-data-driving-growth-ingenuity-and-innovation.pdf. Accessed 15 Apr 2021

Challenges of Mining Twitter Data for Analyzing Service Performance: A Case Study of Transportation Service in Malaysia

Hui Na Chua[1(✉)], Alvin Wei Qiang Liao[1], Yeh Ching Low[1], Angela Siew Hoong Lee[1], and Maizatul Akmar Ismail[2]

[1] Department of Computing and Information Systems, Sunway University, Subang Jay, Malaysia
huinac@sunway.edu.my

[2] Faculty of Computer Science and Information Technology, Universiti Malaya, Kuala Lumpur, Malaysia

Abstract. Literature has shown the prominence of extracting social media data to understand public opinion. However, there are little works on how these opportunities can be realized and the challenges in exploiting the opportunities in the transportation industry. Further, data quality and availability using social media may vary according to different demographics due to population size and languages used. Additionally, most of the related prior studies that show the opportunities of using social media data were conducted in North and South America. With this proposition, we seek to investigate the challenges of using Twitter data with text mining techniques for understanding users' opinions and sentiment through a case study of using the data to assess public transportation service performance specifically in a Malaysian context. Our findings indicate that social media data can only be useful in generating reasonable insights if users could input informative words for forming discussed topics to derive opinion, and incline towards a certain sentiment with adjectives. The findings also identified the need for a more proficient dictionary to classify multilingual tweets. Our research provides original evidence proving the potential of using social media data to assess public transportation services performance which may vary depending on the demographics of the social media users.

Keywords: Text mining · Topic modelling · Latent Dirichlet Allocation (LDA) · Twitter open data · Sentiment analysis · Transportation service performance

1 Introduction

With the rapid growth of internet technology, the development of social networking sites makes it easier for the public to engage and share information. An example of a social media site is Twitter that allows its users to upload opinions and thoughts on a problem or share information [1]. One of the information that can be extracted through analyzing Twitter data is public opinion and trends regarding a topic. The public transportation sector is a key element in the globalization era where economic

W. Abramowicz et al. (Eds.): BIS 2021 Workshops, LNBIP 444, pp. 227–239, 2022.
https://doi.org/10.1007/978-3-031-04216-4_21

opportunities are closely related to the mobility of goods, people, and information. Unfortunately, the quality of the public transportation sector in Malaysia has always not been on par with developed countries due to many reasons, namely poor planning and design of public transportation services [2]. Because of that, the public has drifted away from using public transportation facilities as their main mode of commutation, which results in the increased usage of private vehicles that has led to many accessibility issues as reflected in heavy road congestions and difficulties in looking for a parking spot [3].

There is prior research that exploits social network data to understand users' perspectives, for example, use the data as a resource for sociolinguistic studies as social microscopy [4] and for identifying factors that most influence consumers' scores when composing reviews related to characteristics of hotels [5]. Similarly, through data from social media sites like Twitter, one can identify the sentiment on public transportation services through the capability of sentiment analysis techniques [6, 7]. In addition, text mining techniques, such as topic modelling, can also be applied to Twitter data to uncover hidden topics within the tweets to identify similar topics being discussed [8, 9]. The advantages of measuring sentiment in comparison to using traditional surveys for measuring public transportation users' satisfaction include the relatively low data collection cost plus recent information that can be collected in real-time. According to the prior studies discussed earlier, the potential of a social media platform as means to assess transportation services performance and understand users' sentiment has already been acknowledged. However, there are obstacles to overcome to extract useful information from social media data effectively.

In terms of data quality, social media content is mostly in an unstructured natural language form that cannot be readily interpreted, queried, or aggregated using Markup or Sequential Query languages [10] that support semi or fully structured data formats. Therefore, relevant information must be acquired using text mining techniques [11]. A crucial question is whether social media data is of enough quality to meet the need of transportation service providers in obtaining users' sentiment. Unstructured data might potentially have a bias because there is no guided format or boundary of topics discussed thus such data may be useful, and the processing of textual messages into meaningful data is not straightforward. Furthermore, different countries might have different technical challenges in processing text messages due to different languages being used.

In terms of the data sampling, there is a lack of understanding on the comments given by each individual and if this population is representative of a certain place. If social media are to become supportive platforms in public transportation services planning, a study is needed on understanding the population that is participating and those that may be marginalized [12]. Conversely, the number of tweets available and user input content of Twitter data collected may vary depending on user characteristics. Most of the related prior studies were conducted in North and South America [6–9, 12]. Statista's report showed that 20% of Twitter users are Americans, implying huge data volume is available for analysis on American users compared to other countries [13]. Twitter is considered one of the most popular social media platforms in Malaysia with 49% of internet users between age 16 to 64 use the platform [14].

With the observations from data quality and sampling aspects that may be influenced by user characteristics along with the fact that prior studies are mostly conducted in America, we seek to investigate the challenges of using Twitter data with text mining techniques for understanding users' opinion and sentiment towards public transportation service performance in a Malaysian context. Two research questions (RQ) are formulated to achieve our objective:

(RQ1) How does Twitter data quality influence the discovery of topics discussed to drive user opinion on public transportation service performance in Malaysia?

(RQ2) What are the challenges in mining Twitter data to understand user sentiment towards public transportation system performance in Malaysia?

2 Literature Review

2.1 Public Transportation in the Klang Valley

Klang Valley is an area centered in Kuala Lumpur. Public transportation in the Klang Valley covers a wide variety of different services such as bus services and rail transport. Table 1 states the backgrounds and specifications of six main types of public transportation services in the Klang Valley: The Light Rapid Transit lines (LRT); Mass Rapid Transit line (MRT); Keretapi Tanah Melayu (KTM), Monorail line, Bus Rapid Transit line (BRT), and Public Bus.

Table 1. Background of public transportation services in the Klang Valley

Public transportation service	Line/route/type (opening year)	Daily ridership
LRT	Kelana Jaya Line (1998)	245,569
	Ampang and Sri Petaling Lines (1996)	170,814
MRT	Sungai Buloh – Kajang Line (2016)	158,247
KTM	Port Klang - Seremban Line (1995)	85,120
BRT	BRT Sunway Line (2015)	5,382
Monorail	KL Monorail (2003)	63,778
Public bus	RapidKL by Rapid Bus	400,000 (approx.)

In recent years, traffic congestion and limited parking space problems in the city have become a prominent issue in the Klang Valley [2]. It is also found that most of the services are still lack service quality and performance [15].

2.2 Twitter

Due to Twitter content's characteristics of having short texts, open data available for access, and easy to retrieve through API, it is a commonly used platform for researchers to conduct text mining related analysis. Applications from the internet in the form of social media have the potential to reach a wider range of participants from the most affected stakeholders to the general public [16]. The opportunity to engage with the public through social media is regarded as a means of support for micro-participation for planning, performance analysis, and making decisions [17]. Prior studies used sentiment analysis to identify how the public feels about a public transportation system [4]. Conversely, limitations of this method were also identified [18], for example, the sentiment of tweets that contains sarcasm and irony are well known for being classified poorly by automated computer procedures.

2.3 Topic Modelling on Twitter Using LDA for Sentiment Classification

Topic Modelling is frequently used in text-mining tools to identify latent topics within a group of texts or documents using statistical algorithms. Several topic modeling algorithms have been developed, and Latent Dirichlet Allocation (LDA) is found as a commonly used unsupervised machine learning technique [19] for performing topic modeling. Research was conducted across the years about topic modelling on Twitter data to identify trending events and topics on Twitter using topic modelling [20]. Performing topic modelling on tweets is a complex task due to their irregular and unstructured type of language used [21]. Using LDA to identify topic models within sites, such as Twitter, demonstrated that the distribution for a mixture of topics learned by topic models has certain features that can help in classification problems [22].

Twitter sentiment classification aims to identify the sentiment polarity of short and informal texts (i.e. tweets) if they have a positive or negative sentiment. The majority of the existing studies are streamlined into two main approaches, namely, the lexicon-based approach that uses a dictionary of opinion sentiment words [23, 24], and the machine learning approach [25, 26] that uses a classification model to classify sentiment.

SentiStrength [27] is a lexicon-based sentiment classifier that also uses non-lexical rules and linguistic information to detect the sentiment strength of short and informal English texts such as tweets. For every row of text, the SentiStrength tool outputs two different integers ranging from 1 to 5 for both positive sentiment strength and negative sentiment strength. For this output, 1 represents no sentiment, and 5 represents strong sentiment for positive sentiment strength and negative sentiment strength respectively. For example, a row of text with a classified sentiment strength of (5, 2) would contain a strong positive sentiment and a low negative sentiment. A neutral text will be classified as (1, 1). The goal of SentiStrength is to detect the positive and negative sentiment strength. Besides that, SentiStrength also aims to detect the sentiment expressed within the text rather than the overall polarity of the text [28].

2.4 Related Work on Social Media Data for Public Transportation Analysis

In recent years, studies on social media data and public transportation analysis started to increase in numbers. Collins et al. [6] used Twitter data and a sentiment strength detection algorithm to analyze transit rider's satisfaction in Chicago. Their analysis showed that transit riders tend to use negative sentiments rather than positive sentiments to a situation. On the other hand, Schweitzer [7] also discovered that Twitter users tend to use more negative sentiments when posting a tweet about public transit compared to other public services. Haghighi, et.al [8] proposed a framework that uses LDA topic modelling algorithm and sentiment analysis to evaluate transit riders' opinions and identify the hidden topics regarding their dissatisfaction with the transit service. They found out that the percentage of negative sentiment is higher during weekends as compared to weekdays. Mendez et.al [9] also used a similar approach that uses topic modelling and sentiment analysis on Twitter data to evaluate the satisfaction of users on public transportation services in Santiago, Chile. They concluded that their proposed methodology can effectively, and promptly, diagnoses any problems because it can locate issues and identify trends related to public transportation services. Similarly, Casas & Delmelle [12] used a topic modelling procedure on tweets to filter out important topics relevant to the transit system and performed a structured content analysis to identify perceptions of the public on transit systems in Cali, Columbia.

3 Research Methodology

We conducted a text mining process of extracting relevant Twitter data for deriving users' discussed topics and sentiments towards public transportation services.

3.1 Study Area

Klang Valley was chosen as the study area for this research for two reasons. Firstly, public transportation services are saturated in Klang Valley, with six types of different public transportation services ranging from LRT, MRT, KTM, BRT, Monorail, and public busses [15]. Secondly, a prior study shows that most of the Twitter users in Malaysia originate from more developed states like Selangor, Kuala Lumpur, and Johor in which both Selangor and Kuala Lumpur are part of the Klang Valley [28].

3.2 Data Collection

The data for this research was collected over 5 years and 3 months from 1st of January 2014 to 31st of March 2019 for LRT, KTM, Monorail, and public busses. The rationale behind selecting this timeframe for Twitter data collection is because every all six public transportation services in the Klang Valley have their respective opening years. For example, MRT, being the newest among the other services, started its operations in 2016. On the other hand, public transportation services such as LRT, KTM, and Monorail have been fully operational for more than a decade. Therefore, this timeframe was chosen so that every type of public transportation service used for this study will be treated equally by including more years for older public transportation services as mentioned above. Because of that, the years involving public transportation services that have not to exist will not be used for analysis.

Data for this study was extracted using third-party software called Octoparse [29] was used as a preferred web scraping tool compared to Twitter API because there were certain limitations when using it for this study, for example, only 1500 tweets can be extracted per hour. It is necessary to process the collected tweets using a series of filtering and the removing of stop words, as well as tags and links to produce a finalized structured and cleaned dataset. Twitter data extracted through Octoparse in its raw form consist of several variables which are listed in Table 2, together with their examples for LRT public transportation services in the Klang Valley.

We adopted three filtering approaches in the tweets extraction process to identify relevant data to the public transportation services. Firstly, we used keywords as hashtags to represent the six public transportation services was done using the name of the public transportation service filter (i.e. LRT, MRT, KTM, BRT, Monorail, and Bus) to specify the tweets to be extracted. Secondly, we applied an extraction filter by identifying the hashtags used by users. Therefore, the hashtags that were used for tweet extraction are common hashtags that contain the name of transportation services. Examples of the hashtags are "lrtdelay", "lrtmalaysia", "lrtsucks", and "jomnaiklrt" for filtering LRT-related tweets. The third tweet extraction filter used is the study area, which is the Klang Valley. A location filter (i.e. Kuala Lumpur City, Petaling Jaya, and Klang) was used in the Twitter advance search.

Table 2. Example of Twitter Data for every public transportation service

Twitter variable	Example of LRT
Username	Ganee
ID	Ganee
publishing_date	29/3/2019 21:52
Content	I usually stand in LRT eventhough there is empty seats. Removes the guilt of needing tooffer my seat. Now I see so many young Kids in the LRT. I wonder If the day when a kid will say to me "Pakcik, sila duduk"
Comments	–
Retweet	–
Likes	1
ImageURL	–
VideoURL	–
TweetURL	https://twitter.com/ganee/status/11118537342653

Before data can be analyzed, an iterative series of data preparation was performed to build the final datasets for analysis. Tokenization was performed in the data preparation process to divide individual tweets into parts. Stop words are removed from a tweet that contains common English stop words using the Python R library programming package. Additional stop words not in the default English stop words list is also added and later removed. The stemming process involves obtaining the root or the base of a word by removing suffixes and affixes. The data modelling phase includes running a series of sentiment classification based on polarity score and topic modelling for each tweet using SentiStrength to assign a positive and negative sentiment strength to the text. LDA was used to identify latent topics within public transportation-related tweets.

4 Results and Discussion

4.1 (RQ1) How Does Twitter Data Quality Influence the Discovery of Topics Discussed to Derive User Opinion on Public Transportation Service Performance in Malaysia?

4.1.1 Sample Size of Twitter Data Relating to Public Transportation Service

The total number of tweets collected for this research resulted in 3721 tweets before any data reduction or cleaning methods were applied. The specific number of tweets for each public transportation service and daily ridership is illustrated in Table 3. On average the number of tweets collected daily is less than one tweet for all service providers, and the average monthly tweets are near to zero percent of the daily ridership. This observation implies there is a lack of representation for the daily ridership population of public transportation service users.

Table 3. Daily ridership and total collected twitter data

	LRT	MRT	KTM	BRT	Monorail	Public Bus
Daily Ridership	416,383	58,247	85,120	5,382	63,778	400,000
Total Data Collected (5 years and 3 months)	1,647	354	782	96	274	568
Average daily tweet (from total data collected)	<1	<1	<1	<1	<1	< 1
Average monthly tweet (from total data collected)	<26	<6	<13	<2	<5	< 9

Our results in Table 3 are complemented by [28] which revealed that most of the Malaysian tweets are posted at late night times with their peak is also at late night and that Twitter users are mostly non-active users [30] with less than 2 tweets per day [28]. This finding infers that receiving prompt notification of daytime events that occurred related to public transportation service performance can be hardly feasible. Further, it is observed that Twitter data lacks the understanding of the individual's characteristics who express their opinion and whether this data sample is representative. If social media platforms are to become a mainstay in public participatory assessment for public transportation services, more research is needed on understanding the characteristics of Twitter users who are contributing to the content and those that may be marginalized.

4.1.2 Data Extraction Process to Derive Relevant Tweets

As observed from Table 3, "#lrt", "#mrt" and "#ktm" were not used as a filter to extract tweets. Although "#lrt" was used in certain public transportation-related tweets, the hashtag is also used as an abbreviation for "last retweet" on Twitter. The same applies to "#mrt" and "ktm", whereby "#mrt" also means "Mr T" whom is an American Actor, and "#KTM" is an Austrian motorcycle manufacturer. These hashtags are omitted to prevent the extraction of irrelevant Twitter data. Through the process of extracting relevant content for public transportation services, we learned that defining the filtering criteria can be challenging as it requires a thorough investigation of hashtags, to avoid bias of including those irrelevant hashtags that having the same name.

4.1.3 Words for Discussed Topics that Can Represent Strength and Limitation

Table 4 shows few examples of LDA topic modelling outputs for public transportation service during a specific timeframe. For LRT, words such as "bad" in Topic 1 (implies limitation) and "good" (implies strength) in Topic 2 are examples of adjectives that can describe LRT as a public transportation service performance. Besides that, words such as "delay" in Topic 1 and "wait" in Topic 3 for KTM during the 1st Quarter of 2015 signifies that this public transportation service tends to be delayed or late, hence causing the users to wait. This implies that KTM may not be the most effective for their punctuality hence indicating their limitations in service performance. The words used by the public for MRT-related tweets mostly contain nouns and verbs. For example, "monorail" and "lrt" which are the name of the respective public transportation are considered a noun. Besides that, words like "take" and "launch" are verbs, nouns and verbs generally cannot indicate the strength or limitations of the public transportation service. The results infer that words might not be always representative enough to derive useful public transportation service performance. This drawback is reasonably due to the nature of unstructured data that has no guided boundary of words for a discussed topic. In addition, adjectives are required to suggest the strength or limitations that a public transportation service performance.

Table 4. LDA outputs for LRT, KTM, and MRT tweets during the 1st Quarter of 2015

	LRT			KTM			MRT	
Topic 1	Topic 2	Topic 3	Topic 1	Topic 2	Topic 3	Topic 1	Topic 2	Topic 3
Pic	Take	people	back	time	wait	day	first	Ride
Will	Way	day	home	well	old	now	today	Can
Now	Wait	morn	photo	first	day	like	take	Morn
Come	Still	now	last	post	woot	time	discount	Also
Bad	Good	new	ride	look	later	admin	free	Work
Greet	Petal	home	still	even	earli	shift	monorail	New
Cent	Final	back	delay	dinner	back	part	lrt	Shop
petrona	Mani	need	bass	arriv	need	excit	proud	Razak
Today	Can	first	imol	minut	first	offic	come	Launch
Forev	Laa	let	trebl	fun	let	rememb	friend	Yab

4.2 (RQ2) What are the Challenges in Mining Twitter Data to Understand User Sentiment Towards Public Transportation System Performance in Malaysia?

4.2.1 Classifying Sentiment from Tweets Content

Malaysia is a multiracial and multilingual country. As seen in Fig. 1, for example, the tweets extracted contain a mixture of English and Malay words in the same tweet.

Thing is, with buses and csrs yg autonomous ni, you can repurpose, reprogram the routes or something.

Fig. 1. Example of Bilingual Tweets.

Tweets in the Chinese language have been excluded in the filtering process if they exist.

Due to the absence of a systematic Malay dictionary for text analysis to process non-English words, the sentiment classification tool could not classify the contents of the tweets as Malay words and slangs are not included for processing sentiment classification. This exclusion further reduced the sample data for analysis. This finding implies the need for a more proficient dictionary to classify bilingual tweets. Besides, combining the finding of unhelpful words for identifying topics stated in Sect. 4.1.3, it is necessary to have a different sentiment of words in a different domain which leads to the need for a transportation-specific lexicon. Further, the coverage of existing lexicons can be insufficient for meaningful transportation-related discussed topics analysis, and the possibility of the corpus to be outdated (i.e., there are no updates are being done and word usage can change through time). Another possibility of bias sentiment analysis, that requires further investigation is the fault news, in which peoples' opinions are affected by others opinion, spreading fault news can affect the general opinion.

4.2.2 Identifying Positive and Negative Sentiment Strength

Tables 5 and 6 show the distribution of positive and negative sentiment strengths (scale 1–5 denotes the range of not positive/negative to positive/negative) respectively. The sentiment strength with a scale of 1–2 (towards the intense of not positive/negative) is the largest distribution for both positive and negative sentiment, comprising approximately more than 80% of the distribution for all six public transportation services. The results show that the tweets extracted mostly have a neutral sentiment polarity which is not useful to identify the strengths and limitations of each public transportation service. This also implies that most Twitter users do not incline towards a certain sentiment (i.e. positive or negative) when posting a tweet on Twitter.

Table 5. Frequency of each positive sentiment strength for each public transportation service

Strength	LRT	MRT	KTM	BRT	Monorail	Public Bus	Total tweets found
1	1,156	338	525	86	204	400	2709
2	722	186	384	62	96	262	1712
3	324	150	177	15	60	96	822
4	80	20	24	–	8	16	148
5	10	–	–	–	–	5	15

Table 6. Frequency of each negative sentiment strength for each public transportation service

Strength	LRT	MRT	KTM	BRT	Monorail	Public Bus	Total tweets found
1	1,338	409	657	86	204	400	3094
2	470	116	182	62	96	262	1188
3	135	30	60	15	60	96	396
4	116	32	56	–	8	16	228
5	–	5	–	–	–	5	10

5 Conclusion

5.1 Implications and Recommendations

Compared to prior studies conducted in other countries, the sample data we collected was obviously by far smaller and insufficient to represent the population of public transportation service users. One of the reasons for this low data sample collected can be explained rationally by the exclusion of other languages (such as Malay and Mandarin) besides English. Malaysia is a multiracial and multilingual country. Malay is the national language and tweets in Malay can potentially be the large data sample that we have ignored due to the limitation of system tools for understanding and analyzing the language. Further, our results revealed that most of the Malaysian tweets are posted at late night times with low tweets frequency per day indicating mostly are none-active users. And, most Twitter users incline towards a neutral sentiment when posting a tweet on Twitter. This tendency decreases the potential of social media data as an informative source for user sentiment analysis on public transportation in Malaysia.

Through the evaluation of the usage pattern and characteristics of Malaysian Twitter users in our research study, we learned that to obtain useful feedback based on social media users' comments on service performance requires the participation of active and expressive users in using adjective words. Furthermore, the comparison between our research findings and other studies indicates that the potential of using social media data to assess public transportation services performance may vary depending on the demographics of the social media users. Thus, for organizations or researchers who are planning to conduct a future research study or propose a framework system utilizing social media data to assess transportation services performance, we suggest several considerations:

- Understanding of users' usage patterns is necessary before confirming the use of social media data, which is, to identify if there are enough active users in giving feedback through social media. In this study, only a part of the Malaysian population uses Twitter so the data are not representative of all the users;
- If the study involves multiracial and multilingual population, the capability of system tools to handle different languages is required to achieve a substantial sample size and ensure inclusion of commonly used languages by the population studied;
- Expressiveness of users in terms of using indicative adjectives in their posted message is critical to enable meaningful Sentiment analysis for forming discussed topics to represent an opinion. Prior studies showed that sentiment analysis was possible through social media data which were conducted mainly in North and South America, but our study conducted in Malaysia showed otherwise. This finding motivates us to suggest further research to investigate if different users from different countries have a significant difference in how they construct their posts that reflect the sentiment.
- Due to the nature of social media data that lacks the information of individual's characteristics who express their opinion, we recognized the usage of social media data is currently limited to identifying sentiments and discussed topics along with their evolution instead of replacing the role of surveys that are capable of capturing perception by demographics. Therefore, with the currently limited information provided via social media data, other data sources should also be considered if the objectives of data analysis are beyond identifying sentiments and discussed topics.
- Transportation issues are difficult to analyze, for instance, they are not limited to only to lack of parking lots or traffic congestion, for example, the underground MRT is not affected by road congestion. The issues are probably more linked to the usage increasing of private vehicles, as the reduction of commercial speed, more pollution, noise, accidents and so on. Considering all transport services together may lead to multiple and unrelated results. Although our study shows insufficient useful data using Tweets for analyzing user acceptance, the technique of text mining approach we presented in this paper is suitable to evaluate the performance of the transport systems in terms of user acceptance and other indicators of user perception, but it is not suitable to quantify numeric indicators.

5.2 Future Work

We plan to further analyze social media data incorporating semantic-based modelling techniques that may give other insights into the usefulness of the data for service acceptance analysis. Our research outcomes also acknowledge the need for a more proficient dictionary to classify multilingual tweets, for example, for the Malay language, therefore filling this gap will increase more data samples for analysis. Furthermore, we intend to explore other social media platforms such as Facebook to investigate to compare if different social media platforms' data reflect different characteristics of public opinion on a topic. We plan to investigate if combining multiple platforms' data could provide useful outcomes, this could allow us to have comparisons with results achieved using traditional surveys for measuring public transportation users' satisfaction.

Acknowledgments. This paper was supported by Partnership Grant CR-UM-SST-DCIS-2018–01 and RK004–2017 between Sunway University and the University of Malaya.

References

1. Windasari, I.P., Uzzi, F.N., Satoto, K.I.: Sentiment analysis on Twitter posts: an analysis of positive or negative opinion on GoJek. In: 4th International Conference on Information Technology, Computer, And Electrical Engineering. IEEE 2017, October 2017
2. Almselati, A., Rahmat, R., Jaafar, O.: An overview of urban transport in Malaysia. Soc. Sci. **6**(1), 24–33 (2011)
3. Borhan, M., Syamsunur, D., Mohd Akhir, N., Mat Yazid, M., Ismail, A., Rahmat, R.: Predicting the use of public transportation: a case study from Putrajaya, Malaysia. Sci. World J. **2014**, 1–9 (2014)
4. Hovy, D., Johannsen, A., Søgaard, A.: User review sites as a resource for large-scale sociolinguistic studies. In: Proceedings of the 24th International Conference on World Wide Web, pp. 452–461, May 2015
5. Cozza, V., Petrocchi, M., Spognardi, A.: Mining implicit data association from Tripadvisor Hotel Reviews. In: EDBT/ICDT Workshops, pp. 56–61, March 2018
6. Collins, C., Hasan, S., Ukkusuri, S.: A novel transit rider satisfaction metric: rider sentiments measured from online social media data. J. Public Transp. **16**(2), 21–45 (2013)
7. Schweitzer, L.: Planning and social media: a case study of public transit and stigma on Twitter. J. Am. Plann. Assoc. **80**(3), 218–238 (2014)
8. Haghighi, N.N., Liu, X.C., Wei, R., Li, W., Shao, H.: Using Twitter data for transit performance assessment: a framework for evaluating transit riders' opinions about quality of service. Public Transport **10**(2), 363–377 (2018). https://doi.org/10.1007/s12469-018-0184-4
9. Mendez, J., Lobel, H., Parra, D., Herrera, J.: Using Twitter to infer user satisfaction with public transport: the case of Santiago, Chile. IEEE Access **7**, 60255–60263 (2019)
10. Hoffer, J.A., Venkataraman, R., Topi, H.: Modern database management (Global Edition). ISBN-13: 978–1292263359. Pearson. 2019 (2019)
11. Manning, C.D., Schütze, H., Raghavan, P.: Introduction to Information Retrieval. Cambridge University Press (2008)
12. Casas, I., Delmelle, E.: Tweeting about public transit — Gleaning public perceptions from a socialmedia microblog. Case Stud. TransportPolicy **5**(4), 634–642 (2017)
13. Statista: Twitter: number of monthly active users 2010–2019 (2019). https://www.statista.com/statistics/282087/number-of-monthly-active-twitter-users/. Accessed 5 Jan 2020
14. Datareportal. Digital 2021: Malaysia. https://datareportal.com/reports/digital-2021-malaysia. Accessed 24 May 2021
15. 'Public transport in the Klang Valley (2019). https://en.wikipedia.org/wiki/Public_transport_in_the_Klang_Valley. Accessed 7 Jan 2020
16. Brabham, D.: motivations for participation in a crowdsourcing application to improve public engagement in transit planning. J. Appl. Commun. Res. **40**(3), 307–328 (2012)

17. Evans-Cowley, J., Griffin, G.: Microparticipation with social media for community engagement in transportation planning. Transp. Res. Record J. Transp. Res. Board **2307** (1), 90–98 (2012)
18. Grant-Muller, S., Gal-Tzur, A., Kuflik, T., Minkov, E., Shoor, I., Nocera, S.: Enhancing transport data collection through social media sources: methods, challenges and opportunities for textual data. IET Intel. Transport Syst. **9**(4), 407–417 (2015)
19. Blei, D., Ng, A., Jordan, M.: Latent Dirichlet Allocation. J. Mach. Learn. Res. **3**, 993–1022 (2003)
20. Weng, J., Lim, E.P., Jiang, J., He, Q.: Twitterrank: finding topic-sensitive influential twitterers. In: Proceedings of the Third ACM International Conference on Web Search and Data Mining, pp. 261–270, February 2010
21. Yang, M., Rim, H.: Identifying interesting Twitter contents using topical analysis. Expert Syst. Appl. **41**(9), 4330–4336 (2014)
22. Hong, L., Davison, B.D.: Empirical study of topic modeling in Twitter. In Proceedings of the First Workshop on Social Media Analytics, pp. 80–88, July 2010
23. 'WordNet | A Lexical Database for English', https://wordnet.princeton.edu/. Accessed 7 Jan 2020
24. Ren, Y., Wang, R., Ji, D.: A topic-enhanced word embedding for Twitter sentiment classification. Inf. Sci. **369**, 188–198 (2016)
25. Ali, F., et al.: Transportation sentiment analysis using word embedding and ontology-based topic modeling. Knowl.-Based Syst. **174**, 27–42 (2019)
26. Qi, B., Costin, A., Jia, M.: A framework with efficient extraction and analysis of Twitter data for evaluating public opinions on transportation services. Travel Behav. Soc. **21**, 10–23 (2020)
27. Thelwall, M., Buckley, K., Paltoglou, G., Cai, D., Kappas, A.: Sentiment strength detection in short informal text. J. Am. Soc. Inf. Sci. Technol. **61**(12), 2544–2558 (2010). https://doi.org/10.1002/asi.21416
28. Abu Bakar, M., Mohd Ariff, N., Hui, E.: Exploratory data analysis of Twitter's rhythm in Malaysia. In: AIP Conference Proceedings 2019 (2019)
29. Octoparse. https://www.octoparse.com/. Accessed 25 May 2021
30. Huberman, B.A., Romero, D.M., Wu, F.: Social networks that matter: Twitter under the microscope. arXiv preprint arXiv:0812.1045 (2008)

Data Quality Assessment of Comma Separated Values Using Linked Data Approach

Aparna Nayak(✉), Bojan Božić, and Luca Longo

SFI Centre for Research Training in Machine Learning, School of Computer Science, Technological University Dublin, Dublin, Republic of Ireland
{aparna.nayak,bojan.bozic,luca.longo}@tudublin.ie

Abstract. With an increasing amount of structured data on the web, the need to understand and convert it into linked data is growing. One of the most frequent data formats is Comma Separated Value (CSV). However, it is not easy to describe metadata such as the datatype, data quality and data provenance along with it. Therefore, to publish CSV on the web, it is required to convert CSV into linked data format. Many approaches exist to facilitate the conversion process from structured data to linked data. However, all methods require additional domain knowledge for the conversion process. The goal of this research is to assist publishers in converting CSV files into linked data without human intervention whilst understanding its quality and root causes of data quality violations. The proposed framework consists of two modules. The first module converts the given CSV file into a knowledge graph based on a proposed ontology which is appended with data quality information. In the second module, triples that have violated the data quality constraints are identified. The results show that it is possible to convert a CSV to a knowledge graph by adding its quality information without the help of external mappings.

Keywords: CSV · Data quality · Knowledge graphs · Linked data · Quality assessment · Root cause analysis

1 Introduction

The Semantic Web aims to publish data on the web that can be interpreted by humans and machines. It also helps to reuse, share, and publish data across the web, making it easy to integrate data from different sources. However, significant amounts of data on the web still reside in various formats other than Resource Description Framework (RDF). The rapid growth of data catalogues on the web has led to publishing data in multiple formats. However, the quality of such published data is not known to its users [2,15]. The comma-separated file (tabular data; irrespective of the delimiter used) format is one of the predominant ways to share information that stores data in a two-dimensional array.

W. Abramowicz et al. (Eds.): BIS 2021 Workshops, LNBIP 444, pp. 240–250, 2022.
https://doi.org/10.1007/978-3-031-04216-4_22

The resource format of 37% of the total available dataset in the European data portal[1] is comma-separated (CSV). The remaining 38% of the dataset is in zip format, which includes CSV files. These files contain data in the form of rows and columns as a two-dimensional array. A row represents numerous properties that belong to one instance. A column represents a set of values that belongs to a specific attribute. Converting delimited files to RDF triples that conform to the principles of Semantic Web is an added advantage to the community for reuse and sharing purposes.

Existing approaches require an external mapping to convert any given dataset into RDF triples. This process requires users to have domain knowledge, as they need to be aware of all existing classes in the available ontology to map each column of the CSV file. The proposed method makes use of direct mapping to uplift data to RDF. Direct mapping [1] defines a simple transformation, used to materialize RDF graphs. This method neither requires users to have domain knowledge, nor the dataset should adhere to any ontology/schema. Furthermore, automatic conversion helps to generate RDF triples from datasets that do not adhere to a specific schema or an ontology. On the other hand, mapping languages require the dataset to adhere to an ontology or schema.

CSV to RDF conversion process extracts additional semantic information from the dataset that is added to the triples generated from the dataset. These additional triples later used to assess the quality and it helps to locate the erroneous triples along with the type of violation. It explains users of the dataset to understand which triples have violated a constraint with a precise location in the knowledge graph (KG). The research project aims to determine "To what extent can CSV files be converted to RDF triples with or without human intervention, thus enabling the user to identify root causes of the data quality problems?"

The remainder of this article is structured as follows. Section 2 covers the related work. Section 3 discusses design of the proposed methodology along with use cases. Section 4 summarizes the details of the dataset and results. Finally, Sect. 5 concludes the paper with future work.

2 Related Work

Several solutions have been developed to map non RDF data into RDF. This section briefly introduces some tools that help to map CSV to RDF, and data quality metrics helps to assess the RDF. An automatic conversion framework is implemented to deploy linked data on the Pan-European website [5]. This framework reads CSV headers to generate mappings. In another similar approach, CSV files that belong to the same domain are clustered before mapping it into RDF format [14]. However, one of the drawbacks of such method is that the system requires column must be declared as either a data property or as an object property in their ontology. Another problem is that a typographic error in the column name does not match any properties in the ontology. On the other hand, CSV files are uplifted with the help of associated metadata [11]. The

[1] https://www.europeandataportal.eu/data/eu-international-datasets.

conversion process fails in the absence of metadata. A comprehensive analysis of data mappings helps to uplift any given dataset to RDF mentioned in [8]. Sheet2RDF [6], is a semi-automatic approach to uplift CSV to RDF triples. The graphical user interface of sheet2rdf enables some refinements without the need to explicitly use the underlying mapping specification language. Data uplifting techniques either require an external mapping function or additional software.

Data quality is a multidimensional concept. There are several frameworks for assessing the quality of linked data. A systematic review conducted in [17] identified a number of different data quality dimensions applicable to Linked Data. Luzzu [4] is one of the frameworks that consider most of the quality dimensions defined by [17] to evaluate the linked data. Semquire [10], Databugger [9], LD Sniffer [12] are some other existing data quality assessment tools/frameworks. However, the root causes of the triples that have violated the constraints are not identified in existing data quality assessment frameworks. Luzzu framework is extended to support root causes of quality violations [16] by identifying the erroneous triples based on the failure of the metric. The data quality metrics are associated with corresponding exception class and problematic parts such as subject, object, and predicate. One of the recent works [3], identifies the constraint violated triples using EYE reasoner[2]. However, all the described constraints are domain-dependent. The proposed technique does not require a user to write any external mappings. In addition, it appends data quality information to the RDF triple store along with the identification of root causes of triples that have violated the data quality constraints.

3 Design

This section gives an overview of the system design that includes a data quality assessment framework that assesses the data quality and reports the quality violated triples, if any. One of the best practise suggested by W3C is to publish data on the web along with data quality information [2]. The main goal of the proposed method is to convert a CSV file into RDF triples that also contains data quality information. RDF data is generated with the help of direct mapping which is appended with quality information. Furthermore, these RDF triples are evaluated to identify erroneous triples.

W3C [7] has described two methods to convert any tabular data into RDF triples. Standard mode conversion frames the information gleaned from the cells of the tabular data with details of the rows, tables, and a group of tables within which that information is provided. Minimal mode conversion includes only the information gleaned from the cells of the tabular data. The proposed model follows minimal mode as the dataset considered for the experiment is not an annotated file. Most of the vocabulary terms mentioned by W3C are applicable only for annotated CSV files. However, the standard mode of conversion is used wherever appropriate.

[2] http://eulersharp.sourceforge.net/.

The proposed CSV2RDF ontology conceptually maps all the components of the CSV file to the classes of the ontology as described in Fig. 1. It connects dataset properties to the data arranged in the form of rows and columns. The contents of the CSV file is converted, as shown in Fig. 2. Each row of the CSV is considered to be a class, as mentioned in the CSV2RDF ontology. All the columns of the dataset are considered object properties of the row to maintain the order of data. All the cell values are translated into RDF literals for the given CSV file.

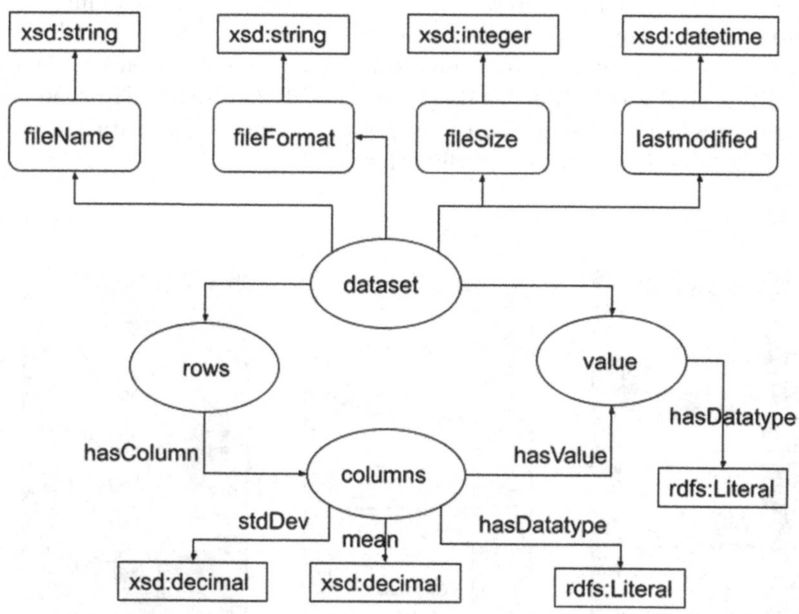

Fig. 1. Ontology describing CSV2RDF data

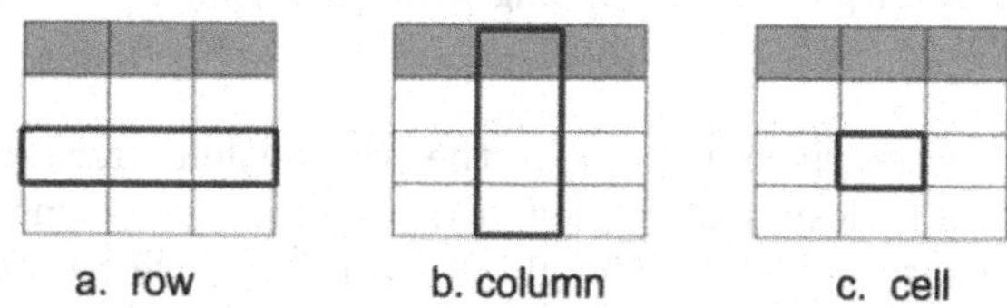

Fig. 2. Row-Column-Cell as RDF triples (a) subject - (b) predicate - (c) object

Data quality assessment pipeline in case of direct mapping is as shown in Fig. 3. The triple modeling module is responsible for uplifting CSV into RDF. The proposed ontology is used to convert CSV into RDF. The data module is responsible for collecting all the required RDF triples, axioms, and constraints.

Axioms contain CSV2RDF ontology. Constraints represent the rules that are used to identify the quality violated triples. The conversion process generated RDF triples along with meta-information such as file name, file format, last modified, and data quality information. The quality metrics used to evaluate the knowledge graph are obtained from the comprehensive list of data quality metrics [17].

Typed literals are created by comparing the lexical token of each cell with basic datatypes. Datatypes such as int, float, date, string and URI are considered in this experiment. RDF:nil is assigned as the datatype for all the empty cells present in the CSV. RDFS:Literal is assigned to all other remaining cells if they do not match any of the afore-mentioned basic datatypes. The datatype is predicted for each column based on values that are present in the column. The lexical form of a cell value is compared with all considered datatype using a regular expression, and matched datatypes are counted. The datatype which has the highest value is assigned to the column.

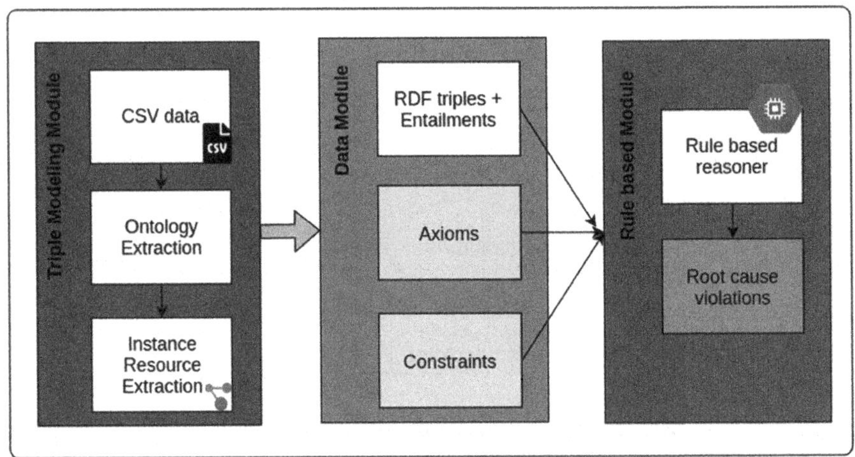

Fig. 3. CSV2RDF direct mapping with data quality assessment

The RDF triples are given as input to the Apache Jena framework. The Jena inference subsystem is designed to allow a range of inference engines or reasoners to be plugged into Jena. Such engines are used to derive additional RDF assertions, which are entailed from some base RDF together with any optional ontology information, the axioms and rules associated with the reasoner. The conventional model makes use of a single reasoner. The proposed model benefits from the cascaded reasoner. RDF triples and OWL axioms are given as input to the OWL reasoner, as shown in Fig. 4. The inferential model generated by the OWL reasoner is given as input to the generic reasoner. The generic reasoner validates the OWL inferential model against the user-defined constraints, and the resulting violations are reported to the user in triple format.

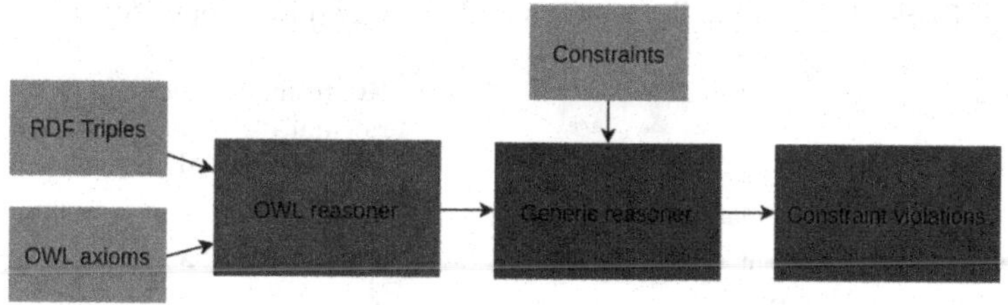

Fig. 4. Cascaded reasoner in Apache Jena

The generic reasoner admits RDF triples as input to validate them against the constraints and axioms. Constraints help to identify erroneous triples in the dataset. Data quality problems such as the presence of outliers and datatype mismatch have been identified using constraints. The explanatory message reported by the reasoner helps the user to understand the exact position of the erroneous triple in the dataset, thereby identifying the root causes of the quality problems. Constraints are written in Datalog as shown in Fig. 5.

The knowledge graph is constructed by considering metrics that are listed in Table 1. Considering the KG is constructed from raw data quality metrics such as deprecated class/property, use of long URIs and some other metrics have been taken care of. Metrics listed under the domain ontology dependent column are assessed with the help of an OWL reasoner. Root cause analysis of ontology dependent metrics is accomplished with the help of the cascaded OWL reasoner. The cascaded reasoner evaluates ontology dependent quality metrics followed by root cause analysis of the metrics mentioned in Table 2.

Table 2 denotes the data quality metrics that are implemented for quality assessment purpose. Assessment refers to the evaluation of the entire dataset for the specific metric. Root cause analysis refers to the detection of triples that have violated a constraint that describes the specific quality metric. All the metrics used for the evaluation are considered from [17]. All columns are assumed to have a header while converting CSV to RDF triple. Semantic Web typically follows the open-world assumption(what is not known to be true or false might be true, or the absence of information is interpreted as unknown information, not as negative information). In this research, the quality metric population completeness is determined by considering the closed world assumption.

Datalog constraints are written to identify the triples that have violated the constraints. Rule R1 in Fig. 5 compares the datatype of the column with the data type of the value. Datatype mismatch error is thrown in case of a mismatch between the column and the value. Rules R2 and R3 in the Fig. 5 helps to locate the outliers based on lower and upper quantiles. Outliers are computed only for columns that have integer datatype.

Table 1. Data quality metrics and corresponding dependency

Metric	KG from raw data	Domain ontology dependent
No use of deprecated class/property	✓	✗
No misreported content types	✓	✗
Detection of long URI	✓	✗
Syntax error	✓	✗
Prolix RDF features	✓	✗
Inverse-functional properties	✗	✓
No use of entities as members of disjoint classes	✗	✓
Correct domain/range definition	✗	✓
Misplaced classes and properties	✗	✓

Table 2. Implemented data quality metrics

Metric	Assessment	Root cause identification	Dimension
Detection of ill typed literals	✓	✓	Syntactic validity
Population completeness	✓	✗	Completeness
No inaccurate values	✓	✗	Semantic accuracy
No outliers	✓	✓	Semantic accuracy
Language used	✓	✗	Interpretability

The users of this framework will be either CSV data consumers who wish to uplift their data to RDF or publishers who wish to keep multiple data source format. The following use cases were listed based on the functionalities that the framework supports.

UC1: Uplifting of CSV: Users who have CSV files can uplift their data to RDF. Direct mapping is a simple technique to map CSV to RDF. It also helps the naive users to get the RDF format of data.

UC2: Assessment of data: Users should be able to assess data. Linked data that is uplifted from CSV is assessed to understand the quality of data.

UC3: View data quality problems: Users will get the data quality problems in the form of a report that is generated using data quality vocabulary [13]. This helps the user to understand the exact cause of the problem.

```
[R1: (?r ?c ?o) (?o ns1:hasValue ?d)
(?c ns1:hasDatatype ?e) notDType(?d, ?e) ->
(?c ns1:constraintType ns1:datatypemismatch)
(?c ns1:expectedDT ?e) (?c ns1:foundDT ?d)
(?c ns1:constraintElement ?r) ]

[R2: (?r ?c ?o) (?o ns1:hasValue ?v)
(?c ns1:stdDev ?std) (?c ns1:mean ?m)
product(?std,3,?threshold) difference(?m ?threshold
?LL) lessThan(?v, ?LL) -> (?c ns1:constraintElement ?v)
(?c ns1:contraintType ns1:OutlierLowerThreshold) ]

[R3: (?r ?c ?o) (?o ns1:hasValue ?v)
(?c ns1:stdDev ?std) (?c ns1:mean ?m)
product(?std,3,?threshold) sum(?m ?threshold ?UL)
greaterThan(?v, ?UL) -> (?c ns1:constraintElement ?v)
(?c ns1:contraintType ns1:OutlierUpperThreshold) ]
```

Fig. 5. Datalog constraints for datatype mismatch and outlier detection

4 Results

The experiment is conducted by considering CSV files from various themes such as agriculture, education, health, house, and synthetic data. Table 3 represents dataset information considered for the experiment. The table gives information about the number of rows, number of columns, and different datatype that the file contains. One of the significant outputs of root cause violations of the datatype mismatch is as shown in the Fig. 6. The exact location of the quality problem is highlighted in error thrown by the program. It includes row number, column name, expected value and found value. The datatype mismatch is found as a result of additional space in the beginning of the email id which was present in the original dataset. The detailed analysis of the problem helps the user to understand data quality in the dataset. For example, in Fig. 6, *ns1:R2* refers to the row number and data property is *ns1:Email.* The expected datatype refers to the datatype of the column, and found datatype indicates the datatype associated with the literal. This is one of the significant outputs found for the datatype

Table 3. Dataset information

Dataset	Rows	Columns	Datatype
Agriculture	18	7	String, URL, Decimal, Integer
Education	13	9	String, Integer, URL
Health	18	16	Integer, String
House	21613	16	Integer, Decimal
Synthetic data	10	9	Integer, String, URL, Decimal

mismatch when the RDF triples are verified against the constraints mentioned in Fig. 5. Additional datatype mismatch triples are the result of the presence of *NULL* values. The algorithm assigns *RDF: nil* to literal values if it finds any empty cells. When such values are compared against a column, a datatype error is thrown.

```
ns1:Email ns1:constraintElement ns1:R2 ;
     ns1:constraintType ns1:datatypemismatch;
     ns1:expectedDT xsd:string ; ns1:foundDT
     " braylib@wocklowcoco.ie"^^xsd:decimal .
```

Fig. 6. Root cause violation : Datatype mismatch

Outlier constraints (R2 and R3 in Fig. 5) were not projected due to the fact that no outliers were present in the dataset. The synthetic data is generated to locate all the root causes that the experiment addresses. The experiment is publicly available on Github[3].

5 Conclusions and Future Work

This research presents an ontology-based framework for constructing the knowledge graph from a CSV file without human intervention. The proposed framework consists of two core modules. The first module deals with converting the representation of knowledge from frames to semantic networks, which are also appended with data quality information. The second part of the work deals with building a knowledge base from production rules to identify triples that have violated the constraints. Creating a knowledge graph from raw data benefits from incorporating multiple quality metrics.

Some of the limitations of this research are i. The annotated CSV files are not considered ii. Semantic enrichment of the knowledge graph. Annotated CSV files contain additional metadata that can be used to identify similar entities on web using natural language processing. Semantic enrichment of the knowledge graph addresses mapping the contents of CSV file to DBpedia classes thus connecting data silos. The future work of this research tries to overcome the limitations of the system along with incorporating more data quality metrics and automatic refinement based on quality violation explanations.

Acknowledgements. This publication has emanated from research conducted with the financial support of Science Foundation Ireland under Grant number 18/CRT/6183. For the purpose of Open Access, the author has applied a CC BY public copyright licence to any Author Accepted Manuscript version arising from this submission.

[3] https://github.com/aparnanayakn/csvdataqualityassessment.

References

1. Arenas, M., Bertails, A., Prud'hommeaux, E., Sequeda, J.: A direct mapping of relational data to RDF. W3C Recommendation **27**, 1–11 (2012)
2. Lóscio, B.F., Caroline Burle, N.C.: Data on the web best practices. W3C Recommendation (2017)
3. De Meester, B., Heyvaert, P., Arndt, D., Dimou, A., Verborgh, R.: RDF graph validation using rule-based reasoning. Semantic Web (Preprint), 1–26 (2020)
4. Debattista, J., Auer, S., Lange, C.: Luzzu-a methodology and framework for linked data quality assessment. J. Data Inf. Qual. (JDIQ) **8**(1), 1–32 (2016)
5. Ermilov, I., Auer, S., Stadler, C.: User-driven semantic mapping of tabular data. In: Proceedings of the 9th International Conference on Semantic Systems, pp. 105–112. Association for Computing Machinery, New York (2013)
6. Fiorelli, M., Lorenzetti, T., Pazienza, M.T., Stellato, A., Turbati, A.: Sheet2RDF: a flexible and dynamic spreadsheet import&lifting framework for RDF. In: Ali, M., Kwon, Y.S., Lee, C.-H., Kim, J., Kim, Y. (eds.) IEA/AIE 2015. LNCS (LNAI), vol. 9101, pp. 131–140. Springer, Cham (2015). https://doi.org/10.1007/978-3-319-19066-2_13
7. Jeremy Tandy, Ivan Herman, G.K.: Generating RDF from tabular data on the web. W3C Recommendation (2015)
8. Junior, A.C., Debruyne, C., Brennan, R., O'Sullivan, D.: FunUL: a method to incorporate functions into uplift mapping languages. In: Proceedings of the 18th International Conference on Information Integration and Web-Based Applications and Services. iiWAS 2016, pp. 267–275. Association for Computing Machinery, New York (2016)
9. Kontokostas, D., Westphal, P., Auer, S., Hellmann, S., Lehmann, J., Cornelissen, R.: Databugger: a test-driven framework for debugging the web of data, pp. 115–118. Association for Computing Machinery, Inc (2014)
10. Langer, A., Siegert, V., Göpfert, C., Gaedke, M.: SemQuire - assessing the data quality of linked open data sources based on DQV. In: Pautasso, C., Sánchez-Figueroa, F., Systä, K., Murillo Rodríguez, J.M. (eds.) ICWE 2018. LNCS, vol. 11153, pp. 163–175. Springer, Cham (2018). https://doi.org/10.1007/978-3-030-03056-8_14
11. Mahmud, S.M.H., Hossin, M.A., Hasan, M.R., Jahan, H., Noori, S.R.H., Ahmed, M.R.: Publishing CSV data as linked data on the web. In: Singh, P.K., Panigrahi, B.K., Suryadevara, N.K., Sharma, S.K., Singh, A.P. (eds.) Proceedings of ICETIT 2019. LNEE, vol. 605, pp. 805–817. Springer, Cham (2020). https://doi.org/10.1007/978-3-030-30577-2_72
12. Mihindukulasooriya, N., García-Castro, R., Gómez-Pérez, A.: LD Sniffer: a quality assessment tool for measuring the accessibility of linked data. In: Ciancarini, P., et al. (eds.) EKAW 2016. LNCS (LNAI), vol. 10180, pp. 149–152. Springer, Cham (2017). https://doi.org/10.1007/978-3-319-58694-6_20
13. Riccardo Albertoni, A.I.: Data on the web best practices: data quality vocabulary. W3C Recommendation (2016)
14. Sharma, K., Marjit, U., Biswas, U.: Automatically converting tabular data to RDF: an ontological approach. Int. J. Web Semant. Technol. **6**, 71–86 (2015)
15. Umbrich, J., Neumaier, S., Polleres, A.: Quality assessment and evolution of open data portals. In: 2015 3rd International Conference on Future Internet of Things and Cloud, pp. 404–411. IEEE (2015)

16. Vaidyambath, R., Debattista, J., Srivatsa, N., Brennan, R.: An intelligent linked data quality dashboard. In: AICS 27th AIAI Irish Conference on Artificial Intelligence and Cognitive Science. pp. 1–12 (2019)
17. Zaveri, A., Rula, A., Maurino, A., Pietrobon, R., Lehmann, J., Auer, S.: Quality assessment for linked data: a survey. Semant. Web **7**(1), 63–93 (2016)

Spatio-temporal Data Sources Integration with Ontology for Road Accidents Analysis

Artem Volkov[1], Nikolay Teslya[2](✉), Georgy Moskvitin[1], Nikolai Brovin[1], and Evgeny Bochkarev[1]

[1] ITMO University, Kronverksky pr., 49 A, 197101 Saint-Petersburg, Russia
[2] SPC RAS, 14th line 39, 199178 Saint-Petersburg, Russia
teslya@iias.spb.su

Abstract. Within a smart city concept, it is possible to combine a large number of information sources that has spatio-temporal characteristics. The complexity of such a combination lies in the high heterogeneity of information, the need to use spatial and temporal characteristics, as well as formats for presenting such information. To date, there are information platforms for smart city sources organization that allow combining heterogeneous data sources, however, in the process of combining, human participation is still required to establish an unambiguous correspondence and process space-time characteristics. The paper proposes an approach based on a microservice architecture, in which each data source is mapped to a microservice, which presents an ontological data model of the source. When forming a query, data is sampled from sources and integrated based on spatial characteristics for subsequent analysis. As an example, the paper provides an analysis of data on road accidents in St. Petersburg, Russia since 2019 in order to determine accident-dangerous sections of roads and the main causes of accidents. The result is accidents clusters obtained by combining accidents data, road types and weather conditions in the accidents area.

Keywords: Smart city system · Spatio-temporal data · Integration · Open data sources · Open weather map · Traffic incidents

1 Introduction

In many tasks related to the development of a smart city, it is required to analyze spatial and temporal data to predict how the decisions made will affect the current situation [5,13]. An example of such an analysis would be the analysis of road network traffic [8] for planning new or renovating existing infrastructure; recommendation of attractions for tourist routes [24], the use of spatial data for the analysis of objects on the record, including outside the visibility of surveillance cameras [12,17]; researching the needs of residents based on their feedback

The reported study was funded by RFBR, project number 20-07-00904.

W. Abramowicz et al. (Eds.): BIS 2021 Workshops, LNBIP 444, pp. 251–262, 2022.
https://doi.org/10.1007/978-3-031-04216-4_23

and messages on social networks [7]. For all the tasks mentioned, it is required to analyze data in relation to space and time from a variety of sources in order to track the dynamics of the situation in city districts.

This paper discusses the integration of data sources for the formation of recommendations to the vehicle driver, aimed at preventing road accidents. For this purpose, the selection and merging of data sources containing the history and description of road accidents in the region, weather conditions, and the state of infrastructure is carried out. Each data source is integrated into a common system through a microservice that provides an ontological description of the source and acts as a mediator between the integration services and data sources.

The method of ontological modeling will be used to describe the sources of open data that are linked to space and time. Its use will make it possible to create an ontological description of the source containing, in addition to the properties of the source itself, data properties with semantic links between them. The presence of semantic links will provide a more complete description of the source, close to the description of the problem area under consideration. A review of existing research has shown that this approach has worked well for all types of data sources. The analysis of accidents in St. Petersburg, Russia was chosen as an example for implementation. Based on the extracted data, clusters were built, combining road accidents depending on the type of road and the number of victims.

The rest of the paper is structured as follows. Section 2 provides an overview of data access methods. Section 3 describes an architecture of proposed approach. Section 4 describes an experiment, including data sources, information model. Section 5 provides results and discussion over the experiment results.

2 Related Works

2.1 Review Open Data Access Methods

Aim of smart city system is to connect heterogeneous parts of city infrastructure for its optimization through data analysis. The analysis effectiveness is directly correlated with quantity and quality of input data. Taking into account that government sources require special access permissions, the best way to build smart city is the usage of open data sources. This method prevents re-creating already existing reference information and helps to give a quick start to any project [16]. However, when using open data, it is important to choose the best way to work with them. The problem is that access to the open data sources as well as their format, are not fully standardized and the presetting phase can take a long time [15].

The smart city system functions can be divided in independent categories based on data sources [11]. For example, the visualization and monitoring tasks require real-time access to data. But historical data better suits for the analytical and predictive tasks. Based on this logic, the data sources can be divided into two types: historical and real-time and data processing should be carried out separately depending on the type.

Data can be characterized by volume and relevance. Accordingly, to evaluate access methods, it is necessary to rely on how this affects the data [14]. Access methods evaluation parameters are listed below.

1. Availability of the whole data set.
2. Possibility to get data slice.
3. Data update frequency.

Common for all access methods parameters are not used in classification as they cannot help in outlining boundaries. These are the parameters: time range, limit of number of requests and spacing. The comparison of the open data access methods is presented in the Table 1.

Table 1. Open data access methods comparison.

Access methods	Availability of the whole data set	Possibility to get data slice	Data update frequency
Direct access to labeled data source	Full access	Yes	Infrequently
Unlabeled source processing	Partial	No	Frequently
Streaming data	Partial	Partial	Real-time
Hybrid method	Full access	Yes	Real-time

A synonym to the unlabeled source processing method is web scraping. The approaches with direct access to the labeled source and with streaming data are opposite and independent in nature. However, they can be combined within a single method. In this case, it is possible to get the entire amount of data and build a slice on it using static tables [21]. For visualization and getting actual values, it would be better to use data in real time.

3 Horizontal Scaling of Data Sources in the Smart City System

3.1 Problem

The effectiveness of a smart city system directly correlates with the amount of input data. With a large number of open data sources, the question arises of how easy it is to add another data source to the system. This task can be defined as horizontal scaling of data sources. There are centralized and distributed ways of storing information during its aggregation [9].

The centralized method is the simplest. It is based on saving of all data in one common database. Data processing is carried out in one application. The method is suitable for small projects. However, as new sources are added, the complexity of the system will begin to grow non-linearly. This leads to an increase in the costs of maintaining the system associated with a variety of options for describing data and their processing [10]. All these options are required to be described in the context of one program and one database. This approach requires a universal processing algorithm for all data sources, which in some cases is not an optimal solution.

The distributed approach implies the division of data storage and processing programs into independent parts. It simplifies the parallel development and connection of new data sources to the system. However, this also increases the complexity of organizing interaction between independent parts, especially in comparison with the centralized approach.

3.2 Architecture

Microservice architecture is one of the variants for implementing the distributed approach. For easy scaling, each data source is separated into a separate microservice along with an application. At the center of the system is a microservice that is aimed at interacting with the user. This microservice can be designated as the system kernel [20].

The kernel is a search engine, a single-entry point for users. Data analysis is not performed in the kernel, since this will happen already in the microservices. Each microservice is a standalone application. The microservice does not redirect the request to another microservice.

A distributed approach to the aggregation, storage and processing of data from heterogeneous sources is preferable for the smart city system. This allows clear separation of data and makes it easier to add new sources. With the further development of the system, if new sources of geospatial data appear, it will only require setting up interaction with the custom module.

4 Road Accidents Data Retrieval in the Area of Saint Petersburg, Russia

Recommendation systems such as an advanced driver-assistance system (ADAS) use information of the condition of the driver e.g., trying to catch drowsiness and show a recommendation about places to rest. The system can detect dangerous behavior via analyzing the trajectory of vehicles movement. There are also systems which scan the outside environment and raise alarm when leaving line or car is to close to the leading transport.

Such systems have been already implemented in modern cars which can provide information about current road signs and displaying this information on the windshield. Road signs can be both regulating road traffic and having a recommendation nature. As examples of such combining tasks can be the sign with a recommended speed in rainy weather or a warning about approaching a dangerous section of the road [6]. Usually, these signs are installed if a major accident has occurred on the road or if an accident of the same type occurs frequently. Unfortunately, these road signs are not installed at all dangerous places for many reasons from bureaucracy to inaction of road services.

The open data source of road accidents allows to avoid all barriers for determining problem areas on the roads and delivering the recommendations to drivers [23]. Actual current problem places can be detected by analyzing the road accidents and presented to the user through an application on a mobile phone or navigator.

4.1 Data Retrieval

The task is to get information from different resources. Combining all data sources together, we can get complete information about road accidents. Data sources are enumerated below.

1. Main Directorate for Traffic Safety of the Ministry of Internal Affairs of Russia statistics website.
2. Portal “Safe Roads”.
3. Map “OpenStreetMap” (Provide information about road types).
4. Weather service “OpenWeatherMap”.

Accident card information is obtained from the open source data service [1]. This web-service was analyzed to find out the REST API endpoints which allow to obtain the necessary information. The data is returned in not well-formed JSON format, thus additional operations with data are required to cast them to valid JSON after which all fields can be extracted and stored in database. Due to the fact that in the system there is a high probability of a human factor the data should be validated before the extraction process.

Weather information is taken from the service [2] via their Historical API. It allows to get weather data by coordinates and time. During the extraction of information about the data on the accident cards, additional information about the weather is requested from this resource according to the received coordinates and time. The output is returned in a JSON format in the body of the REST API response. Data schema of traffic accident cards is described below (Fig. 1).

The data obtained from the traffic police statistics portal is not guaranteed to be correct. Therefore, it is necessary to validate, clean and correct it before analysis. The following errors were most often encountered in the received data: blank fields, information inappropriate to the field (for instance, text information in the house number field), incorrect coordinates. The validator program checks data at the stage of decomposition of the received accident cards into separate fields, the correspondence of data types to their fields and subsequent correction. For instance, missing information is supplemented to the baseline values which based on the analysis of other road accident cards and completely inappropriate to the field is deleted. Incorrect coordinates are not corrected but they are not used for the data analysis as the geozone is specified. It is planned to integrate with the Google Geocoding API service. The service can correct the address

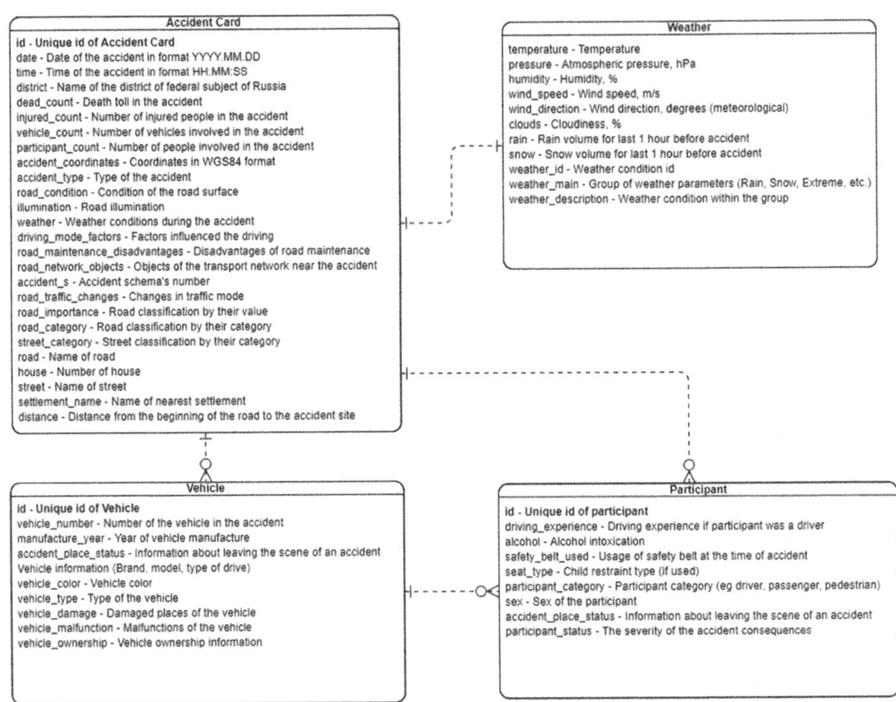

Fig. 1. Entity-relationship diagram with fields descriptions

information based on the available information about the address of the accident site. If it is possible to obtain reliable coordinates, the system compare them with those in the accident card.

4.2 Special Data Characteristics

Spatio-temporal (ST) data has their own characteristics and ignoring the nature of it could lead to poor results [4]. There are two properties which make a ST data domain differ from other data domains: autocorrelation and heterogeneity between data instances. Presence of autocorrelation in ST Data domains is caused by the fact that data instances could be not independent to each other. Observations which are carried out in nearby spatial or temporal or both dimensions can correlate to each other. The second property is that ST data instances can belong to different plurality of instances.

ST datasets can show heterogeneity (or non-stationarity) both in space and time in varying ways and levels. For instance, satellite measurements of vegetation at a location on Earth show a cyclical pattern in time due to the presence of seasonal cycles. But the classical data-mining approach assumes that data instances are homogeneous (stationary). Therefore, trying to apply to ST data the classic data-mining approach, such as clustering, could lead to poor results.

A way of representation of ST is another issue for analysis. Contrary to classical data which are described via well-defined set of features the ST data can be described in variety ways. For instance, in the ST dataset spatial values which can be defined as distinct locations could be represented as time objects. Or otherwise, the time slices could be taken as objects and spatial information as object features. As well as other data the ST data can contain non-ST features in addition to time and location [18].

The non-ST variables provide additional information about each data instance: types (e.g. type of disease or crime), quantitative characteristics, such as citizens' population in the particular region and so on. These features allow us to present data in a variety of ways. The finite decision how to represent data depends on the field of the data and the final goal of the research.

The provided data is a set of facts of road accidents which can be characterized as events with time and location where and when accidents occur. As it mentioned, the data provided as events in particular location and time with many additional features. Analyzing all sets of features is complicated and hard to interpret results. Therefore, in this research we focus on accumulations of road accidents. Thus, the result of the study will be the places of occurrence of types of accidents, depending on the condition of the road surface.

4.3 Ontology Description

In order to aggregate different data sources, it is necessary to have an idea of the internal structure and what information contained in each element of sources. A solution is to present the source metadata as an RDF graph [19]. The road accident has an event nature and represents a point in space and time. In this way the initial graph has two vertices: a location and a timestamp. The graph is extended when new data sources have new information about the road accident domain, or remains unchanged in case there is no to add. The merging process is based on searching the closest hyponym vertices. In this work ontology is manually created for road accident cards (Fig. 2). In the future, the process of creating and expanding an ontology will be automated. And the expert's work will be to check the system operation.

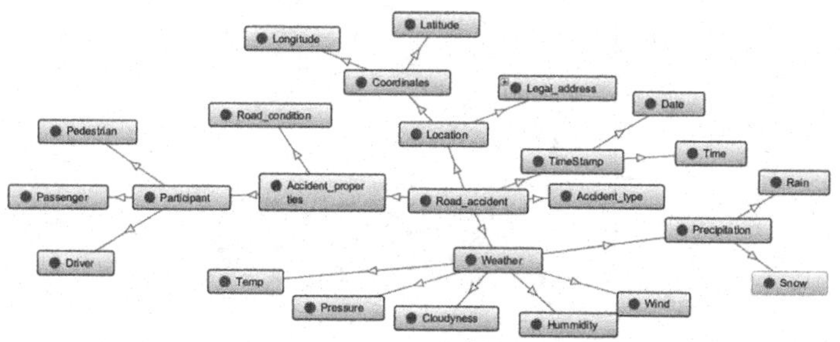

Fig. 2. RDF graph of road accident domain

5 Road Accidents Data Analysis

Data scrapper and data provider for another microservices in the system are combined together into one web-service. This service is written in Java using Spring framework and following libraries: Hibernate, Gson, Lombok. Data is stored in PostgreSQL database with the PostGis plugin. All interactions with service are performed via REST API.

As a starting dataset, accident cards were received for the last 2 years for Saint-Petersburg and Leningrad region. Number of received cards is approximately 12 000.

In software implementation QGIS for points visualization and PostgreSQL for data storage are used. The task was divided into the development of four microservices: a microservice for obtaining weather information, a microservice for obtaining traffic incidents cards, clustering microservice and a visualization microservice.

Clustering algorithm is written in Python using K-means algorithm. The algorithm minimize the total squared deviation of cluster points from the centers of these clusters. The algorithm stops when there is no change in the intracluster distance on some iteration. So we need to minimize lost function:

$$J(C) = \sum_{k=1}^{K} \sum_{i\,\in\,C_k} ||x_i - \mu_k||^2 \rightarrow \min_{C} \tag{1}$$

where C - set of clusters with power K, μ_k- center of cluster C_k.

Firstly we have tried to create the space clustering using coordinates [22]. But we have to know a number of clusters. We chose the number of clusters from which the loss function hardly changes. Figure 3 shows us the loss function and we decided that number of clusters is 20.

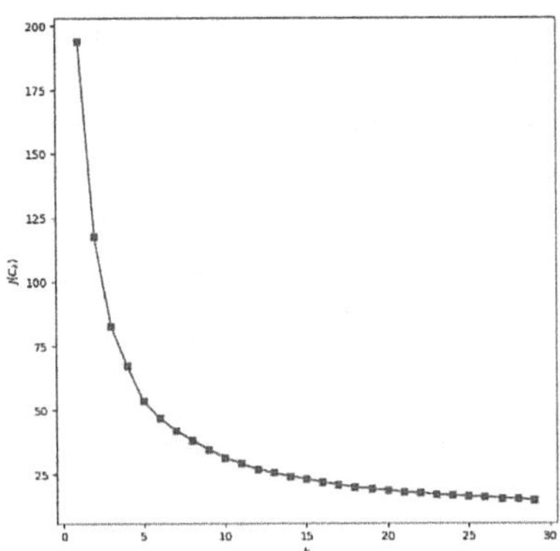

Fig. 3. Loss function $J(C)$

Space based clustering is presented on Fig. 4. Space cluster areas duplicate state districts. For rural areas clusters are mostly concentrated near federal routes. The more branched route net the more distributed cluster.

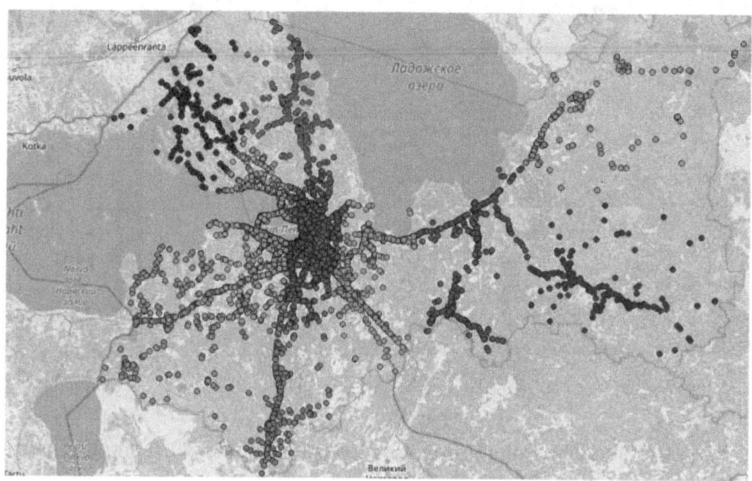

Fig. 4. Traffic incidents clustering map based on space with big clusters

Also we tried to increase the amount of clusters (Fig. 5).

You can see the result of vehicle based clustering of Saint-Petersburg accident cards in the following Fig. 6. Red dots represents one vehicle, blue - two vehicles, yellow - three vehicles.

Concentrations of one vehicle accidents can be explained as bad road conditions or this accident is a hitting a pedestrian. Two vehicle accidents can be explained as low organization of oncoming traffic. It is possible to reduce such accidents by dividing traffic. Also we can see a difficult crossroads, streets and a highway exit.

During the experiment it was discovered that some of the points are bad geoencoded. Particularly longitude and altitude are mixed up. As the result in Saint-Petersburg geozone there are points from India and other Russia regions.

Data analysis can be performed on a huge number of accident card parameters including such parameters as: weather and road conditions, brand and condition of a car, traffic rule violations. The system is developed as an open-source project [3]. It has been partially implemented and is under an active development. Data analysis is carried out by working with the database directly and there is no open access to the data due to the risk of overloading the system with queries. User web-interface is in the development plan.

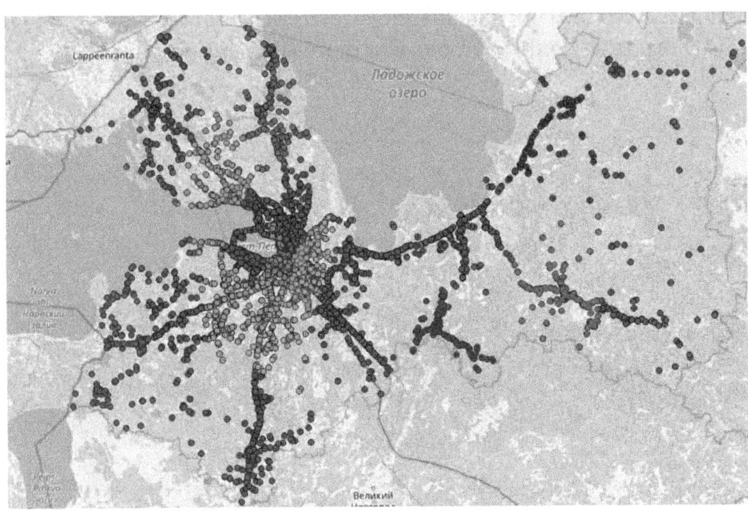

Fig. 5. Traffic incidents clustering map based on space with small clusters

Fig. 6. Traffic incidents clustering map based on involved vehicles (Color figure online)

6 Conclusion

The main methods of access to open data are considered, their advantages and disadvantages are given. The access method affects the processing and storage of data, as well as the possible range of tasks to be solved. Therefore, before choosing a data access method, it is important to determine what problem needs to be solved. Then, how relevant and how much data is needed. The frequency of data updates also plays an important role. All this refers to the requirements for solving the problem.

Without working out the questions of the final result in the middle of the implementation, it may turn out that the chosen method is not suitable. The choice of a particular data access method affects the architecture of the system. This means that changing the method entails high costs starting from the first step of the implementation.

The direction of further work is to connect more data sources and use more parameters for clustering. Accident cards contain information about the condition of the road surface, the condition of the driver, the number of passengers, the type and make of the vehicle, the type of accident, which can also be the basis for new clusters and show interesting results.

References

1. Main directorate for traffic safety of the ministry of internal affairs of Russia statistics homepage. http://stat.gibdd.ru. Accessed 16 Apr 2021
2. Openweather homepage. https://openweathermap.org/. Accessed 16 Apr 2021
3. Road accident analysis source code. https://cais.iias.spb.su/gitlab/smartcity/geosources/road-accident-analysis. Accessed 16 Apr 2021
4. Atluri, G., Karpatne, A., Kumar, V.: Spatio-temporal data mining: a survey of problems and methods. ACM Comput. Surv. **51**(4), 1–41 (2018). https://doi.org/10.1145/3161602
5. Babar, M., Arif, F.: Smart urban planning using Big Data analytics to contend with the interoperability in Internet of Things. Future Gener. Comput. Syst. **77**, 65–76 (2017). https://doi.org/10.1016/j.future.2017.07.029
6. Barrachina, J., et al.: VEACON: a vehicular accident ontology designed to improve safety on the roads. J. Netw. Comput. Appl. **35** (2012). https://doi.org/10.1016/j.jnca.2012.07.013
7. Dilawar, N., et al.: Understanding citizen issues through reviews: a step towards data informed planning in smart cities. Appl. Sci. **8**(9), 1589 (2018). https://doi.org/10.3390/app8091589
8. France-Mensah, J., O'Brien, W.J.: A shared ontology for integrated highway planning. Adv. Eng. Inform. **41**, 100929 (2019). https://doi.org/10.1016/j.aei.2019.100929
9. Gaur, A., Scotney, B., Parr, G., McClean, S.: Smart city architecture and its applications based on IoT. Procedia Comput. Sci. **52**, 1089–1094 (2015). https://doi.org/10.1016/j.procs.2015.05.122, The 6th International Conference on Ambient Systems, Networks and Technologies (ANT-2015), the 5th International Conference on Sustainable Energy Information Technology (SEIT-2015)

10. Gomes, V.C.F., Queiroz, G.R., Ferreira, K.R.: An overview of platforms for big earth observation data management and analysis. Remote Sens. **12**(8), 1253 (2020). https://doi.org/10.3390/rs12081253
11. Haslam, P.A.: Bigger data and quantitative methods in the study of socio-environmental conflicts. Sustainability **12**(18), 7673 (2020). https://doi.org/10.3390/su12187673
12. Jin, W., Zhao, Z., Li, Y., Li, J.I.E., Xiao, J.U.N., Zhuang, Y.: Video question answering via knowledge-based progressive spatial-temporal attention network. ACM Trans. Multimedia Comput. Commun. Appl. **15**(2s), 1–22 (2019). https://doi.org/10.1145/3321505
13. Khan, M., Babar, M., Ahmed, S.H., Shah, S.C., Han, K.: Smart city designing and planning based on big data analytics. Sustain. Cities Soc. **35**, 271–279 (2017). https://doi.org/10.1016/j.scs.2017.07.012
14. Kucera, J., Chlapek, D., Klimek, J., Necasky, M.: Methodologies and best practices for open data publication. CEUR Workshop Proc. **1343**, 52–64 (2015)
15. Malik, K., Sam, Y., Hussain, M., Abuarqoub, A.: A methodology for real-time data sustainability in smart city: Towards inferencing and analytics for big-data. Sustain. Cities Soc. **39**, 548–556 (2017). https://doi.org/10.1016/j.scs.2017.11.031
16. Ojo, A., Curry, E., Zeleti, F.A.: A tale of open data innovations in five smart cities, pp. 2326–2335, January 2015. https://doi.org/10.1109/HICSS.2015.280
17. Olszewska, J.I.: Detecting hidden objects using efficient spatio-temporal knowledge. Representation **10162**, 302–313 (2017). https://doi.org/10.1007/978-3-319-53354-417
18. Po, L., Bikakis, N., Desimoni, F., Papastefanatos, G.: Linked data visualization: techniques, tools, and big data. Synth. Lect. Semant. Web Theory Technol. **10**(1), 1–157 (2020). https://doi.org/10.2200/s00967ed1v01y201911wbe019, https://www.morganclaypool.com/doi/10.2200/S00967ED1V01Y201911WBE019
19. Ramar, K., Mirnalinee, T.: A semantic web for weather forecasting systems, pp. 1–6 (2014). https://doi.org/10.1109/ICRTIT.2014.6996127
20. Rodriguez, J.A., Fernandez, F.J., Arboleya, P.: Study of the architecture of a smart city. Proceedings **2**(23), 1485 (2018). https://doi.org/10.3390/proceedings2231485
21. Schauppenlehner, T., Muhar, A.: Theoretical availability versus practical accessibility: the critical role of metadata management in open data portals. Sustainability **10**(2), 545 (2018). https://doi.org/10.3390/su10020545
22. Sun, Y., Wang, Y., Yuan, K., Chan, T.O., Huang, Y.: Discovering spatio-temporal clusters of road collisions using the method of fast Bayesian model-based cluster detection. Sustainability **12**(20), 8681 (2020). https://doi.org/10.3390/su12208681, https://www.mdpi.com/2071-1050/12/20/8681
23. Wu, H., Zhong, B., Medjdoub, B., Xing, X., Jiao, L.: An ontological metro accident case retrieval using CBR and NLP. Appl. Sci. **10**(15), 5298 (2020). https://doi.org/10.3390/app10155298, https://www.mdpi.com/2076-3417/10/15/5298
24. Zhang, W., Gu, T., Sun, W., Phatpicha, Y., Chang, L., Bin, C.: Travel attractions recommendation with travel spatial-temporal knowledge graphs. In: Zhou, Q., Miao, Q., Wang, H., Xie, W., Wang, Y., Lu, Z. (eds.) ICPCSEE 2018. CCIS, vol. 902, pp. 213–226. Springer, Singapore (2018). https://doi.org/10.1007/978-981-13-2206-8_19

BITA Workshop

BITA 2021 Workshop Chairs' Message

A contemporary challenge for enterprises is to keep up with the pace of changing business demands imposed on them in different ways and especially through digitalization. There is today an obvious demand for continuous improvement and alignment in enterprises, but many organizations lack proper instruments (methods, tools, patterns, best practices, etc.) to achieve this. Enterprise modeling, enterprise architecture, and business process management are three areas belonging to traditions where the mission is to improve business practice and business and IT alignment (BITA). BITA is often manifested through the transition of enterprises from one state (AS-IS) into an improved state (TO-BE), i.e. a transformation of the enterprise with integrated digital solutions and platforms to meet future needs. A challenge with BITA is to move beyond a narrow focus on one tradition or technology into multidisciplinary ecosystems. There is a need to be aware of and able to deal with a number of dimensions of the enterprise architecture and their relations in order to create alignment. Examples of such dimensions are organizational structures, strategies, business models, work practices, processes, and IS/IT structures. Among the concepts that deserve special attention in this context is enterprise architecture management (EAM). An effective EAM aligns IT investments with overall business priorities, determines who makes the IT decisions, and assigns accountability for the outcomes. IT governance is also a dimension that traditionally has had a strong impact on BITA. There are ordinarily three governance mechanisms that an enterprise needs to have in place, 1) decision-making structures, 2) alignment process, and 3) formal communications.

This workshop aimed to bring together people who have an interest in BITA. We invited researchers and practitioners from both industry and academia to submit original results from their completed or ongoing projects. We encouraged broad understanding of possible approaches and solutions for BITA, including EAM and IT governance subjects. Specific focus was on practices of business and IT alignment, i.e. we encouraged submission of case study and experiences papers.

The workshop received eight submissions. The Program Committee selected four submissions for presentation at the workshop.

We thank all members of the Program Committee, authors, and local organizers for their efforts and support.

Ulf Seigerroth
Kurt Sandkuhl

Organization

Chairs

Ulf Seigerroth	Jönköping University, Sweden
Kurt Sandkuhl	Rostock University, Germany

Program Committee

Alexander Smirnov	SPIIRAS, Russia
Henderik Proper	Luxembourg Institute of Science and Technology, Luxembourg
Hasan Koç	Berlin International University of Applied Sciences, Germany
Birger Lantow	University of Rostock, Germany
Michael Fellmann	University of Rostock, Germany
Hans-Georg Fill	University of Fribourg, Switzerland
Jānis Grabis	Riga Technical University, Latvia
Dominik Bork	TU Wien, Austria
Stijn Hoppenbrouwers	HAN University of Applied Sciences, The Netherlands
Janis Stirna	Stockholm University, Sweden
Nikolay Shilov	SPIIRAS, Russia
Marite Kirikova	Riga Technical University, Latvia

IT-Service Value Modeling: A Systematic Literature Analysis

Henning Richter(✉) and Birger Lantow

University of Rostock, 18051 Rostock, Germany
{Henning.Richter,Birger.Lantow}@uni-rostock.de
http://wirtschaftsinformatik.uni-rostock.de

Abstract. In 2020 the new ITIL v4 standard was introduced. ITIL standardization had and still has a big influence on how IT-Service Management is seen and performed in practice. Thus, the new standard is expected to have a high impact as well. A key element of ITIL v4 is the strong focus on Stakeholder Value in Service Design. Yet apart from ITIL, stakeholder orientation is a current trend in business analysis. This work provides a Systematic Literature Analysis with regard to approaches that allow value modeling for IT-Services. As a result, no approach that fits all requirements inherent in the ITIL v4 standard and IT-Service Design could be identified. However, a set of requirements that should be considered when developing methods, notations, and tools for IT-Service Value modeling is derived.

Keywords: ITIL v4 · Service modeling · Service value · Value stream modeling · Stakeholder value · Service Blueprinting · IT-Service Management

1 Introduction

In 2020 ITIL v4 was released. It primarily focuses on enabling IT-Service providers to respond to new stakeholder demands quickly and simply. According to [1], a company's purpose is to create value for its stakeholders. Everything a company does must serve (directly or implicitly) creating value for their stakeholders. ITIL has a strong industrial background, and it is likely that many companies will adopt the new version in order to improve their IT-Service Management capabilities. While ITIL v4 generally describes these capabilities and their integration, a concrete method or toolset for the integration of stakeholder value in service design is not provided. Even if an enterprise does not intend to implement ITIL v4, considering Stakeholder Value in IT-Service Management can improve demand orientation. This study investigates the current state of research with regard to modeling and analyzing stakeholder value for IT-Services based on a Systematic Literature Analysis. We defined two Research Questions:

RQ1: From an ITIL v4 perspective, which potential methods and notations for IT-Service value modeling and analysis are defined in the scientific literature?

W. Abramowicz et al. (Eds.): BIS 2021 Workshops, LNBIP 444, pp. 267–278, 2022.
https://doi.org/10.1007/978-3-031-04216-4_24

RQ2: What is the potential of found notations to serve value modeling for IT-Services from an ITIL v4 perspective?

RQ1 collects already existing methods in the area. Methods can be described at various granularities. Since ITIL v4 already describes the approach to a value-oriented IT-Service Management, future implementers benefit from approaches that are operationalized down to a level that allows tool support for modeling and analysis. Service Blueprinting, Stakeholder Value modeling, and Value Stream modeling have been selected as base modeling concepts because they are explicitly mentioned as such in the ITIL v4 standard [1]. RQ2 emphasizes on the fit of found notations (either as part of a method or not) to the requirements of IT-Service Management or ITIL v4, respectively.

Before describing the Systematic Literature Analysis in Sect. 3, we define important concepts with regard to value and service management from the perspective of ITIL v4 in Sect. 2. These concepts are also used to provide a structure for the analysis of found literature, where answers to the research questions are provided in the Sects. 4 and 5. This leads to a collection modeling requirements for IT-Service value modeling as it is intended by ITIL v4, provided by a summary and outlook in Sect. 6.

2 Important Concepts

Though there are other definitions of the following terms, this study uses the ITIL v4 concepts as a common ground because this might help practitioners to link the results to the demands of ITIL v4.

Stakeholder Value. According to [1,2], the term *value* is a set of perceived usefulness, importance, and benefits of something. Moreover, value is highly dependent on each relevant stakeholder individually. [1] specifies that value is created by active collaboration between the service provider, the service consumer, and other stakeholders. Each one of them can have one or more different and unique points of view for what they actually value. According to ITIL v4 [1], value can be material (e.g., money, saving costs, materials or products), or immaterial (e.g., good user experience), or as a special category - lower risks. In the past, value analysis was mostly focused on saving costs [2].

Value Stream. A Value Stream is a series of steps carried out by a company to create and deliver products or services to their consumers [1,2]. When structuring a company's activities as Value Streams, a clear overview is created, showing what the company actually delivers. Services can be analyzed for elements hindering the workflow and activities not adding any value. Such activities are commonly referred to as *"waste"* and should be eliminated. According to [1,2], Value Streams focus on the end-to-end flow of activity from demand to value. Value Streams are not processes, but processes can be referenced in a Value Stream. The processes are units of work (at different granularities or contexts). Value Streams use information provided by stakeholders as inputs or other Value Streams and use resources of service providers and service consumers to generate outputs required to create outcomes demanded by the stakeholders. Key objectives are maximizing value generation and minimizing waste.

Service Blueprinting. With [3], ITIL v4 recommends Service Blueprinting to model and understand the customer journey. A customer journey is the complete end-to-end experience customers have with service providers and their products offered through touchpoints and service interactions. A touchpoint of a customer journey is an event where a (potential) service consumer interacts with the service provider's products or resources. Service Blueprints can help to understand these touchpoints and interactions by visualizing them. Service Blueprints are architectural drawings depicting how the service should look like and what is required for service provision. The key elements of Service Blueprints are the line of interaction (pinpointing the direct interactions between the consumer and provider), the line of visibility (separating visible and invisible activities from the consumer's perspective), and the line of internal interaction (separating employees that are either directly or not directly supporting customer interactions).

3 Paper Selection for Analysis

A Systematic Literature Analysis has been performed in order to assess the state of research regarding methods for IT-Service Value Modeling. The analysis process is oriented along with the guidelines for a Systematic Literature Analysis presented by Kitchenham [4] and Webster [5]. The review process is divided into four different parts (see Fig. 1). The first activity is to identify conference series, journals, and catalogs that are likely to represent the state of the art of research on the topic of interest. Here, a base set of papers for review is extracted by keyword search. The second step is the exclusion/inclusion of papers based on title and abstract. Then, the remaining papers have to be classified, and data has to be extracted with regard to the research questions. The classification is based on the concepts that have been described in Sect. 2. The fourth and last step is to analyze the extracted data. The first two steps are described in this section. Data extraction and analysis will be part of the following sections that answer the Research Questions.

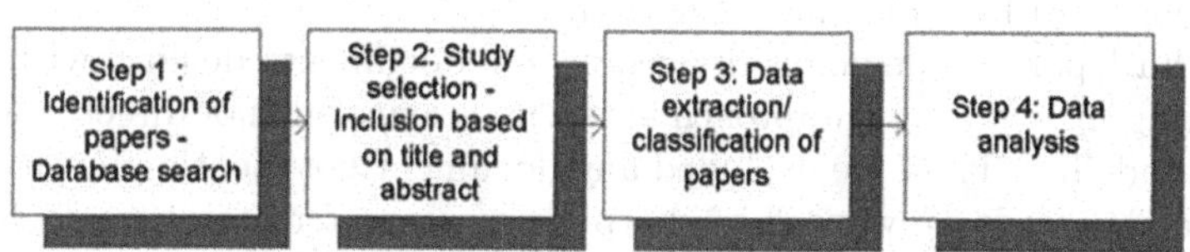

Fig. 1. Systematic literature analysis process [6].

The first step is the identification of papers dealing with methods for value and service modeling. The well-known portal for high-quality scientific literature indexing Scopus has been selected as literature source. The query is based on the main concepts described in Sect. 2. It contains four parts:

(1) *((TITLE ("value" W/1 ("modeling" OR "modelling" OR "designing")) AND ALL ("notation"))*

OR

(2) *TITLE ("value stream" AND ("modeling" OR "modelling" OR "designing" OR ("design*" W/3 "method")))*

OR

(3) *((TITLE ("service blueprinting") OR (TITLE("service blueprint*" AND ("modeling" OR "modeling" OR "designing")))) AND ALL ("value")))*

AND

(4) *PUBYEAR> 1999 AND (LIMIT-TO(SUBJAREA, "ENGI") OR LIMIT-TO(SUBJAREA, "COMP") OR LIMIT-TO(SUBJAREA, "BUSI"))*

The first part aims at notations in the area of value modeling and designing. Therefore we looked for paper titles that have the term "value" close to one of the terms "model(l)ing" and "designing". Hence, important literature like [7] with titles including *"value delivery modeling"* were found as well. The focus on notations is met by searching for the term "notation" somewhere in the document. Thus, all documents that refer to some notation for modeling or designing are fetched even when this is not explicitly mentioned in the title, abstract, or keywords.

The second part of the query is directly dedicated to value stream design methods. Since modeling is a part of the design process and sometimes authors do not clearly distinguish both, we also included modeling in this part of the query. Although we knew about the synonymous and very common term of Value Stream Mapping, Value Stream Mapping seems just to be a general approach and does not represent an actual modeling method with a specific notation. Thus, we stuck to documents explicitly mentioning *"value stream"*. Additionally, we considered documents introducing a design method where both of these terms are not separated by more than three words.

The third part implements the search for documents dealing with Service Blueprinting. It is a general approach in service science and not directly dedicated to value modeling. Thus, we excluded any document not mentioning the term of *value* somewhere, as we wanted to sharpen our scope for literature dealing with value.

The last part just limits the literature to the 21st century (to exclude documents presenting outdated solutions) and reasonable subject areas (to exclude documents, e.g. dealing with modeling mathematical values).

The results of the query have been verified the last time on April 23th 2021. In total, 42 possibly relevant papers have been identified (Part (1): 9, Part (2):

Table 1. Overview on the method classification and the literature associated

Domain	Method	Literature
Stakeholder Value Modeling	e3value	[14,15]
	Dynamic Value Description	[14]
	Possession, Ownership, Availability	[7]
	Value Delivery Modeling Language	[7]
Value Stream Modeling	Value Stream Mapping	[16–25]
	Risk Value Stream Mapping	[26]
	Dynamic Value Stream Mapping	[16]
	Value Stream Design 4.0/Value Stream Method 4.0	[27,28]
	Extended Value Stream Method	[29]
Service Blueprint Modeling	"traditional" SB	[30–32]
	IT Self-service Blueprint	[33,34]
	Industrial Service Blueprint	[35]
	Product Service Blueprint	[36]
	Flowthing Model	[37]

23, Part (3): 11). Six of these papers were not accessible for us due to missing licenses: [8–13]. Systematic Literature Analysis draws a sample of scientific literature. Therefore, missing out on some sources reduces the sample size and thus the potential significance of the study. Out of the remaining 36 papers, nine have been classified as not relevant to the topic, and two were not available in English. We carefully investigated the title and abstract of each accessible document. If title and abstract showed that the particular document is in a different domain, we classified it as not relevant (e.g., focusing on modeling chemical expressions or just focusing on business processes instead of an entire Value Stream). Furthermore, we scanned the remaining documents regarding their potential contributions to the Research Questions. We assumed any document suiting this purpose as relevant. In the end, 25 papers have been considered relevant and used for data extraction and analysis.

4 RQ1: Modeling Methods and Notations

As mentioned before, identified modeling methods and notations found by the different parts of the query are mapped to the concepts defined in Sect. 2. Table 1 provides an overview. The *Value Stream Mapping* method is by far the most often presented one for modeling Value Streams compared to the others. However, as already stated in Sect. 3, Value *Stream Mapping* turned out to be a very general approach that subsumes a variety of methods and notations.

In the following, we concentrate on the expressivity of the notations connected to the found methods. We used a classification based on the concepts presented in Sect. 2. For reasons of brevity, the visual notations are not shown. However, we refer to the abstract notation, meaning the concepts that are available in the different approaches and their semantics.

Notations for **Stakeholder Value Modeling**, as found by the first part of the query, mainly focus on the exchange of value between different stakeholders. Thus, provision and receipt of value is modeled for each stakeholder. Generally, value is considered as an economic value. Thus, it can be measured in monetary terms. For example, in the *e3value* notation, economically independent actors transfer value through value ports and interfaces by performing activities. Value interfaces group value ports providing or requesting value objects to or from actors or market segments. A market segment is a group of actors sharing a set of common properties (see e.g. [14]). Several authors make a reference to *e3value* in their approaches. Johannesson and Bergholz [15] extended the *e3value* notation with concepts that allow modeling rights, claims, or custody in addition to tangibles that can be exchanged. Souza et al. [14] define in their *Dynamic Value Description Method* a special view presenting value exchange stakeholder-centric using the same concepts as *e3value*. Souza et al. claim that this approach performs better than *e3value* in terms of effectiveness, efficiency, the perceived ease of use, usefulness, and intention to use. Another important addition is the introduction of value level agreements which emphasizes the immaterial value that is attributed to quality rather than the exchange of goods. However, there is no special concept for these quality aspects. They are described by comments in the model. The *Possession, Ownership, Availability* model by Scheller and Hruby [7] is similar to the approach by Johannesson and Bergholz [15]. There is more expressivity regarding the nature of exchange compared to *e3value*. Furthermore, concepts for modeling processes, resources, and events are added. In consequence, more information on how the value is created and consumed can be provided in a model. However, the focus is still on the value exchange. Value Delivery Modeling Language (*VDML*)[1] is a standard defined by the Open Management Group (OMG). Scheller and Hruby only mention it in [7]. Therefore, we used the official standard documents for analysis. *VDML* integrates several approaches of value modeling and connected domains, such as *e3value* and business models. It also addresses value stream modeling. It provides a general modeling framework that is intended for a high level of abstraction. Thus, the idea of non-economic value is supported by providing a general measure concept for value, but the definition of concrete value categories and their measurement is left open. In consequence, *VDML* can be used for an integrated model that addresses stakeholder value and value stream modeling. Still, it is less specific in terms of used modeling concepts and views than most of the analyzed approaches. An important addition of *VDML* stemming from the integrated Structured Metrics Metamodel[2] is the possibility of different measures to influence each other positively or negatively. Thus, different value propositions can influence each other positively or negatively as well in *VDML* models.

[1] https://www.omg.org/spec/VDML/About-VDML/.

[2] https://www.omg.org/spec/SMM/1.2.

Value Stream Modeling has its origins in lean production. As mentioned earlier, *Value Stream Mapping* is a general approach in this domain. Traditionally, there is a focus on cycle times in the process of value creation in the value stream. *Value Stream mapping* visualizes all required steps from demand over raw materials to product delivery. It aims at optimization in terms of reduced cycle times, idle times, transport times, and so on. Many authors [17–20,22–25] base their models on exactly these concepts - a value (good) that is created and the sequence of steps or processes for value creation containing information on timing. Extensions are suggested by Dotoli et al. [22], that used UML models for specifying single steps inside the Value Stream, and by Schoenemann et al. [21], who proposed to focus on the linkage between products and processes. Furthermore, Noto and Cosenz [16] combine *Value Stream Mapping* with Systems Dynamics in order to detect bottlenecks.

Besides these "traditional" *Value Stream Mapping* approaches, there are approaches that also consider information flows connected with the Value Stream and provide means to optimize them as well. The *Risk Value Stream Mapping* framework by Willumsen et al. [26] focuses on the information flow regarding risk and uncertainty. It defines key decision points and required information, allowing to assess the value of these risk-related activities but also to detect waste in terms of such activities that do not create value. A more general approach is the *Value Stream Method 4.0* for analyzing and designing Value Streams in industry 4.0 by Hartmann et al. [27,28]. It integrates product and process information flows. Furthermore, IT systems are explicitly modeled in the Value Stream. For optimization, information flow-related Key Performance Indicators (KPI) are defined. An example KPI is the Digitization Rate. The *Extended Value Stream Method* developed by Lewin et al. [29] provides additional expressivity with regard to the used IT-Infrastructure. It uses additional concepts dealing with IT infrastructure. The fundamental extensions are swim lanes assigned to different storage media or types of data usage, arrows showing flow direction, and data boxes with frequency or type of data acquisition. Additionally, it supports the concepts of raw data (e.g., data detected by a sensor), information (i.e., data which has already been processed and has information as content and a higher value), and implications (i.e., information that requires a response in the system).

Service Blueprinting is an approach for service modeling from management sciences. It focuses on service delivery as a sequence of interactions between involved stakeholders. Originally, neither value nor IT systems are considered. However, a *Service Blueprint* can help to pinpoint problems with perceived service quality and their causes. Generally, a *"traditional" Service Blueprint* contains customer actions, front-office/visible contact employee actions, back-office/invisible contact employee actions, support processes, and the physical evidence (service provider resources in the interaction) [30–32].

IT Self-Service Blueprints focus on the special domain of self-services. Suzianti and Chairunnisa [34] introduced the idea of *Service Blueprints* for customer self-service in the area of transportation. There is no notation provided by these authors. The adaption and extension to IT-Services has been developed by Bär et al. [33]. They introduced a model for different stakeholders and their capabilities. Depending on these capabilities, a different service quality can be expected. Problems with service quality are indicated by Fail Points. Examples for Fail Points are ambiguous information or forbidden actions. Further classification is possible with regard to the type of service (Communication/Information Seeking/...) and solutions to identified Fail Points (Training/Cashback/Authorization/...). *Industrial Service Blueprints* as introduced by Biege at al. [35] use a layer for each stakeholder instead of the five predefined layers of *"traditional" Service Blueprints.* Furthermore, there is a stronger focus on processes. Therefore, alternative and parallel control flows can be modeled using gateways, and multiple start- and endpoints are possible while *"traditional" Service Blueprints* just allow a sequence flow. The *Product Service Blueprint* approach was developed by [36] and addresses the trend of servitizing products. It consist of three parts – a product area, a service area, and a supporting area. The product area provides the flow of product-related activities, the service area describes the flow of service provision activities and the supporting area provides the underlying activities required to provide the product-service system at all. The *Flowthing Model* method was introduced by Al-Fedaghi [37] as an alternative approach to *Service Blueprinting* in general. It extends the *Service Blueprint* approach by concepts of BPMN which has some commonalities with the *Industrial Service Blueprint.* A major concept are so-called *"flow things"* and can be e.g. goods, money, information, or data. These things flow in a so-called *"flow system"* consisting of specific sub-areas dependent on the use case. The term *flow* refers to the transformation of a flow thing passing (flowing) through six states (also called stages) in a flow system. These stages are creating, transferring (input/output), processing, releasing, arriving, and accepting. Thus, there are some commonalities with the approaches discussed with regard to *Stakeholder Value Modeling.*

5 RQ2: Potential for IT-Service Value Modeling

Although *e3value* seems to be a well-known and commonly used method for modeling value, it does not suit the requirements of ITIL v4 perfectly. In *e3value*, the term of value is interpreted as a material good that can be transferred from one actor to another. However, ITIL v4 requires value to be an immaterial good (e.g. customer satisfaction) as well. Moreover, *e3value* does not highlight the perspectives of each relevant stakeholder involved (which is required by ITIL v4 as well). However, *Dynamic Value Description* delivers these perspectives by stakeholder-specific views. Still, the only possibilities in *Dynamic Value Description* to model immaterial value are the value level agreements and comments. Consequently, not all requirements of ITIL v4 are met. As mentioned in Sect. 4 *VDML* supports the required concepts and also introduces the modeling of conflicts in value delivery by the possibility of negative influences between value propositions. However, the concrete concepts and stakeholder specific views need to be defined in addition to *VDML*. In general, it is too abstract for analyzing the value delivery of a concrete IT-Service.

Value Stream Mapping does not provide any concepts of IT systems or information flows. Thus, it does not suit modeling Value Streams for IT-Services in compliance with ITIL v4. In extension, the *Risk Value Stream Mapping* method introduces very specific information flows for the field of Risk Management. It allows modeling Value Streams with risk information flows. Thus, the influence of risk on value creation as discussed in Sect. 2 is considered here. However, risk information is only one of many possible information flows. *Value Stream Method 4.0* provides integration of concepts for IT systems and information flows. Nevertheless, it does not provide concepts for modeling cycle times. Lastly, the *Extended Value Stream Method* shows similarities with the *Value Stream Method 4.0*. Both describe concepts for IT systems and their links to the other components of the Value Stream by using swim lanes.

The *"traditional" Service Blueprint* approach already seems to be a helpful tool to understand the relationships between the customers and the service components. It complies with the definition of a *Service Blueprint* by ITIL v4. However, the *IT Self-Service Blueprint* method seems to be an even better fit for ITIL v4, as it was explicitly developed for IT (self-)services. Thus, it already contains concepts required for IT-Services and thus also for ITIL v4 (e.g., concepts for IT-systems, actor capabilities or fail point solutions). In contrast, the *Industrial Service Blueprint* approach focuses on the industrial domain explicitly. Thus, it does not provide concepts for IT. Moreover, *Industrial Service Blueprints* are presented visually more like a process and not like a traditional blueprint overview. Thus, they do not provide a compact overview of the relationships between the customer and the service components. Consequently, *Industrial Service Blueprints* do not seem to suit ITIL v4 as well as previous approaches do. The *Product Service Blueprint* approach seems to have similar issues as well, as it primarily focuses on connecting products with services. Further, *Product Service Blueprints* do not seem to be as comprehensible as usual *Service Blueprints* are due to the additional product area. However, in IT-Services, the service itself is mostly the product sold, and a differentiation between product and service does not seem to be necessary. Still, the servitization of products is mostly based on a combination of products with IT-Services. In *Flowthing*, many new concepts were introduced to visualize the flow of things. These new concepts drastically increase the complexity of the *Flowthing* method compared to *"traditional" Service Blueprints.*

6 Summary and Outlook

This Systematic Literature Analysis drew a sample on the literature with regard to modeling Stakeholder Value, Value Streams, and Service Blueprints. Considering the analyzed literature, there is no approach that completely fits the requirements for IT-Service Value modeling. Each of the approaches provides well-justified concepts based on modeling requirements. Summarizing, the following requirements apart from the "baseline" of modeling can be identified:

Value for Different Stakeholders. The *Dynamic Value Description* and *IT-Self Service Blueprinting* provide stakeholder-specific models.

Quality Dependent Value. This concept has also been introduced by *Dynamic Value Description*. *Service Blueprinting*, with its focus on interactions and physical evidence, also supports this aspect by considering the perceived quality of service.

Conflicting Value Propositions. Emphasizing the optimization of a certain value or value category might diminish other values. This effect is considered in *VDML*, for example.

Value Based on Information Flows. While *Risk Value Stream Mapping* emphasises on risk-related information flows, *Value Stream Method 4.0* and *Extended Value Stream Method* generally introduce information flows.

Risk Dependent Value. *Risk Value Stream Mapping* allows the explicit modeling of risks and the analysis of value with regard to risk-targeted information flows as well as activities.

Alternatives in the Control Flow. Approaches like *Industrial Service Blueprints* and *Flowthing* use control flow gateways known, e.g. from BPMN.

Value Creation Through Multiple Processes and Resources. *"traditional" Service Blueprinting* defines five standard layers that describe different actors and their processes involved in service delivery. *Industrial Service Blueprints* extend this concept for an arbitrary number of actors. However, the *Possession, Ownership, Availability* approach includes resources as well. *Value Stream Method 4.0* and *Extended Value Stream Method* use swim-lanes to identify different actors.

IT Resources. *Value Stream Method 4.0*, *Extended Value Stream Method*, and *IT-Self Service Blueprinting* provide concepts to include IT resources in Value Stream and IT-Service modeling.

Considering all mentioned aspects, IT-Service Value models tend to be complex. Mechanisms for complexity management, as for example suggested by Moody [38] or the definition of modeling methods that adapt to the specific context by selecting appropriate method components and hence a subset of modeling concepts should be investigated. A study on the importance of each found modeling requirement for IT-Service Value modeling might also help to define a set of mandatory concepts. All this is future work as well as the provision of a modeling tool. Since none of the found approaches fits all the requirements, it is likely that also no appropriate modeling tool yet exists.

References

1. AXELOS: ITIL Foundation, ITIL 4 edn (German edn). The Stationery Office (2019)
2. AXELOS: ITIL 4 Managing Professional Create, Deliver and Support. The Stationery Office (2019)
3. AXELOS: ITIL 4 Managing Professional Drive Stakeholder Value. The Stationery Office (2019)
4. Kitchenham, B.: Procedures for Performing Systematic Reviews, vol. 33, pp. 1–26. Keele University, Keele (2004)

5. Webster, J., Watson, R.T.: Analyzing the past to prepare for the future: writing a literature review. MIS Quart. xiii–xxiii (2002)
6. Ivarsson, M., Gorschek, T.: Technology transfer decision support in requirements engineering research: a systematic review of REJ. Requir. Eng. **14**(3), 155–175 (2009)
7. Scheller, C., Hruby, P.: Business process and value delivery modeling using possession, ownership, and availability (POA) in enterprises and business networks. J. Inf. Syst. **30**(2), 5–47 (2016)
8. Tortorella, G., Pradhan, N., Macias de Anda, E., Trevino Martinez, S., Sawhney, R., Kumar, M.: Designing lean value streams in the fourth industrial revolution era: proposition of technology-integrated guidelines. Int. J. Prod. Res. **58**(16), 5020–5033 (2020)
9. Lim, T., et al.: An industrial case study on discrete event modelling of value stream mapping for industry 4.0. Int. J. Mechatron. Manuf. Syst. **13**(1), 90–110 (2020)
10. Stadnicka, D., Litwin, P.: Value stream mapping and system dynamics integration for manufacturing line modelling and analysis. Int. J. Prod. Econ. **208**, 400–411 (2019)
11. Becker, J., Beverungen, D., Knackstedt, R., Winkelmann, A.: E-Services in Retail - An Extended Service Blueprinting Approach, vol. 1, p. 335 (2009)
12. Polonsky, M., Garma, R.: Service blueprinting: a potential tool for improving cause-donor exchanges. J. Nonprofit Publ. Sect. Mark. **16**(1–2), 1–20 (2006)
13. Lian, Y.H., Van Landegheni, H.: Simulation-based value stream mapping the formal modeling procedure. EUROSIS 79–84 (2005)
14. Souza, E., Moreira, A., Araújo, J., Abrahão, S., Insfran, E., Silveira, D.: Comparing business value modeling methods: a family of experiments. Inf. Softw. Technol. **104**, 179–193 (2018)
15. Johannesson, P., Bergholtz, M.: Rights and Intentions in Value Modeling. Springer, Heidelberg (2010)
16. Noto, G., Cosenz, F.: Introducing a strategic perspective in lean thinking applications through system dynamics modelling: the dynamic value stream map. Bus. Process Manag. J. **27**, 306–327 (2020)
17. Mohammed Faisal, A., Ravi, A.: Simulation modeling and analysis for the implementation of total quality management using value stream mapping in labour-intensive small and medium-sized enterprises, pp. 1442–1444. Institute of Electrical and Electronics Engineers Inc. (2018)
18. Faisal, A.: Predictive simulation modeling and analytics of value stream mapping for the implementation of lean manufacturing: a case study of small and medium-sized enterprises (SMES), pp. 582–585. Institute of Electrical and Electronics Engineers Inc. (2018)
19. Faisal, A.: Simulation modeling and analysis of value stream mapping for the implementation of lean manufacturing in labour-intensive small and medium-sized enterprises, pp. 3567–3569. Institute of Electrical and Electronics Engineers Inc. (2016)
20. Azizi, A., Manoharan, T.: Designing a future value stream mapping to reduce lead time using SMED-a case study. Procedia Manuf. **2**, 153–158 (2015)
21. Schoenemann, M., Thiede, S., Herrmann, C.: Integrating Product Characteristics into Extended Value Stream Modeling, vol. 17, pp. 368–373. Elsevier B.V. (2014)
22. Dotoli, M., Fanti, M., Iacobellis, G., Rotunno, G.: A Lean Manufacturing Strategy Using Value Stream Mapping, the Unified Modeling Language, and Discrete Event Simulation, pp. 668–673 (2012)

23. Agyapong-Kodua, K., Ajaefobi, J., Weston, R., Ratchev, S.: Development of a multi-product cost and value stream modelling methodology. Int. J. Prod. Res. **50**(22), 6431–6456 (2012)
24. Xie, Y., Peng, Q.: Integration of value stream mapping and agent-based modeling for or improvement. Bus. Process Manag. J. **18**(4), 585–599 (2012)
25. Agyapong-Kodua, K., Ajaefobi, J., Weston, R.: Modelling dynamic value streams in support of process design and evaluation. Int. J. Comput. Integrat. Manuf. **22**(5), 411–427 (2009)
26. Willumsen, P., Oehmen, J., Rossi, M.: Designing Risk Management: Applying Value Stream Mapping to Risk Management, vol. 2019, pp. 2229–2238. Cambridge University Press (2019)
27. Hartmann, L., Meudt, T., Seifermann, S., Metternich, J.: Value Stream Method 4.0: Holistic Method to Analyse and Design Value Streams in the Digital Age, vol. 78, pp. 249–254. Elsevier B.V. (2018)
28. Hartmann, L., Meudt, T., Seifermann, S., Metternich, J.: Value stream design 4.0: designing lean value streams in times of digitalization and industrie 4.0 [wertstromdesign 4.0: Gestaltung schlanker wertstroeme im zeitalter von digitalisierung und industrie 4.0]. ZWF Zeitschrift fuer Wirtschaftlichen Fabrikbetrieb **113**(6), 393–397 (2018)
29. Lewin, M., Voigtlander, S., Fay, A.: Method for Process Modelling and Analysis with Regard to the Requirements of Industry 4.0: An Extension of the Value Stream Method, vol. 2017, pp. 3957–3962. Institute of Electrical and Electronics Engineers Inc. (2017)
30. Milton, S., Johnson, L.: Service blueprinting and BPMN: a comparison. Manag. Serv. Qual. **22**(6), 606–621 (2012)
31. Kostopoulos, G., Gounaris, S., Boukis, A.: Service blueprinting effectiveness: drivers of success. Manag. Serv. Qual. **22**(6), 580–591 (2012)
32. Bitner, M., Ostrom, A., Morgan, F.: Service blueprinting: a practical technique for service innovation. California Manag. Rev. **50**(3), 66–94 (2008)
33. Schoenwaelder, M., Szilagyi, T., Baer, F., Lantow, B., Sandkuhl, K.: It Self-Service Blueprinting a Visual Notation for Designing it Self-services (CEUR-WS), vol. 2238, pp. 88–99 (2018)
34. Suzianti, A., Chairunnisa, A.: Designing Service Blueprint of Self-service Technology (SST) Based Public Transportation Service in Indonesia Using SSTQual, Kano Model, and QFD, EDP Sciences, vol. 237 (2018)
35. Biege, S., Lay, G., Buschak, D.: Mapping service processes in manufacturing companies: industrial service blueprinting. Int. J. Oper. Prod. Manag. **32**(8), 932–957 (2012)
36. Geum, Y., Park, Y.: Designing the sustainable product-service integration: a product-service blueprint approach. J. Clean. Prod. **19**(14), 1601–1614 (2011)
37. Al-Fedaghi, S.: Alternative Approach to Service Blueprinting, vol. 2015, pp. 54–61. IEEE Computer Society (2015)
38. Moody, D.: The "physics" of notations: toward a scientific basis for constructing visual notations in software engineering. IEEE Trans. Softw. Eng. **35**(6), 756–779 (2009)

Consolidating Academic and Practical Guidelines for Digital Transformation

Ziboud Van Veldhoven(✉) and Jan Vanthienen

KU Leuven, 3000 Leuven, Belgium
ziboud.vanveldhoven@kuleuven.be

Abstract. Digital transformation is of paramount importance for companies nowadays but successfully doing so proves to be a challenging task. Researchers have proposed a myriad of guidelines on how to tackle digital transformation projects, but most of these are quite general and limited work has been conducted to compile these. In this paper, we aim to consolidate the numerous guidelines into one framework and expand upon each guideline with practical examples. The preliminary framework was validated by expert questionnaires and expanded with their recommendations on how to implement these guidelines in practice. In total, we list 78 guidelines structured in three levels: abstract, general, and practical. In future work, the preliminary framework will be further improved through surveys with special attention to real-life practical examples. This work can aid researchers and practitioners dealing with digital transformation.

Keywords: Digital transformation · Guidelines · Best practices · Digitalization

1 Introduction

Digital transformation (DT) is a fast-growing topic in information systems (IS), management, and business research in the past five years. The shifting paradigm towards a digital-first world brings forward a myriad of novel business opportunities but also numerous problems and challenges. Following this shift has quickly become one of the top priorities across many industries [1]. DT has many definitions but is commonly defined in the business scope as a process in which combinations of technologies aim to significantly improve an entity [2]. In a wider scope, DT can be defined as the increasing interaction between technology, business, and society which has transformational effects [3]. In DT, companies make the most out of the current technological capabilities by digitalizing their business processes and internal workings, improving the customer experience across all customer touch-points, and rethinking the business model [4, 5].

Planning and rolling out a DT plan is not easy [6–8]. Digital leaders are often looked at as the prime example of agile, collaborative, and digital-first thinkers. They rely on superior business and IT alignment (BITA) to embrace value, to improve customer experience, and to offer unique business models. Many companies aim to mimic their tactics, but this is neither always possible nor desirable. However, current research has brought forward numerous guidelines for businesses and managers to

W. Abramowicz et al. (Eds.): BIS 2021 Workshops, LNBIP 444, pp. 279–290, 2022.
https://doi.org/10.1007/978-3-031-04216-4_25

follow to successfully deal with DT. These guidelines differ widely, both on the level of generality and scope. Most of these guidelines rest on a certain level of abstraction. For example, a common recommendation for companies is to become more *agile* but how and whether to implement these general guidelines into practice are still insufficiently explored [9]. Furthermore, the proposed guidelines and best practices are spread among a wide range of papers, books, and reports. As a result, the available collections are often incomplete, overlapping, and inconsistently structured.

In this paper, we shine more light on the DT guidelines and their practical implementation. We aim to give a structured overview of the guidelines in three levels: abstract, general, and practical. As such, we attempt to bridge the gap between academic guidelines and practical implementation. The contribution of this paper is twofold. First, we consolidate the different guidelines for DT into one framework and secondly, we collaborate on how these guidelines can be implemented in practice. The obtained framework is beneficial for both researchers and practitioners active in DT.

We proceed with the background in which we go deeper into the commonly agreed upon guidelines for DT projects found in the literature. In Sect. 3, we explain our action-design research into more detail. Next, we give the preliminary framework in which the guidelines we found in the literature are combined with the findings from the interviews. We discuss the contributions, limitations, and future work in Sect. 5, and conclude the paper in Sect. 6.

2 Background

Many researchers have studied the concept of DT and derived recommendations for companies to follow. From qualitative research such as case studies, interviews, and surveys, research have proposed a myriad of guidelines. It is possible to structure these guidelines into three levels: abstract, general, and practical. The abstract level contains the suggested business areas a company must pay attention to when venturing on a DT process. On the general level, the business areas are given more concrete points of attention. How companies can implement these points of attention in practice is given on the practical level. The commonly agreed upon abstract guidelines suggested for companies are having a digital strategy, improve business agility, become innovative, have a modern organizational structure, have a digital and agile culture, have support from the top management, and have an adequate IT infrastructure. These form the abstract level, and the foundation of the framework.

The digital strategy often considered as a starting point for any DT efforts [10]. In general, the digital strategy must *align the IT and business strategy* and outline how the company will create value as digital technologies change the market [11]. In other words, the digital strategy formulates a *digital business model* [12]. Practical guidelines for the digital business model include customer engagement, or digitized solutions strategy [13]. The digital strategy must be *continuously reassessed* because the technological capabilities, the market, and customer demands are changing at a rapid pace [14]. Furthermore, it is generally agreed upon that a digital strategy must be *customer-driven* [15]. In practice, companies can do this by delivering personal, relevant, and timely information. In addition, effort must be made to integrate all customer touchpoints while some companies can think about co-creation projects with their customers. Finally, many researchers suggest the importance of *collaboration* with strategic partners in the digital world. One way to do so is to be active on digital *platforms*.

Business agility stems from many aspects of a company. Being able to quickly adapt to the changing customer requirements and competitive landscape has become of paramount importance. Therefore, common guidelines for business agility consist of *outsourcing* areas that are too slow to change [16], to empower employees to make more decisions [17], *agile development methods* such as DevOps and Scrum, and *cloud computing* to quickly adapt the IT capacity to the dynamic requirements of the market [18].

In a similar vein, innovation is becoming a key competitive differentiator. Companies must constantly experiment, assess, and innovate to stay competitive. At the same time, the standard value chains must be ensured. This is often referred to as business *ambidexterity*, i.e. establishing a balance between innovation and current processes [19]. To stimulate innovation, Matzler et al. [20] recommend investing or co-operating with *start-ups* to secure long-term innovation without interrupting the current business. Another strategy to increase innovation is to create a new division for innovation [21]. Lastly, some companies experiment with product innovation days. These can take form of hackathons, innovation jams, or offsite days for brainstorming [22].

Part of the business agility and innovation comes from the right business structure and culture. Modern organization structures that are focused on openness, collaboration, and agility are making their way to firms. For example, more than 70% of digitally mature companies rely on *cross-functional teams* to organize work [5]. These are groups of employees with different expertise working together towards clear goals. Several organizational models exist to implement these such as the Spotify model. To increase agility and innovation, common recommendations are to flatten the hierarchy so that decisions can be taken faster [22]. For example, Zappos is experimenting with holacracy: a business structure in which there are no titles and managers. To stimulate agility, business structures can be changed so that *collaboration* is more common by for example adopting open hours and free-roaming desks [23]. Finally, many researchers recommend firms to adopt a *digital division* that is tightly integrated with all divisions and leads the DT efforts [24].

A culture that stimulates innovation and agility, has several differences with traditional corporate culture. First, *change* must become accepted as an inherent part of everyday life [25]. Reward systems that offer monetary or intrinsic rewards to promote change are being evaluated. Another difference is that companies are moving away from rigid-rules to value-based rules [26]. Furthermore, researchers recommend employing a *digital mindset* in which novel technologies are embraced such as data analytics [2]. Often, this requires offering extra digital training. Finally, companies must give employees the chance for *continuous personal development* [27]. A practical guideline to achieve this is offering time-off hours for personal development.

The *support* of top management is especially important for DT initiatives [27]. Sing and Hess [28] recommend that companies create a *chief digital officer (CDO)* that leads the DT initiatives. It has been suggested that top management must lead in a sympathetic, supportive, and proactive way [6, 23]. The CDO, or other DT leader, must have sufficient authority and work in close collaboration with the CEO. One of the tasks of the top management is to clearly *create and communicate the business' vision* on digitalization efforts [29]. Finally, several studies have urged managers to create a *sense of urgency* to initiate DT [30].

Finally, DT relies on an adequate IT infrastructure. Previous studies have demonstrated the importance of an *operational backbone* that ensures the efficiency and reliability of core operations [13, 31], and of a *digital services platform* that enables the rapid development, implementation, and rollout of digital innovations [30, 31]. In addition, several studies have urged companies to create a *unified database* in which all the corporate and customer data is centralized [32]. Finally, the IT capabilities ought to support novel technologies that are beneficial for the company's digital transformation such as social media, mobiles, analytics, cloud computing, Internet of Things (SMA-CIT), 3D printing, and more. At the same time, security, privacy, and safety must be guaranteed by the IT infrastructure [2].

3 Methodology

To consolidate the various DT guidelines in a structured way, an action design research (ADR) approach was applied in five steps [33]. In the first step, we drew on the literature to establish the common guidelines for DT. The groundwork of the framework is based on the works of [2, 5, 34–36] and extended through backward and forward referencing [37] and a general literature search as seen in Sect. 2. From this, we constructed a preliminary framework in step two. In the validation phase, we distributed a questionnaire to seven DT experts asking about the completeness, correctness, and level of importance of the identified guidelines. After each questionnaire, the framework was iteratively updated. For future work, we will conduct a survey with companies from various sectors to find out how they implement these guidelines in practice, and whether these guidelines were found useful. In the final phase, a thorough analysis and interpretation of the survey results will take place and used to finalize the framework. The methodology is shown in Fig. 1.

Fig. 1. Five-step methodology. Step 4 and 5 are future work.

In detail, the validation with domain experts was done using a semi-structured questionnaire with a focus on deriving the validation and completeness of our framework and possible modifications [38]. The semi-structured questionnaire was distributed in the summer of 2020 by the head researcher. Of the 20 experts who were sent invitations to participate, seven responded. Applicants were required to be active in a domain related with DT with at least five years of experience. An overview of the questioned experts is shown in Table 1.

In the first part of the questionnaire, we asked about their position, experience, and key tasks related to their DT efforts. In the second part, experts were asked for their opinion on each guideline and whether the related guidelines were relevant and logical. The different general guidelines were also ordered based on perceived importance by the specialists. Most importantly, we asked to share how these guidelines can be put into practice. In the end, we asked the interviewees to elaborate on the framework's completeness with room for own suggestions.

Table 1. Overview of questioned experts.

	Firm	Position	Current DT tasks
E1	Public employment service	Director of architecture and innovation	Managing AI team, leading DT projects
E2	Management consulting	CIO	Guide companies in their DT projects
E3	Major supermarket	Project manager	Implementation of enterprise resource planning technologies, and document management systems (DMS)
E4	SaaS product consultancy	CEO	Guide companies in their DT projects
E5	Major Belgian bank	CDO	Manage DT projects, and data architecture
E6	Major technology hardware producer	Partner business manager	B2B strategic sales manager, strategy manager
E7	Business consultancy	Business architect and business owner	Managing digital products with cross-functional teams

4 Results

Following the literature review, an initial framework with the guidelines from the literature review was constructed. To assess the guidelines' completeness and correctness, seven DT specialists were asked to validate them. The validation was used to iteratively update and improve the framework, of which the end-result is shown in Table 2. The validation step revealed a general level of support for the framework. Most of the abstract and general guidelines from the literature were considered as 'important' to 'very important' by the experts. Several guidelines received mixed feedback. The perception of importance was used to rank the general guidelines from top to bot accordingly. Moreover, the experts provided additional guidelines focused on the practical level.

Table 2. The preliminary framework of guidelines for DT.

Abstract	General	Practical
Create a Digital strategy	Customer-driven	Give personal, relevant, and timely information
		Integrate all customer touchpoints
		Co-creation projects
	Continuous business and IT alignment (BITA)	Product focused mindset (E4, E7)
	Create a Digital business model	Customer engagement model
		Digitized solutions model
	Continuous reassessment	Bi-annual board meetings to reassess the roadmap (E5)
		Step-by-step implementation (E5)
	Strategic collaborations	Be part of platforms, ecosystem-thinking (E3)
Improve Business Agility	Use Agile development methods	DevOps, Scrum
	Utilize cloud computing	Step-by-step switching over (E5)
	Outsource inflexible areas	Rightsourcing (E2)
	Stakeholder management	Employee empowerment (limited to their domain) (E2, E5)
		Involve all stakeholders (customers, partners, suppliers) in DT projects (E4)
	Reduce administrative barriers (E6, E7)	Employee empowerment (E2, E5)
Increase Innovation and ambidexterity	Create a new division for innovation	Must be close aligned with the core business (E5)
	Invest in start-ups	Corporate venturing mindset (E2, E4)
	Be part of innovation networks (E2)	Follow social networks and innovation blogs (E2)
	Plan product innovation days	Hackathons, innovation jams, offsite days for innovation, innovation workshops (E2)
Modern Organizational Structure	Cross-functional teams	Spotify model
		Product- and feature-oriented (E4, E7)
	Stimulate collaboration	Open hours for divisions
		No assigned desks
	Create a flatter hierarchy	Holacracy (Zappos)
	A digital division to lead DT	
Digital and Agile Culture	Collaboration part of the culture	Reward system that promotes collaboration
	Change part of the culture	Reward system that promotes change
	Create value-based rules	Look at the result, not at hours worked (E6)
		Work-life-blend (E6)
	Create a digital mindset	Digital training
	Offer continuous personal development	Time-off hours for personal development
		Different generations in the same team to reduce the digital skill gap (E6)

(continued)

Table 2. (*continued*)

Abstract	General	Practical
Top management support	Create and communicate a digital vision	Stakeholder management (E7)
	Establish a sense of urgency	
	Supportive leadership	Lead in a participatory way
		Servant leadership (E1)
	Create a DT leader	CDO, CIO, or CEO responsibility
Adequate IT Infrastructure	Support for novel technologies	SMACIT, 3D printing
	An operational backbone	
	A digital services platform	IT as a product and revenue stream (E4)
	Data privacy and safety (E3)	Personal devices must be kept secure (E6)
	A unified Data Structure	Data ownership (E1)

The digital strategy was considered especially important. The most important guideline was found to be customer-driven followed by aligning the IT and business goals. E2 mentions that the digital strategy must be flexible "so that it can quickly be altered when new opportunities or challenges arrive". Putting the digital strategy into practice is not easy. Several experts said that a digital strategy must be implemented step-by-step and "that means not to change everything at once" (E5). Furthermore, businesses think too often that DT means automating the existing business processes (E1). What is important is to be pragmatic and value-thinking (E2). To reassess the digital strategy, one interviewee recommends having bi-annual board meetings in which the roadmap is evaluated and changed if needed (E5). The continuous evaluation of the results of the DT strategy is of paramount importance given the large uncertainty surrounding DT (E1).

Regarding business agility, agile development methods were sought highly important. One remark is that the agile methodology must be aligned with a long-term goal and the right architecture. "If the architecture is not right, a lot of value will be lost in integration-problem solving and rollout of products" (E2). Outsourcing, cloud computing, and employee empowerment have mixed responses. Regarding outsourcing, E2 recommends "rightsourcing". Only those capabilities that cannot be executed efficiently internally must be outsourced, and preferably as close as possible. Albeit the employee empowerment was considered beneficial for agility, E2 note that it must be limited to their domain and not at the decision level, otherwise this can lead to chaos. This was backed by E5 who said that empowerment must only be for decisions that must be taken fast and that do not have a big impact on the long term. For cloud computing, E2 says it can offer huge opportunities regarding scaling and rollout, but it is not always the cheapest option, especially not for large firms. Nevertheless, the added benefits of security, flexibility, and disaster recovery make cloud computing potentially interesting. Another important recommendation for business agility is to intensely involve all employees and stakeholders (E2). Secondly, companies should start step-by-step in their agile transformation and validate their efforts with clear key performance indicators (KPI) to see if agile has the right business effects. E6 and E7 add the general guideline of reducing administrative barriers. It is important to note that business agility is also highly dependent on the innovation measures, the business structure, and culture, explained below.

On the question of innovation and ambidexterity, the three proposed guidelines from the literature received mixed responses. First of all, it is important to establish that innovation is a continuous process just as DT. As a result, product innovation days were not marked very important. A new division of innovation can be useful. However, there is the risk that this will create resistance in the company (E6) and is in contrast to the cross-functional teams (E7). Investing in startups can be a good strategy. A starting point for this is introducing a corporate venturing mindset by investing in startups or scale-ups, becoming member or advisor of the board, collaborating by offering facilities and putting aside corporate venture capital budgets (E2, E4). Finally, another guideline is to follow and be part of social networks, blogs, and innovation networks that specialize in innovation in your sector (E2).

For business structure, all experts highlighted that cross-functional teams are very important. This guideline was marked the most important across the entire questionnaire. These teams must have clear goals, or *north stars*, rather than specific requirements (E4). Flattening the business hierarchy and stimulating collaboration were both considered important. One way to stimulate collaboration is to have a product- and feature-oriented mindset across the entire company with a clear role distribution (E4, E7). A product has a strategic, long-term objective, and is necessarily connected with multiple divisions. Furthermore, management and HR must work actively on modernizing the business structure using change management techniques (E2). A separate division to lead DT has mixed results. Two experts say it is not important at all (E2, E4), while the others indicate a medium level of importance. One of the experts with a negative attitude for a separate division explains that "DT does not have to be the role of an elite division but of the entire company" (E2).

All of the listed guidelines for digital culture were considered highly important. By including DT in the business strategy, it will naturally become an evident part of the business culture (E1). Collaboration was the most important guideline. This is in line with the cross-functional teams mentioned previously. In the second place is continuous development for employees in which they learn new skills and keep up with trends. When the cross-functional teams are responsible for their products, they must be given the freedom to learn the necessary skills to continuously navigate towards the *north star* of their product (cf. product-thinking mindset above, E4). A practical way to do this is by putting different generations together in cross-functional teams. This can reduce the digital skill gap between employees (E6). Regarding the digital-first mindset, one expert notes to not be overly confident in digital technologies: "a digital-where-suitable mindset is a better name" (E2). The real value comes from both human and digital contact. Value-based rules can help in a digital culture; managers must look at the result instead of the number of hours worked (E6). Some companies move away from strict office hours towards work-life-blend: being able to freely plan your day, mixing work, and private (E6).

As far as the top management is concerned, there is some debate about who should assume the role of DT leader. While the literature is generally in favor of establishing a CDO function, the expert interviews indicate a low level of importance for the function. E2 explains that there are already many chief executive functions, one of which can take the extra role of DT leader. Furthermore, the experts state that DT is the responsibility of the entire company (E7). Top management should, however, create and communicate the digital vision to all stakeholders. E6 notes that communication is not a one-time thing but must happen continually throughout the project. Besides, nearly all respondents agreed that creating a feeling of urgency is extremely important. Regarding the leadership philosophy, proactive and supportive leadership is generally considered important. E1 advises servant leadership, in which the goal of the leader is to serve in contrast to the traditional leadership where the goal is to lead and decide. In a similar vein, E7 stresses that leaders must not only talk about it but be an example themselves.

Lastly, the right IT infrastructure has been deemed as an essential requirement for DT. All the experts recommend an operational backbone and service backbone. Having a central database received mixed opinions. E1 states that "a central database is not realistic. The real question is about data ownership. In addition, the data platform must support the needs of all stakeholders such as the analytics and reporting teams". Furthermore, change for the sake of change is not needed. "Legacy software is working software", says E4, who continues by saying that moving everything to the cloud will create its own problems. The needed capabilities and faced challenges change all the time so it is often not worth the investment to change what works. In contrast, E2 recommends having a cloud-first strategy for the IT infrastructure because cloud software follows a lean and efficient IT structure. A practical guideline to keep in mind when changing the IT infrastructure is to take small steps. A large step requires a higher budget and is often slower in practice (E4). Another practical guideline is to move away from IT as a service to IT as a product and revenue stream. The IT infrastructure must also ensure security, data safety, and privacy (E3). E6 states that private devices of employees form the biggest risk and must be regulated.

5 Discussion

The contribution of this paper is twofold. First, we consolidate a wide range of guidelines in a structured framework. Practitioners and researchers can be informed through our framework about the general best practices recommended by researchers for DT projects. As such, the framework can be used as a checklist or as inspiration for DT practitioners. In addition, the structured framework can be useful for future research. Secondly, we received interesting feedback from DT experts regarding the applicability of the framework. In general, the common DT guidelines found in the literature are considered important but always require a degree of nuance. Investigating the generality of these guidelines is another area that can be explored by other researchers. The guidelines suggestions by the experts also highlight the discrepancy between the literature and practice. More research must be done in this area.

It is important to note the limitations of this framework. First, the framework forms a high-level generalization of common guidelines for DT. The exact implementation is firm-specific and might differ significantly between various industries. Due to the dynamic and complex nature of DT, there exists some overlap between the different guidelines and levels, and with other research areas such as project management. Secondly, the practical guidelines were only suggested by seven experts and require further investigations. Furthermore, DT is a complex process making it is hard to measure whether these best practices contribute to the successful implementation of any DT project. Investigating the effect of DT guidelines on DT success could be a fruitful area for further research. One interesting question to further investigate is whether these guidelines must be followed in a particular order and what alternative instruments can be used.

In future work, we plan to expand this framework by means of a large-scale survey across all types of DT practitioners. This way, we want to get a good overview of how these guidelines can be implemented in practice. We believe the practical side of DT is often underrepresented in the literature. In addition, the survey will be used to further update the framework with new guidelines, restructuring and reordering guidelines, and more examples.

6 Conclusion

In this paper, we consolidated various guidelines for DT found in the literature in a novel framework that consists of three levels: abstract, general, and practical guidelines. The framework was validated using expert interviews who expressed a general acceptance. Several changes were made to the original framework, together with several extensions regarding the practical guidelines. This framework is one of the first that consolidates the DT guidelines found in the literature with a focus on the practical implementation and can be used both by researchers and practitioners as explorative means when embarking on DT projects. In future work, a large-scale survey will be conducted to further investigate and enhance the framework.

References

1. Chanias, S., Myers, M.D., Hess, T.: Digital transformation strategy making in pre-digital organizations: the case of a financial services provider. J. Strateg. Inf. Syst. **28**, 17–33 (2019). https://doi.org/10.1016/j.jsis.2018.11.003
2. Vial, G.: Understanding digital transformation: a review and a research agenda. J. Strateg. Inf. Syst. **28**, 1–27 (2019). https://doi.org/10.1016/j.jsis.2019.01.003
3. Van Veldhoven, Z., Vanthienen, J.: Digital transformation as an interaction-driven perspective between business, society, and technology. Electron. Mark. **1**, 16 (2021). https://doi.org/10.1007/s12525-021-00464-5
4. Hess, T., Benlian, A., Matt, C., Wiesböck, F.: Options for formulating a digital transformation strategy. MIS Q. Exec. **15**, 123–139 (2016)
5. Kane, G., Palmer, D., Phillips, A.N., Kiron, D., Buckly, N.: Achieving digital maturity. MIT Sloan Manag. Rev. **59**, 1–29 (2017)

6. Ebert, C., Duarte, C.H.C.: Digital transformation. IEEE Softw. **35**, 16–21 (2018). https://doi.org/10.1109/MS.2018.2801537
7. Besson, P., Rowe, F.: Strategizing information systems-enabled organizational transformation: a transdisciplinary review and new directions. J. Strateg. Inf. Syst. **21**, 103–124 (2012)
8. Heavin, C., Power, D.J.: Challenges for digital transformation–towards a conceptual decision support guide for managers. J. Decis. Syst. **27**, 38–45 (2018). https://doi.org/10.1080/12460125.2018.1468697
9. Warner, K.S.R., Wäger, M.: Building dynamic capabilities for digital transformation: an ongoing process of strategic renewal. Long Range Plann. **52**, 326–349 (2019). https://doi.org/10.1016/j.lrp.2018.12.001
10. Goerzig, D., Bauernhansl, T.: Enterprise architectures for the digital transformation in small and medium-sized enterprises. In: Procedia CIRP. pp. 540–545 (2018). https://doi.org/10.1016/j.procir.2017.12.257
11. Kane, G., Palmer, D., Philips, A.N., Kiron, D., Buckley, N.: Strategy, not technology, drives digital transformation. MIT Sloan Manag. Rev. Deloitte. **27** (2015)
12. Remane, G., Hanelt, A., Nickerson, R.C., Kolbe, L.M.: Discovering digital business models in traditional industries. J. Bus. Strategy. **38**, 41–51 (2017)
13. Sebastian, I.M., Ross, J.W., Beath, C., Mocker, M., Moloney, K.G., Fonstad, N.O.: How big old companies navigate digital transformation. MIS Q. Exec. **16**, 197–213 (2017). https://doi.org/10.1017/S0021859600058731
14. Matt, C., Hess, T., Benlian, A.: Digital transformation strategies. Bus. Inf. Syst. Eng. **57**(5), 339–343 (2015). https://doi.org/10.1007/s12599-015-0401-5
15. von Leipzig, T., et al.: Initialising customer-orientated digital transformation in enterprises. Procedia Manuf. **8**, 517–524 (2017). https://doi.org/10.1016/j.promfg.2017.02.066
16. Ebrahimpur, G., Jacob, M.: Restructuring for agility at volvo car technical service (VCTS). Eur. J. Innov. Manag. **4**, 64–72 (2001). https://doi.org/10.1108/14601060110390558
17. Kotter, J.P.: Leading change: why transformation efforts fail. Harv. Bus. Rev. **12** (1995). https://doi.org/10.1016/0029-1021(73)90084-4
18. Bharadwaj, A., El Sawy, O.A., Pavlou, P.A., Venkatraman, N.: Digital business strategy: toward a next generation of insights. MIS Q. **37**, 471–482 (2013)
19. Schuchmann, D., Seufert, S.: Corporate learning in times of digital transformation: a conceptual framework and service portfolio for the learning function in banking organisations. Int. J. Adv. Corp. Learn. **8**, 31–39 (2015). https://doi.org/10.3991/ijac.v8i1.4440
20. Matzler, K., Friedrich von den Eichen, S., Anschober, M., Kohler, T.: The crusade of digital disruption. J. Bus. Strategy. **39**, 13–20 (2018)
21. Christensen, C.M., Raynor, M.E., McDonald, R.: What is disruptive innovation. Harv. Bus. Rev. 1–16 (2015)
22. Berghaus, S., Back, A.: Disentangling the fuzzy front end of digital transformation: activities and approaches. In: Thirty Eight International Conference on Information Systems. p. 17 (2017)
23. Guest, M.: Building your digital DNA: Lessons from digital leaders Contents The digital organisation. Deloitte. **51** (2014)
24. Neugebauer, R. (ed.): Digital Transformation. Springer, Heidelberg (2019). https://doi.org/10.1007/978-3-662-58134-6
25. Kane, G.: The technology fallacy: people are the real key to digital transformation. Res. Technol. Manag. **62**, 44–49 (2019). https://doi.org/10.1080/08956308.2019.1661079
26. Berman, S.J.: Digital transformation: opportunities to create new business models. Strateg. Leadersh. **40**, 16–24 (2012). https://doi.org/10.1108/10878571211209314

27. Kane, G., Palmer, D., Phillips, A.N., Kiron, D.: Winning the war for talent. MIT Sloan Manag. Rev. **58**, 17–19 (2017)
28. Singh, A., Hess, T.: How chief digital officers promote the digital transformation of their companies. MIS Q. Exec. **16**, 1–17 (2017)
29. Mugge, P., Abbu, H., Michaelis, T.L., Kwiatkowski, A., Gudergan, G.: Patterns of digitization: a practical guide to digital transformation. Res. Technol. Manag. **63**, 27–35 (2020)
30. Westerman, G., Bonnet, D., McAfee, A.: The digital capabilities your company needs. MIT Sloan Manag. Rev. (2012)
31. Ross, J.W., Sebastian, I.M., Beath, C., Mocker, M., Moloney, K.G., Fonstad, N.O.: Designing and executing digital strategies. In: 2016 Int. Conf. Inf. Syst. ICIS 2016. pp. 1–17 (2016)
32. Brock, J.-U., von Wangenheim, F.: Demystifying AI: what digital transformation leaders can teach you about realistic artificial intelligence. California Manag. Rev. **61**(4), 110–134 (2019). https://doi.org/10.1177/1536504219865226
33. Peffers, K., Tuunanen, T., Rothenberger, M.A., Chatterjee, S.: A design science research methodology for information systems research. J. Manag. Inf. Syst. **24**, 45–77 (2008). https://doi.org/10.2753/mis0742-1222240302
34. Morgan, R.E., Page, K.: Managing business transformation to deliver strategic agility. Strateg. Chang. **17**, 155–168 (2008)
35. Muehlburger, M., Rueckel, D., Koch, S.: A Framework of Factors Enabling Digital Transformation. In: Twenty-fifth Americas Conference on Information Systems. pp. 1–10 (2019)
36. Westerman, G., Calméjane, C., Bonnet, D., Ferraris, P., McAfee, A.: Digital Transformation: A Roadmap for Billion-Dollar Organizations. Capgemini Consulting (2011)
37. Webster, J., Watson, R.T.: Analyzing the past to prepare for the future: writing a literature review. MIS Q. **26**, xiii–xxiii (2002). 10.1.1.104.6570
38. Myers, M.D., Newman, M.: The qualitative interview in IS research: examining the craft. Inf. Organ. **17**, 2–26 (2007)

Integrated Security Management of Public and Private Sector for Critical Infrastructures – Problem Investigation

Thomas Rehbohm[1], Kurt Sandkuhl[1,4](✉), Clemens H. Cap[1], and Thomas Kemmerich[2,3]

[1] Institute of Computer Science, University of Rostock, Rostock, Germany
kurt.sandkuhl@uni-rostock.de
[2] Technologie-Zentrum Informatik und Informationstechnik, University of Bremen, Bremen, Germany
[3] Norwegian University of Science and Technology, Trondheim, Norway
[4] Jönköping University, Jönköping, Sweden

Abstract. The interaction between security management in public and private organisations includes complex challenges. In particular in critical infrastructure sectors, there is a need for instruments that enable the holistic and overarching management of private and public providers. Cross-organisational structures and processes should be defined, but are difficult to establish in federal governmental structures due to different legislative levels and scopes. The paper investigates this challenge using Germany and the Free Hanseatic City of Bremen as example.

The study proposes the development of an "Enterprise Architecture Framework" integrating and overarching the organizational structurers for both, a federal state, its municipalities and the (private) critical infrastructure providers in these municipalities. The main contributions of this paper are based on the results of an interview study. The interview partners were representatives of enterprises and public bodies covered by the federal IT security regulations. The contribution of the paper is the identification of security management challenges for services of general interest and how to increase the resilience of public service providers. Cybersecurity management in the context of public institutions is in focus.

Keywords: Cybersecurity · Security management · Services of general interest

1 Introduction

Cyber- and information security has become even more important in pandemic times. The current crisis impressively shows how systemically relevant the availability of digital services and information and communication technology (ICT) infrastructures are, even for countries or federal states such as Germany's Free Hanseatic City of Bremen, which serves as motivating case and subject of investigation in this paper. Despite this all-German case the findings of this paper are relevant for other states and contexts, as the challenge of cooperation between governmental structures (in our case in a federal state) and private companies being part of critical infrastructure sectors are similar in many democratic countries. Not only is much more data being processed - including sensitive

W. Abramowicz et al. (Eds.): BIS 2021 Workshops, LNBIP 444, pp. 291–303, 2022.
https://doi.org/10.1007/978-3-031-04216-4_26

data - but digital services and secure data networks are the backbone of society in the age of contactless working. Information security and data protection must meet these increased demands. Therefore, the basic values of information security, such as availability, integrity and confidentiality, are becoming increasingly important.

Using the example of the Bremen, the research work presented here aims to gain a better understanding of organisational structures and processes relevant to services of general interest, which particularly concerns the interface between organisations responsible in the public sector and private operators of critical infrastructures. For the Bremen federal state administration, not only its own infrastructures are of elementary importance, but also the functioning of the ICT infrastructures that enable a common life in the whole region. This includes in particular the "system-relevant" providers of services in the municipal environment. Such service providers were partially addressed in Germany by the Act to Increase the Security of Information Technology Systems (cf. [15]). With the following BSI Kritis[1] Ordinance, companies classified as critical infrastructure sectors and ensure supply for a certain size of population were obliged to implement requirements for the security of information technology systems [16].

However, there is no established procedure, best practice, public body, organisational structure or other instrument to support implementation of cyber-security for critical infrastructures when it comes to the cooperation between federal state, municipality and private sector companies. This gap has been addressed by research on Business and IT-alignment (BITA) [14], for example in [21], but is not fully covered yet. BITA is a continuous process aiming at aligning strategic and operational objectives and ways to implement them between the business divisions of an organization and the organization's information technology division. The concrete contribution of the research work is an empirical study among the operators of critical infrastructures in the Bremen in order to collect information on the security management of infrastructures relevant to the provision of public services that has been difficult to access for research to date. After a brief description of the theoretical background (Sect. 2), research methodology (Sect. 3) and state of research (Sect. 4), the hypothesis-driven survey is described in Sect. 5 and analysed in Sect. 6. Section 7 summarizes conclusions.

2 Theoretical Background

2.1 Enterprise Architecture Management and Business/IT-Alignment

In general, an enterprise architecture (EA) captures and structures all relevant components for describing an enterprise, including the processes used for development of the EA as such [1]. Research activities in EAM are manifold. The literature analysis included in [7] shows that elements of EAM [2], process and principles [3], and implementation drivers and strategies [8] are among the frequently researched subjects. Furthermore, there is work on architecture analysis [4], decision making based on architectures [5] and IT governance [6]. However, there is no specific focus on EAM use in the German public sector.

[1] BSI = Federal Office for Information Security. Kritis = Critical Infrastructure Protection.

The existing research on EAM and IT governance for the public sector to a large extent is specific to certain countries. This is obviously caused by the strong influence on national regulations and laws on governing structures, decision and implementation procedures, and policies. Much work for the public sector has been done in the USA (see, for example, [9]) which is due to the fact that the Clinger-Cohen act already in 1996 made it mandatory for public agencies to show how planned investments in information technology would improve efficiency and effectiveness. Similar regulations and related research can be observed in Australia (see [10]). Examples for EA in public administration in Europe can be found in Denmark, the Netherlands [11] and Finland [12]. However, due to the specifics of the German governmental system with its federal structure and a combination of state-level and federation-level regulations, the work from other European countries is not easily applicable in Germany.

Business and IT-alignment (BITA) in general is a continuous process aiming at aligning strategic and operational objectives and ways to implement them between the business divisions of an organization and the organization's information technology division [14]. Many challenges are linked to BITA since the business environment continuously changes and so does the IT in an enterprise, but the pace of change and the time frame needed to implement changes are different. Cyber-security management is considered as major driver of changes in both, business environment and IT. As a consequence; BITA is of high relevance in implementation of cyber-security. In this context, the integration of security management in public and private sector can serve as contributions to structuring the BITA.

2.2 Legal framework for Critical Infrastructure Sectors in Germany

With the Article Law on Increasing the Security of Information Technology Systems (IT Security Law) [15] Germany goes much further in the scope of application than the European regulations envisaged (cf. [20]). The legal framework is based on the findings and experiences from Germany's cyber security strategy from 2011, which was first published in 2016 and is currently being revised and updated. Due to the federal structures in Germany, the federal legislation for the federal state is only partially possible, as there is only limited regulatory competence for this.

For different sectors of critical infrastructure (Kritis) the Federal Government defined requirements for companies of general interest. The Federal Office for Information Security (BSI) was assigned the mandate to supervise the fulfilment of the requirements of the companies affected by the IT Security Act. The BSI provides for rights and obligations, in particular obligations to provide evidence, of the operators. Public administration in the federal states and municipalities are also belongs to the critical infrastructures, but could not be addressed by the federal legislator for legal reasons of the federal state structure. The companies identified in the Bremen are active in various Kritis sectors. Thereby, the number of companies is significantly smaller than the identified Kritis company parts. For instance, municipal water treatment and disposal are bundled in one company, whereas water distribution, due to regulation by the Federal Network Agency (BNetzA), is located in other companies. For the Bremen, there are currently less than 10 identified companies of general interest that fall under the federal law. This approach, based on self-disclosure, is not expedient for a federal state and falls short of increasing the resilience of critical infrastructure operators in the Bremen. The federal government and also the federal states must decide how systemically important companies are identified and addressed. Qualifying characteristics for federal states are preferable to a purely quantitative assessment.

If a company does not fall under federal regulation, there is currently no legislation that would oblige a company to report to a federal state. For operators of critical infrastructures, the BSI negotiates individual verification obligations with the respective company (for instance, introduction of management in accordance with ISO27001) or facilitates a sector-specific security standard (B3S) of the respective sector working group of a critical sector. The federal regulation of IT security for operators of critical infrastructures is currently under revision and has identified challenges in the federal states and municipalities that were not conclusively regulated by the law itself.

3 Research Approach

The work presented in this paper is part of a research project aimed at method and tool support for cross-organizational structures and processes between the public and the private sector supporting systematic implementation of cyber-security management in critical infrastructure sectors. The core of the method and tool support is expected to be an EA-based reference model for how to organize the cooperation between the different levels of public administration (i.e., federal state, municipality) and the private sector (i.e., enterprise in the critical infrastructure sector). The overall research project follows the paradigm of design science research [13]. This study concerns a step towards the explication of problems and elicitation of requirements for the envisioned design artifact: EA based reference model for integrated caber-security management. The work presented in this paper started from the following research question: *RQ: In the context of cyber-security in critical infrastructure sectors, what challenges are visible in the cooperation of public administration and private sector and what problems have to be addressed in method support?*

The research method used for working on this research question is a combination of literature study, expert interview and argumentative-deductive work. Based on the research question, we started identifying research areas with relevant work for this question and analysed the literature in these areas. The purpose of the analysis was to find existing approaches or theories supporting cyber-security integration in public/private sectors allowing us to study the problem relevance in detail. Since the literature study showed a lack of established approaches (see Sect. 4), we decided to focus on an expert interview in a real-world case for investigating the problem (see Sect. 5). The case is used to explore the existence and shape of the challenge of public/private sector integration.

4 State of Research

For a few years now, the number of research papers in the field of cyber and information security has been growing steadily. Due to the transposition of the Directive of the European Parliament and of the Council on measures to ensure a high common level of security of network and information systems [20] into national law, this topic is of high interest in the member states. In the context of the digitalisation of public administration and society, the results of the colloquium "Do we need a new art of statecraft" - challenges for state and administrative action due to digital development- are processed in a socio-political, state-theoretical, constitutional and administrative discourse [18]. A comprehensive presentation of IT security law and its current technical basis is provided in the legal handbook "Cybersecurity" [17]. In particular, it shows how responsible corporate management and supervision (corporate governance) and compliance with laws and regulations (corporate compliance) are to be interpreted in this context. IT governance comprises management and organisational models, which are supplemented by the aspects of conformity with rules on processes and law. The intended research of this work is to close the gap that arises in the federal design of the ambitious digitisation projects from the perspective of the federal state administration.

4.1 Literature Analysis

In addition to the book publications mentioned above, an examination of conference and journal publications is also necessary to analyse the state of research. To this end, a literature analysis was conducted in relevant German and English-language literature databases. Specifically, SSOAR (administrative sciences), EconLit and WISO (public service, business administration) as well as Scopus (various disciplines, mainly English) were selected. The following table shows the search terms, database and number of hits. The search was conducted in the title, abstract and keywords of the publication. The search was conducted at the End of January 2021.

Search term	Literature	Hit	Relevant
Daseinsvorsorge AND Cybersicherheit	Scopus	0	0
services for the public AND cybersecurity		0	0
Public services AND cybersecurity		12	0
Public services AND IT-securtiy		3	0
Public services AND information securtiy		43	0
Daseinsvorsorge AND (Cybersicherheit OR IT-Sicherheit OR Informationssicherheit OR IT-Sicherheitsmanagement)	EBSCO EconLit	0	0
(Kommune OR Bundesland) AND (Cybersicherheit OR IT-Sicherheit OR Informationssicherheit OR IT-Sicherheitsmanagement)		0	0
Daseinsvorsorge	SSOAR	59	0
Daseinsvorsorge UND Sicherheit	WISO	2	0

The 12 hits in Scopus on public services and cybersecurity concerned military aspects of cybersecurity or case studies in emerging countries. The 3 and 43 hits on IT security or information security in the public sector were also not relevant to our work, as they dealt with studies on the state of affairs in specific countries (e.g. Indonesia, Cameroon, Nigeria), specific aspects of IT security (e.g. access control or continuity), or aspects of continuing education or e-learning. None of the 59 hits in SSOAR for the search term services of general interest dealt with a topic in the context of IT security. The two articles in WISO focused only on security in healthcare.

Overall, no publication on cyber security or IT security and services of general interest could be found through the literature search.

4.2 Summary

Current cybersecurity research is very wide-ranging, dealing in particular with technologies, digital sovereignty and liability issues in the context of the European Union, regulation and national implementation. Cybersecurity research cannot be limited to one specific field only. In contrast, the critical infrastructure "public administration", together with the responsibility for municipal services of general interest, is little researched. Organisational structures should be researched, should be the focus of further development, in order to ensure the interaction of federal state government, municipalities and companies for the common good. It is about the interaction between law and information technology.

5 Interview Study

5.1 Methodological Approach

For the research in the field of cyber security, a study was conducted in the form of a guideline-based interview. The participants were identified through the notification of the BSI, to the coordinating bodies of the federal state (single point of contact), in this case the Bremen. The head offices for IT, the management and also the CISOs (Chief Information Security Officers) of the companies concerned were interviewed with defined and open questions. A total of nine interviews were conducted with the organisations concerned in a period of between one and four hours. The obtained and transcribed results were subjected to a qualitative content analysis according to [19]. The same text passages of identical main categories were summarised and refined by forming subcategories. The formation of subcategories corresponds to the induction process. In addition to the category-based evaluation, individual case interpretations were created along the main categories. The central research question is further defined by the interview study. Organisational structures or organisational architectures, as well as the accompanying legal design, should promote cooperation between the public administration and the business community in such a way that an exchange of content is made possible and legitimised, obligations are created where necessary and joint IT security status reports can be created.

5.2 Hypotheses/Theses

In order to be able to examine the central research question of this work in the interview study, theses were first developed which served as a basis for the definition of interview questions. The following 9 statements form an important input for the qualitative content analysis of the interviews.

1. "The companies have not been informed that they are affected!" The BSI Kritis Regulation did not lead to the automatic identification of the addressed companies according to coverage. The more time passed, the more companies found out in their compliance management that they were affected only by chance.
2. "The companies do not consider themselves to be part of the municipal utilities and are therefore not willing to recognise the potential impact of services of general interest!" Affected companies do not see their own service as elementary for the common good. Distribution effects in the region can compensate for any shortfalls at any time.
3. "The municipality is not seen as a stakeholder by business enterprises!" The federal state government or the municipality is not seen as an institution to be involved with business enterprises, rather it is seen as a disruptive factor. The more regulations are created, the more dependent business decisions are.
4. "Companies are of the opinion that IT is part of the normal core business that can be afforded by the company itself, this also applies to IT security (networking idea underdeveloped)!" IT infrastructures and services are part of the company's own value chain and must be located within the company. IT and information security can be managed within the company. The more complex the requirements become, the more there is a willingness to outsource.
5. "The development and transfer (stabilisation) of know-how can be achieved without outsourcing partners!" If there are requirements that force a qualification of employees, this is granted within the company. External consultations are only helpful to a limited extent.
6. "The risk in networking office IT and OT is underestimated by companies!" Networking (connecting) operational and office IT has more benefits for the organisation. Cybersecurity risks can be managed. In particular, the focus here is on the "creeping" networking of systems of operational IT, office IT and possibly further IoT systems.
7. "The choice of information security and data protection frameworks are inconsistent and influenced by several factors!" If there are external requirements, for instance from the Basic Data Protection Regulation, management systems are introduced. Implementing management systems voluntarily are only cost factors and have no added value. Are the classic functions of planning, organisation, personnel, management and control available in a form that is conducive to cyber security?
8. "Interdependencies were not taken into account!" The own service is not dependent on other service providers. A failure of suppliers or providers of ICT services is unacceptable and has a fundamental impact on value creation.

9. "The companies do not see the public authorities as partners, but as a source of information. They do not see any regulatory deficits on the part of the legislator!" The more regulatory requirements there are, the less advantageous this is for service provision. Non-binding networking with public administration institutions, including the exchange of current threat situations, can be beneficial.

The prepared 16 questions, which were to be used to investigate the hypotheses, were asked of all participants in the course of the discussion - but not chronologically. The results were written down during the interviews.

5.3 Interviewed Companies and Objectives of the Interviews

The operators of the critical infrastructures located in the Bremen, which were identified by the Federal IT Security Act, are not yet dependent on or even obliged to cooperate with the federal states and municipalities in terms of content due to the existing legal regulations. Operators of a critical infrastructure within the meaning of the law must, in particular, name a contact point to the BSI within six months of the BSI KritisV coming into force, via which they can be reached at any time. However, in the view of the federal state government, this circumstance is not to be considered appropriate, since the federal states were not informed or were only partially informed about the addressed companies. In order to better understand the needs of the operators and also of the federal state government or the supervisory authorities in municipalities within the administration, structured interviews were conducted with the entrepreneurs currently notified. These companies reported to the state are critical infrastructures that did not fall under existing special legal regulations. In particular, companies that fall under the Energy Industry Act, the Telecommunications Act or the Telemedia Act were not among the interviewees.

The companies surveyed are operators in Bremen and Bremerhaven that provide services in the following facility categories: Wastewater disposal, facilities for ordering, distribution and sale of food and raw materials, oil and product storage, water distribution network, control facilities, sewerage, wastewater treatment plants, logistics centres and hospitals.

These companies in the Bremen belong to the Kritis sectors: sewage treatment plants, food supply, fuel and heating oil supply, drinking water supply, sewerage, passenger and goods transport and stationary medical care. These twelve notified operators belong to fewer than 8 companies in the Bremen. In addition, there are operators who fall under the IT Security Act and are to be designated as municipal.

Telephone appointments were arranged with all operators identified in the Bremen and face-to-face appointments with individual operators. The telephone interview partners were, depending on the organisation and size of the respective company, the managing director, the IT security officer and/or the IT manager. The crisis/emergency management officer, the information security officer, the head of network operations and a member of the management were present at the face-to-face appointment. The survey was conducted in 2020.

Questions such as "Which regulatory authorities (local, state or federal state) do you have?", or "Are there established processes and reporting channels?" helped to understand the context of the organisation. Questions that approached the topic of critical infrastructure on a technical level (especially the operation of ICT systems) were intended to reveal whether the respective company operates particularly high-risk infrastructures (e.g. "Are operational information technology and office IT separate in your company?").

The aim of the interviews was to find out where comparable challenges exist among the operators, how the operators can imagine cooperation with the supervisory authorities in terms of content, and what regulatory deficits, training needs or further training requirements exist. From the perspective of the central information security management of the Bremen, the consideration of interdependencies in the context of service provision is also of elementary importance.

6 Results

The 9 hypotheses/theses are evaluated in detail below:

H1 The examination of whether the company is affected by the IT Security Act was carried out exclusively internally; in the broadest sense, compliance management is in place at larger companies. It can be assumed that many companies in the general interest are not yet aware of this. Extensive laws, regulations and controls are common in other subject areas (environment [water and emission protection], trade supervision). More extensive regulations are not advocated by any company. For instance, one interviewee replied: "It is not clear to us how the threshold values were arrived at. Although we are not a big supplier, in our company the annual production thresholds are reached on a weekly basis."

H2 The own understanding of service provision - in the context of services of general interest - is not very well developed. Only once was a comprehensive emergency management system found that also addressed cyber security risks. Although the ongoing pandemic did not begin until after the interviews, one company also had concepts for the operation of systems separated in terms of personnel and space. For instance, one interviewee replied meaningfully: "Our company is a global player that does not operate according to rules in Germany."

H3 Companies that perform public tasks in the broadest sense (e.g. water supply) have a much better understanding of cooperation with municipal public authorities and the relevance for services of general interest. Other sectors are only committed to their original business objective and do not see their service in the context of services of general interest. For instance, one interviewee replied: "We are in contact with Bremerhaven's municipal authorities and the federal state Criminal Police Office regarding threatening situations."

H4 Depending on the size of the company, the IT equipment and the associated IT security are developed very differently. Large companies rely on partners in the IT environment and have awareness-raising concepts for their employees. Networking with industry associations is well developed. Centralised (corporate) IT was also cited in one company to make it clear that necessary decisions are not made in Germany.

Smaller companies are not very networked, operate traditional IT environments and see themselves as committed to data protection goals at most. For instance, one interviewee replied: "The combination of regional and central partners is effective. A Security Operation Centre (SOC) is at the corporate level already established."

H5 Few companies had a training concept or tracked their training in a management system. When training is provided, it tends to be on an annual basis and often as an electronic and optional learning opportunity. On-, off- and change boarding processes for employees are established in the companies' HR management. For instance, one interviewee replied: "We have a network of internal and external partners, we are a global player! But it also varies from country to country. We have a global safety policy, as well as e-learning offers."

H6 The separation of operational and office IT varies greatly; the smaller the company and the resources available for IT, the higher the degree of networking and thus the risk appetite. In larger companies, complete separation of the IT systems can be found, up to and including regulated technical transitions. The danger of networking these two segments is perceived as a high risk. For instance, one interviewee replied: "Production software is integrated, but subject to special regulations with high technical requirements" or "There is only one IT, not connected in a separable way."

H7 There are few industry-specific security standards (B3S) published at the time of the interview. The BSI has often recognised ISO 27001 (native) as a framework, as evidence for the BSI KritisV (Guidance on evidence). Data protection management, on the other hand, is consistently established. For instance, meaningfully, one interviewee replied, "It is well known that you can participate or contribute to industry or thematic working groups. It's a resource issue."

H8 Public authorities are expected to provide a framework. No further explicit requirements or obligations are expected. Rather, voluntary cooperation is advocated. In the meantime, the Bremen federal state Criminal Police Office has contacted the operators affected by the IT Security Act and offered to exchange information. For instance, one interviewee replied: "Further regulations are not really desired. Companies like these, are active in all federal states (of Germany), networking would be overregulated. Too many contact persons."

H9 Very few companies have taken precautions to make themselves less dependent on supplier services in the ICT environment. Nor are most entrepreneurs aware of whom they are dependent on for performance (internet and infrastructures) in the first place. One single company (as mentioned above) is aware of almost all aspects of business continuity management (BCM), thanks to its well-developed emergency and disaster recovery management. Contractual arrangements have also been concluded with replacement service suppliers in this example. For instance, one interviewee replied: "There is complete dependency on ICT. A technical failure in Africa meant that we were no longer able to operate there for a week. ICT can only survive without partners at the head office in Denmark."

7 Conclusions and Future Work

The validity of the analysis is generally given by the use of several evaluators and multiple qualitative content analysis of the interview transcripts (data). For the existing data sets of the nine interviewed companies, this was postponed, as further surveys, especially supra-regional-municipal studies, will follow, which should also verify the present result. The transferability and possibility of generalisation already exists in the authors' view, although there is the possibility that the answers of the interview partners depended on the interview partners' "role" in the company. The very heterogeneous structure of the interviewed companies (local company to part of a group) shows that there are no uniform or comparable approaches or architectures in the organisation of the companies to adequately counter cyber security risks.

Technically and organisationally, the measures taken are mostly not effective. The larger and more decentralised the IT, the more agile and also sensitive the companies are with regard to the required cyber security. Smaller companies can be said to have an underdeveloped understanding of resilience issues. Companies currently affected by the IT Security Act are not fully covered, as the IT Security Act aims at "self-disclosure". The BSI has not reported several large companies in Bremen known to the public administration to the Senator for the Interior. A deficiency that can only be countered by identifying system-relevant enterprises oneself. The required screening of critical infrastructures must map such enterprises of the region or the federal state dependent services (interdependencies) must be fully worked through.

It should be noted that Bremen, like the other federal states, has not yet been able to create any binding structures to promote cooperation between the relevant stakeholders in the common good. The evaluation also shows that most of the Bremen companies reported so far do not have well-developed resilience structures. A common organisational architecture must be developed. Other federal states (such as Saarland, Lower Saxony and Saxony) have adopted legal foundations for information security in the federal state administration. In this context, cooperation with the public organisations can only be targeted on a voluntary basis.

Future Work: Based on the present study, supplemented by further municipal and country-specific studies, an enterprise architecture model is to be developed, which can be integrated into the existing management systems. The architecture must consider the following aspects: The public administration is to network the providers of services of general interest identified based on qualitative characteristics in, for instance, a "country-based alliance for cyber security". Rights and obligations of the participants are defined; in particular, the automated exchange of information, e.g. on cyber security risks and current threat situations must be made possible. The establishment of a cyber-security control centre, modelled on a Computer Emergency Response Team, should be established between stakeholders. At the same time, legal foundations must be created to oblige the providers of services of general interest in the federal state, especially because the tasks of disaster control, crisis management and hazard prevention are the original responsibility of a federal state.

References

1. Ahlemann, F., Stettiner, E., Messerschmidt, M., Legner, C. (eds.): Strategic Enterprise Architecture Management. Springer, Heidelberg (2012). https://doi.org/10.1007/978-3-642-24223-6
2. Buckl, S., Dierl, T., Matthes, F., Schweda, C.M.: Building blocks for enterprise architecture management solutions. In: Practice-Driven Research on Enterprise Transformation, Lecture Notes in Business Information Processing, vol. 69, pp. 17–46 (2011)
3. Glissmann, S., Sanz, J.: An approach to building effective enterprise architectures. In: HICSS 2011, IEEE Computer Society 2011, Washington, D.C., pp. 1–10 (2011)
4. Johnson, P., Lagerström, R., Närman, P., Simonsson, M.: Enterprise architecture analysis with extended influence diagrams. Inf. Syst. Front. **9**(2–3), 163–180 (2007)
5. Johnson, P., Ekstedt, M., Silva, E., Plazaola, L.: Using enterprise architecture for CIO decision-making. In: Second Annual Conference on Systems Engineering Research (2004)
6. Simonsson, M., Johnson, P., Ekstedt, M.: The effect of IT governance maturity on IT governance performance. Inf. Syst. Manag. **27**(1), 10–24 (2010)
7. Wißotzki, M., Sandkuhl, K.: Elements and characteristics of enterprise architecture capabilities. In: Matulevičius, R., Dumas, M. (eds.) BIR 2015. LNBIP, vol. 229, pp. 82–96. Springer, Cham (2015). https://doi.org/10.1007/978-3-319-21915-8_6
8. Szilagyi, T.: Enterprise Architecture Management and Digitization in the German Public Sector. Literature Analysis. Universität Rostock, Rostock (2018)
9. Pang, M.S.: IT governance and business value in the public sector organizations - the role of elected representatives in IT governance and its impact on IT value in US state governments. Dec. Supp. Syst. **59**, 274–285 (2014)
10. Ali, S., Green, P.: IT governance mechanisms in public sector organisations: an Australian context. J. Glob. Inf. Manag. **15**, 41–63 (2007)
11. Janssen, M., Hjort-Madsen, K.: Analyzing enterprise architecture in national governments: the cases of Denmark and the Netherlands. In: HICSS 2007, IEEE Computer Society 2007, Washington, D.C. (2007)
12. Valtonen, M.K.: Management structure based government enterprise architecture framework adaption in situ. In: Poels, G., Gailly, F., Serral Asensio, E., Snoeck, M. (eds.) PoEM 2017. LNBIP, vol. 305, pp. 267–282. Springer, Cham (2017). https://doi.org/10.1007/978-3-319-70241-4_18
14. Johannesson, P., Perjons, E.: An Introduction to Design Science. Springer, Cham (2014). https://doi.org/10.1007/978-3-319-10632-8
14. Seigerroth, U.: Enterprise modeling and enterprise architecture: constituents of transformation and alignment of business and IT. IJITBAG **2**(1), 16–34 (2011)
15. Bundesgesetzblatt. Gesetz zur Erhöhung der Sicherheit Informationstechnischer Systeme, 17 July 2015. Nordrhein Westfalen, Bonn, Deuschland (2015)
16. Katastrophenhilfe, B.F.: Handlungsempfehlungen für Unternehmen, insbesondere Betreiber kritischer Infrastrukturen (2021). https://www.kritis.bund.de/SharedDocs/Downloads/Kritis/DE/200302_HinweisePandemie.pdf?__blob=publicationFile
17. Wichum, R.: Cybersecurity. In: Kasprowicz, D., Rieger, S. (eds.) Handbuch Virtualität, pp. 669–680. Springer, Wiesbaden (2020). https://doi.org/10.1007/978-3-658-16342-6_36
18. Lühr, H.: Brauchen wir eine neue Staatskunst. Kellner Verlag, Bremen (2019)
19. Mayring, P.: Qualitative Inhaltsanalyse, Grundlagen und Techniken. Beltz Verlag, Weinheim und Basel (2008)

20. Rates, R.: Maßnahmen zur Gewährleistung eines hohen gemeinsamen Sicherheitsniveaus von Netz- und Informationssystemen, 06 July 2021. Europäische Union, Straßburg, Frankreich (2016)
21. Rehbohm, T., Sandkuhl, K., Kemmerich, T.: On challenges of cyber and information security management in federal structures-the example of German Public administration. In: BIR Workshops, pp. 1–13 (2019). http://ceur-ws.org/Vol-2443/paper01.pdf

Market Launch and Regulative Assessment of ICT-Based Medical Devices: Case Study and Problem Definition

Maciej Piwowarczyk vel Dabrowski[1,2] and Kurt Sandkuhl[1(✉)]

[1] Institute of Computer Science, University of Rostock, Rostock, Germany
maciej.piwowarczyk@fokus.fraunhofer.de,
kurt.sandkuhl@uni-rostock.de

[2] Fraunhofer FOKUS, Berlin, Germany

Abstract. The market launch and regulative assessment of ICT-based medical devices in Europe is very complex due to a multitude of regulations to be considered during requirements engineering and product management. Additionally, there are no established standards, best practices or support tools how to launch medical devices on the market. The paper is part of a project aiming for methodical support for medical device launch and assessment, and is dedicated to investigating problem relevance. To understand the processes and requirements three case studies were analysed and the necessary processes and requirements were matched towards enterprise architectures (EA). Based on this finding, we argue that EA could be a suitable way to visualize and recommend required processes and structures for medical device management. The main contributions of our work are (a) a literature analysis of EA use in health care and especially for telemedicine, (b) results from use case analysis investigating the business perspective from inside three health tech companies and (c) the analysis of problems of telemedicine integration into EA.

Keywords: Medical device · Market launch · Assessment process · Enterprise architecture

1 Introduction

Medical devices have been an increasingly important part of patient care for the practicing physician and the whole health system since many decades. Medical devices (MD) in general are instruments intended for application in diagnosis, treatment, cure, or prevention of disease (cf. Sect. 2.1) – ranging from simple stethoscopes to pacemakers and complex medical imaging or even drug-device combinations, like coronary stents containing antibiotics. Medical devices traditionally are subject to both national and international regulation, including premarket evaluation and approval processes, surveillance during the operation phase, and post-market evaluation. The focus of regulation essentially is on safety and effectiveness of the medical devices.

Of particular interest for our research are medical devices with a value proposition based on the use of information and communication technologies (see also Sect. 2.1). Examples are remote monitoring of health functions for patients at home, the use of

W. Abramowicz et al. (Eds.): BIS 2021 Workshops, LNBIP 444, pp. 304–315, 2022.
https://doi.org/10.1007/978-3-031-04216-4_27

mobile devices and software-apps supporting the treatment of chronic diseases, or wearables for the care of persons with dementia. Such medical devices cause additional challenges to enterprises planning, their development and market introduction. One challenge is additional regulation, for example from a data protection perspective, as the processing of personal patient information is required. Other challenges are caused from the business model and the technical architecture. Usually, not only the single devices are part of the product but even operations of the technical infrastructure for information transfer and processing, and the actual users often are not the customers but the public health system or health insurances are covering the costs.

In this context, we consider systematic and integral management of regulative compliance, business model and IT implementation as essential for successful medical devices. However, existing recommendations and best practices only cover specific aspects of medical product management, such as the technology assessment, privacy protection or business development (see Sect. 4). Based on observations in industrial cases and focus groups, and judging from an analysis of existing work, this area has not attracted much research so far. The long-term objective of our research is to develop a recommendation for processes, organizational structure and IT support in the field of medical device development. In this paper, the purpose of our research is– as a first step – to investigate the state of research and the relevance of the topic from a business perspective. Guiding question is "*In market launch and assessment of medical devises, what are specific problems and needs of enterprises with ICT-based medical devices?*".

The main contributions of our work are (a) a literature analysis of EA use in health care and especially for telemedicine, (b) results from use case analysis investigating the business perspective from inside three health tech companies and (c) the analysis of problems of telemedicine integration into EA. The rest of the paper is structured as follows: Sect. 2 summarizes the background and related work on telemedicine and EAM. Section 3 introduces the research methods used in the paper. Section 4 is focused on the expert interview and use cases we performed as part of our study with three health tech companies in the field of telemedicine. Section 5 shows the reason why EAM integration of requirements for telemedicine manufacturers is recommendable. Section 6 summarizes our findings and discusses future work.

2 Background and Related Work

2.1 Telemedicine and Medical Devices

As shown below, there are numerous definitions of the term "telemedicine". One of them is: *"Telemedicine is a collective term for various medical care concepts that have in common the principle approach of providing medical services for the health care of the population in the areas of diagnostics, therapy and rehabilitation, as well as in medical decision-making consultation over long distances (or time offsets). Information and communication technologies are used for this purpose."* [1, transl. by the author].

By summarizing other definitions, the term "telemedicine" can be defined as the use of audio-visual communication technologies for the purpose of diagnostics, conducting consultations, and providing emergency medical services when the parties involved are separated by a physical distance [2, 3].

The term "medical device" is defined in the EU Regulation 2017/745 also called Medical Device Regulation (MDR): *"'medical device' means any instrument, apparatus, appliance, software, implant, reagent, material or other article intended by the manufacturer to be used, alone or in combination, for human beings for one or more of the following specific medical purposes: - diagnosis, prevention, monitoring, prediction, prognosis, treatment or alleviation of disease, [...]"* (Art. 2 Para. 1 MDR).

It can be stated that the definitions of a medical device and telemedicine applications overlap, especially with regard to diagnostics, therapy and rehabilitation. Telemedicine applications can consist of various components and combinations of non-medical devices and medical devices. For telemedicine systems, software seems to be excluded as such according to the above-mentioned definition of accessories, but Recital (Rec.) 19, Sentence 2 MDR clarifies that software can also be considered as an accessory. Furthermore, guidance is provided by means of guidance and MedDev documents of the European Commission for the delimitation of the medical device properties of software: "*Medical device software is software that is intended to be used, alone or in combination, for a purpose as specified in the definition of a 'medical device' in the MDR or IVDR, regardless of whether the software is independent or driving or influencing the use of a device.*" (MDCG 2019-11). In recital 19 sentence 1 MDR repeatedly emphasizes that software intended for medical purposes is a medical device. It simultaneously restricts that "*software for general purposes, even when used in a healthcare setting, or software intended for life-style and well-being purposes is not a medical device.*"

Furthermore, software or apps without a medical purpose that are used purely for sports, fitness, wellness or nutrition can generally be assumed not to be medical devices [4, 5]. Although the guidance of the Federal Institute for Drugs and Medical Devices (Bundesinstitut für Arzneimittel und Medizinprodukte – BfArM) [4] was written before the MDR was published, it is still legitimate today, as there are only marginal differences between the English definition of a medical device in the MDR and in Directive 93/42/EEC, on which the German "Medizinproduktegesetz" (MPG) is based. A further aid to differentiating between software-based medical devices is a judgement by the ECJ. This declares software or a software module in which "one of the functionalities makes it possible to use patient data in order to determine, among other things, contraindications, drug interactions and overdoses" to be a medical device. This also applies if this software does not act directly in or on the human body [6].

On the basis of the facts on which the aforementioned judgment is based, it is evident, on the one hand, that the term "medical device" is to be interpreted very broadly. On the other hand, the judgement indirectly shows that the manufacturer himself can influence the distinction between products and medical devices by means of the intended purpose formulated by him. The *"'intended purpose' means the use for which a device is intended according to the data supplied by the manufacturer on the label, in the instructions for use or in promotional or sales materials or statements and as specified by the manufacturer in the clinical evaluation;"* (Art. 2 No. 12 MDR). If an official delimitation decision is made, usage information, the website and information from app stores are also used [4].

2.2 Complexity of the Market Launch of Medical Devices in Germany

The complexity of the European healthcare market is a barrier to market entry for numerous companies, because although there are numerous telemedicine solutions, they are not available nationwide or established in the provision of care due to the particular structure of the healthcare market. In addition, there are no universal ways for manufacturers to enter the healthcare sector [7].

Taking Germany as example, this statement is also in line with the conclusions based on a survey by Bitkom[1] in cooperation with BfArM[2] on the market launch of digital health products and market knowledge of the participants. This revealed that "*participants at all levels [of market launch] would like to see improved support for digital health products and thus also for their path to certification and reimbursement*" [8].

The interdisciplinary nature of telemedicine can be cited as an initial obstacle to market introduction. This leads to increased complexity in the implementation of telemedicine projects and requires close cooperation between the individual players, as different requirements are imposed from the medical, technical and regulatory sides [9, 10]. On the other hand, in addition to the usual business risks, manufacturers of medical devices are threatened with confiscation of the products by state authorities, up to and including imprisonment, if the products are not placed on the market in conformity with the law despite the complex legal situation (cf. § 93–95 MPDG and § 40–43 MPG). It should be noted that telemedicine systems are subject to constant changes and further developments due to rapid technical development, which represents a particular hurdle in the given requirements for digital medical devices [11, 12].

Table 1. Stakeholder and requirement levels of medical devices in telemedicine [18, 19]

Determinants	Influencing factors and barriers
Technology	
Acceptance	Patients
	Insurers
	Physicians
Financing	Business model
	Reimbursement
Organisation	Organisation
Policy and Legislation	Regulators
	Medical institutions

In order to speed up how medical device manufacturers can get into reimbursement models in Germany, the Federal Ministry of Health issued the "Digital Care Act" (Digitale-Versorgung-Gesetz - DVG), which has been in force since 01.01.2020 [13]. It

[1] Bitkom = Germany's association of Information and Communication Technology companies.

[2] BfArM = Federal Institute for Drugs and Medical Devices in Germany.

remains questionable to what extent manufacturers will succeed in overcoming the additional regulatory requirements of this law [14]. In fact, the success of this law can only be measured in a few years' time, as the market launch process of a digital health application in standard healthcare, including the preparation and application processes, can take longer than a year [15, 16]. Tanriverdi and Iacono summarise barriers to the diffusion of telemedicine systems based on use cases. They found that successful telemedicine implementations required consideration of organisation, remuneration and involvement of healthcare professionals, patients, insurers and regulation. Similar parameters on determinants of successful telemedical implementations are found by Broens et al. with the categories: Technology, Acceptance, Financing, Organisation and Policy and Legislation [18, 19].

The domains analysed are also provided for in Health Technology Assessments (HTA). In addition to medical devices, the HTA also evaluates medicinal products, medical procedures and analyses. The use of an HTA is recommended during the development of new technologies, as this enables active innovation management to determine the benefits and quality of the health system and to be able to introduce effective technologies [17]. According to international consensus, the following aspects are the subject of an HTA: safety, efficacy, cost and economic evaluation, ethical analysis, organisational aspects, social aspects, legal aspects [17].

3 Research Approach

The work presented in this paper is part of a research aimed at technological and methodical tool support for medical device launch, assessment and monitoring. We envision the development of a (normative) recommendation how to integrate the required management processes into an organisation, including the business perspective and supporting information systems and technologies. An enterprise architecture (EA) based representation of such a recommendation seems possible. The work follows the paradigm of design science research [23]. This study concerns a step towards the explication of problems and elicitation of requirements for the envisioned design artifact: EA reference model for medical device launch, assessment and management. The work presented in this paper started from the following research question: *RQ: In the context of medical device launch and assessment, what regulative and business challenges are visible in industrial practice and what problems have to be addressed in method support?*

The research method used for working on this research question is a combination of literature study, case study and argumentative-deductive work. Based on the research question, we started identifying research areas with relevant work for this question and analysed the literature in these areas. The purpose of the analysis was to find existing approaches or practices supporting medical device launch and assessment and real-world case studies allowing us to study the problem relevance in detail. Since the literature study showed a lack of established approaches and many open challenges (see Sect. 2), we decided to focus on case studies from companies working with medical device development and market introduction for investigating the problem (see Sect. 4). The cases are used to explore the existence and shape of the challenge of medical device launch and assessment.

4 Case Study on Medical Device Assessment

4.1 Case Study Design

In order to investigate challenges in market launch and regulative assessment of ICT-based medical devices, we had the possibility to study three start-ups. The researchers collected information about requirements, work processes, technologies used and practices by taking part and organising the stakeholder meetings, analysing the law and creating the documentation and requirements list. This allowed to study the requirements engineering (RE) process. All founders of the case studies where in touch with insurers, therefore the researchers had access to the results and the feedback of HTAs.

The company in Case 1 was newly founded with the goal of enabling a telemedicine system of sensor-based location-independent care of rehabilitation patients. This was realized by an optical sensor, software with motion analysis algorithms, a backend for synchronisation and a clinical interface to allow the physician and therapists to adapt the therapy. The employees of the company are specialists in software development, integration of sensor data, medical device and data privacy regulation and in consulting of companies.

Case 2 deals with an app to remind glaucoma patients to register daily eye drops. This project already existed as a working app, but had not yet been established as a company at the time of the analysis. The potential founders were ophthalmologist and a professional app designer, both with years of experience in their field.

Case 3 describes a company founded in May 2018 that developed a chatbot to answer health-related questions from patients with qualified content. Furthermore, other app solutions were planned at the time of the analysis, for example as an interface between patients and ECG measuring devices with the aim of documenting and simplifying parameter output for patients. The founders were experts in informatics and business administration.

4.2 Summary of Case Study Data

In all cases we studied the core work processes and the tasks and responsibilities of all involved roles. The data sources used are documents provided and generated by the company and notes taken by the personnel involved. Additionally, in Case 1 and 2 we had access to internal negotiations and their results and we could take notes during the meetings. Furthermore, in Case 1 one researcher worked 1,5 years as CEO of this company and as "person responsible for regulatory compliance" (Art. 15 MDR). Case 3 was an external customer of one of the researchers and generated the least data and documents due to a strict non-disclosure agreement (NDA) and a short research period of only 3 months. Case 1 was analysed over the last 7 years and case 2 over the last 3 years.

Case 1 originated from a cooperation between industry and science aimed at implementing process innovations in the healthcare sector by using this innovative telemedicine system. The fact that it was newly founded resulted in the rare opportunity to study the implementation of processes and documentation inside the company and the process of finalising the prototype from a research prototype to a market ready product. This case was analysed with requirements engineering processes by the researcher and his co-workers in the company and with HTA by insurance companies and academia projects. The HTA showed a marked need and high degree of maturity of the solution due to all projects and clinical studies the founders had done in the research. The founders were preparing the process according to the new MDR. This required to establish a quality management system according to ISO 13485 which is audited regularly. The requirements engineering showed the technology was not mature enough to be released on market, additionally there was a long way to comply with the medical device regulation and to set up the necessary technical documentation and the clinical evaluation based on the previous studies. Based on this result the founders planned to switch to the Council Directive 93/42/EEC (Medical Device Directive - MDD), which is the precursor of the MDR, which classifies their product into a lower risk class and therefore they do not have to implement a quality management system.

Case 2 originated from a cooperation between an ophthalmologist and an app designer, who came up with an idea to improve the compulsory compliance of the patients by programming the app. According to HTA this project was in an early stage because no clinical studies to prove the increase of compliance had been done by this app. Also, a lot of requirements weren't clear to the founders, especially the requirements of data privacy and the medical device regulation. Therefore, they were trying to find experts and to build up a team around their start up idea. They did a lot of user experience tests which have proven a high quality of the user experience, but due to a lack of documentation and regulatory knowledge, quality of the backend regarding the security and data privacy presupposed by law stayed unclear. The founders tried to bring the product on the market based on the MDD but the preparation and documentation processes took too long, which resulted in the necessity to follow the MDR.

Case 3 originated from the research where the founders wanted to improve the information flow to the patients. Additionally, they were studying chatbots. The analysis in the requirements engineering has shown this case had a high technological maturity, but compliance with regulatory security and data privacy requirements stayed unclear. The HTA showed a high degree of maturity of the solution due to a clinical trial but the market need was unclear. Case 3 is the only analysed project to finalize the go-to- market process by Declaration of Conformity according to MDD. The status of reimbursement and negotiations with insurance companies is unknown.

4.3 Case Study Discussion

As shown above, all three start-ups followed the determinants and influencing factors listed in Table 1. During the requirements engineering processes additionally two approaches were studied that explicitly deal with Ambient Assisted Living (AAL) and telemedicine-assisted patient monitoring in the field of requirements engineering. These two approaches only focus on the technical implementation without mentioning the legal requirements [20, 21]. However, the use cases have shown, that it is precisely these regulatory requirements and laws that are of fundamental importance for medical devices. None of the start-ups was prepared for the huge number of legal requirements, regarding the MDD and especially the MDR to fulfil marketing authorization. What further made things more complicated were the frequent publications of new guidelines, laws and requirements by local authorities, changes in the interpretation and all this during the reform of the marketing authorization of medical products. Additionally, all products affect several areas of law where product liability and data privacy create the most requirements.

Consequently, it became apparent, that classical requirements engineering is insufficient to deal with medical devices or telemedicine products and it is necessary to develop an optimised approach for the RE of telemedicine products that goes beyond the existing methods. A second conclusion was the exclusive application of an HTA to determine requirements for the market introduction of telemedicine products is insufficient and must be supplemented with further approaches. An HTA is made for assessing the maturity and benefit from a working solution. Even though the companies can use HTAs to find requirements to prepare for a negotiation with insurance companies. But the HTA was not helpful in the use cases to find the necessary steps for a go-to-market. However, the HTA was very helpful for the insurance companies to determine if a solution is mature enough.

During the use cases the researchers were faced with a complex legal starting point regarding the fulfilment of data privacy, MDR and reimbursement requirements and frequent changes in the legal requirements which lead to strategic changes in the start-ups. This outcome confirms a particularly complex situation for manufacturers of telemedicine systems. Therefore, a requirement engineering process is not sufficient and a management system has to be created and implemented.

5 Towards EAM-Based Support of Medical Device Assessment

Enterprise architecture management (EAM) is an established discipline in many companies and aims at a coordinated and long-term development of the business and IT-aspects of an enterprise [22]. Enterprise Architecture aims to model, align and understand important interactions between business and IT to set a prerequisite for a well-adjusted and strategically oriented decision-making framework for both digital business and digital technologies [25]. Although EAM has been acknowledged as a relevant approach for the health sector as well, there is a lack of research on how new medical device development, launch, assessment and monitoring should be integrated into the enterprise architecture of companies offering such products. So far, guidance or proposals on how to best integrate medical device development, compliance management and monitoring into existing EA are rare.

Enterprise Architecture Management, as defined by several standards such as ArchiMate [26] and TOGAF [27] today, uses a relatively large set of different views and perspectives for managing and documenting the business-IT-alignment (BITA) [24]. EAM represents a management approach that establishes, maintains and uses a coherent set of guidelines, architecture principles and governance regimes that offer direction and support in the design and development of an architecture to realize the enterprise's transformation objectives.

TOGAF, a widely used industrial approach to EAM, divides EA into four partial architectures [27]:

- Business Architecture defines the business strategy, governance, organization and the most important business processes.
- Data Architecture describes the structure of logical and physical data elements and data management resources in the organization.
- Application Architecture defines a design for the individually used application system and its relationship to the core business processes of the organization.
- Technology Architecture describes the software and hardware functions required to support the development of business, data and application services. This includes IT infrastructure, middleware, networks and communication.

Table 2. Analysis of EA deficits and requirements in case study 1

TOGAF partial architecture	Missing elements for MD market launch in case study	Affected existing process/ structures in case study
Business architecture	– Reimbursement model – Quality management system – Risk management – Roles: Data protection Person responsible for regulatory compliance (PRRC), Medical Device consultant (MDC) – Data protection compliant delivery process	– Clinical trial had to be defined for market launch as digital health application according to Digital Care Act – Processes had to be established for: QM, RM – Roles were defined – Depending on the customer, the refurbishment process had to be changed
Data architecture	– Clinical evaluation – Technical documentation/file – Data Protection Concept – Post-Market Surveillance Process and Vigilance	– Process for continuous clinical evaluation – Developers had to be involved in the documentation process and edit the documentation directly during development
Application architecture	– Security testing process of applications necessary (automatic documentation and alarming optional) – Application for technical documentation, document versioning and control	– Security testing process had to be established, application for packet inspection, automated documentation and alarming was designed and the implementation had begun – Storage of technical documentation in Gitlab, enabling versioning and controlling of documents
Technology architecture	– Data protection and security compliant database and server infrastructure (and separation of patient data depending on clinic)	Disposition of the planned database and server structure in clusters depending on the federal states and clinics, partial outsourcing of the infrastructure to clinic data centres was necessary

Analysis of the case studies in Sect. 4 shows that medical device assessment requires elements and structures in all partial architectures, and – more important – that companies lack the ability to implement proper structures and integrate them with the existing business in an efficient way. Table 2 shows for all partial architectures examples of missing elements in the case 1 and for affected parts of the existing enterprise.

Based on this finding we argue that EA could be a suitable way to visualize and recommend required processes and structures for medical device management. Table 2 shows only initial findings from the case and has to be elaborated into a more fine-granular model, preferably represented in a modeling language, such as ArchiMate. Similar models for other cases and identification of commonalities and improvement potential could form the basis for recommendations of elements and structures in each architecture, and how to implement them. This will be part of future work.

6 Conclusions and Future Work

Based on an analysis of existing regulations and practices, the paper examined cases of medical device launch and assessment to investigate business needs in this area. The cases confirmed that market launch and regulative assessment of ICT-based medical devices in Europe is very complex. Additionally, there are no tools, strategies or best practices how to launch medical devices on market.

Based on this finding, we consider systematic and integral management of regulative compliance, business model and IT implementation as essential for successful medical devices. However, existing recommendations and best practices only cover specific aspects of medical product management, such as the technology assessment, privacy protection or business development (see Sect. 4). The long-term objective of our research is to develop a recommendation for processes, organizational structure and IT support in the field of medical device development.

References

1. Bundesärztekammer. Hinweise und Erläuterungen zu § 7 Absatz 4 MBO-Ä (Fernbehandlung). Bundesärztekammer (2015). http://www.bundesaerztekammer.de/fileadmin/user_upload/downloads/pdf-Ordner/Recht/2015-12-11_Hinweise_und_Erlaeuterungen_zur_Fernbehandlung.pdf. Accessed 22 Nov 2019
2. Field, M.J.: Telemedicine – a guide to assessing telecommunications in health care. In: Institute of Medicine (U.S.) – Committee on Evaluating Clinical Applications of Telemedicine. Journal of Digital Imaging, 2nd edn, vol. 10, pp. 26–27 & 248. National Academy of Science, Washington D.C. (1996). https://doi.org/10.17226/5296
3. Meystre, S.: The current state of telemonitoring: a comment on the literature. In: Telemedicine and e-Health, vol. 11, no. 1, p. 63. Mary Ann Liebert, Inc., Utah (2005). https://doi.org/10.1089/tmj.2005.11.63
4. BfArM. Orientierungshilfe Medical Apps. BfArM (2015). https://www.bfarm.de/DE/Medizinprodukte/Abgrenzung/MedicalApps/_node.html. Accessed 05 Nov 2020

5. Johner, C.: Software als Medizinprodukt – Software as Medical Device. Johner Institut (2019). https://www.johner-institut.de/blog/regulatory-affairs/software-als-medizinprodukt-definition/. Accessed 31 Mar 2019
6. ECJ. Judgment of 07.12.2017, case C-329/16 (2017). http://curia.europa.eu/juris/document/document_print.jsf?doclang=DE&text=&pageIndex=0&part=1&mode=DOC&docid=197527&occ=first&dir=&cid=658263. Accessed 05 Nov 2020
7. Lehmann, B., Bitzer, E.-M.: Vom Projekt in die Versorgung – Wie gelangen telemedizinische Anwendungen (nicht) in den Versorgungsalltag? In: Pfannstiel, M.A., Da-Cruz, P., Mehlich, H. (eds.) Digitale Transformation von Dienstleistungen im Gesundheitswesen VI, pp. 91–116. Springer, Wiesbaden (2019). https://doi.org/10.1007/978-3-658-25461-2_6
8. Hagen, J., Lauer, W.: Reiseführer gesucht – Ergebnis einer Umfrage bei E-Health-Start-ups. In: Bundesgesundheitsblatt, vol. 61, pp. 291–297. Springer, Deutschland (2018).https://doi.org/10.1007/s00103-018-2692-4
9. Brauns, H.-J., Loos, W.: Telemedizin in Deutschland - Stand Hemmnisse Perspektiven. In: Bundesgesundheitsblatt - Gesundheitsforschung – Gesundheitsschutz, vol. 10, pp. 1068–1073. Springer, Heidelberg (2015).https://doi.org/10.1007/s00103-015-2223-5
10. BMBF. Studie und Expertengespräch zu Umsetzungshemmnissen telemedizinischer Anwendungen, written for AGENON by: Bianca Lehmann, Eva-Maria Bitzer, Steffen Bohm, Ulrich Reinacher, Heinz-Werner Priess und Anne de Vries and for Fraunhofer-Institut FOKUS by: Michael John and Johannes Einhaus (2019). https://www.bmbf.de/files/Telemedizin-Endbericht_barrierefrei.pdf. Accessed 03 Sept 2020
11. Henke, K.-D., Troppens, S., Braeseke, G., Dreher, B., Merda, M.: Innovationsimpulse der Gesundheitswirtschaft – Auswirkungen auf Krankheitskosten, Wettbewerbsfähigkeit und Beschäftigung – Forschungsprojekt im Auftrag des BMWi. In: Zentrum für innovative Gesundheitstechnologie an der TU Berlin (2011)
12. Ratzel, R., Lippert, H.D.: Kommentar zur Musterberufsordnung der deutschen Ärzte (MBO), 6th edn, para. 1–4 & 75–77. Springer, Heidelberg (2015). https://doi.org/10.1007/978-3-642-54413-2
13. Bundesgesetzesblatt. Part 1, vol. 2019, no. 49 from 18 December 2019, pp. 2562–2584 (2019). http://www.bgbl.de/xaver/bgbl/start.xav?startbk=Bundesanzeiger_BGBl&jumpTo=bgbl119s2562.pdf. Accessed 10 Sept 2020
14. Johner, C.: Übergangsfristen der MDR. Johner-Institut (2020). https://www.johner-institut.de/blog/johner-institut/uebergangsfristen-mdr/. Accessed 23 Nov 2020
15. BfArM. Digitale Gesundheitsanwendungen (DiGA) (2020). https://www.bfarm.de/DE/Medizinprodukte/DVG/_node.html. Accessed 24 Aug 2020
16. BfArM. Das Fast Track Verfahren für digitale Gesundheitsanwendungen (DiGA) nach § 139e SGB V - Ein Leitfaden für Hersteller, Leistungserbringer und Anwender (2020). https://www.bfarm.de/SharedDocs/Downloads/DE/Service/Beratungsverfahren/DiGA-Leitfaden.pdf?__blob=publicationFile. Accessed 28 Aug 2020
17. Perleth, M., et al.: Health Technology Assessment – Konzepte, Methoden, Praxis für Wissenschaft und Entscheidungsfindung, 2nd edn, p. 5, 56, 142 & 161. MWV Medizinisch Wissenschaftliche Verlagsgesellschaft, Berlin (2014)
18. Broens, T., Huis in't Veld, R., Vollenbroek-Hutten, M., Hermens, H., van Halteren, A., Nieuwenhuis, L.: Determinants of successful telemedicine implementations: a literature study. In: Telemed Telecare 1 September 2007, vol. 13, no. 6, pp. 303–309 (2007). https://doi.org/10.1258/135763307781644951
19. Tanriverdi, H., Suzanne Iacono, C.: Knowledge barriers to diffusion of telemedicine. In: Hirschheim, R., Newman, M., DeGross, J.I. (eds.) Proceedings of the International Conference on Information Systems (ICIS), Atlanta, 1998, pp. 39–50 (1998). https://doi.org/10.5555/353053.353057

20. Ruiz-López, T., Noguera, M., José Rodríguez, M., Luis Garrido, J., Chung, L.: REUBI - a requirements engineering method for ubiquitous systems. In: Science of Computer Programming, vol. 78, no. 10, pp. 1895–1911 (2012)
21. Cheng, B.H.C., Sawyer, P., Bencomo, N., Whittle, J.: A goal-based modeling approach to develop requirements of an adaptive system with environmental uncertainty. In: Schürr, A., Selic, B. (eds.) MODELS 2009. LNCS, vol. 5795, pp. 468–483. Springer, Heidelberg (2009). https://doi.org/10.1007/978-3-642-04425-0_36
22. Lankhorst, M.: Enterprise Architecture at Work. Modelling, Communication and Analysis. The Enterprise Engineering Series. Springer, Heidelberg (2017). https://doi.org/10.1007/978-3-642-29651-2
23. Johannesson, P., Perjons, E.: An Introduction to Design Science. Springer, Cham (2014). https://doi.org/10.1007/978-3-319-10632-8
24. Simon, D., Fischbach, K., Schoder, D.: Enterprise architecture management and its role in corporate strategic management. IseB **12**(1), 5–42 (2013). https://doi.org/10.1007/s10257-013-0213-4
25. Niemi, E., Pekkola, S.: Using enterprise architecture artefacts in an organisation. Enterp. Inf. Syst. **11**(3), 313–338 (2017). https://doi.org/10.1080/17517575.2015.1048831
26. The Open Group. ArchiMate® 3.0.1 Specification, 1st edn. Van Haren Publishing, Zaltbommel (2017)
27. The Open Group. The TOGAF Standard, Version 9.2. TOGAF Series. Van Haren Publishing, Zaltbommel (2018)

BSCT Workshop

BSCT 2019 Workshop Chairs' Message

The 4th Workshop on Blockchain and Smart Contract Technologies (BSCT 2021), organized in conjunction with the 24th International Conference on Business Information Systems (BIS 2021), took place at the Königlicher Pferdestall at Leibniz University Hannover, Germany. Because of the COVID-19 pandemic, the workshop was organized online. The presented volume contains the selected and revised papers prepared for the workshop.

There were 11 papers submitted and the Program Committee decided to accept five papers (an acceptance rate of 45%). There were 17 members in the Program Committee, representing 17 institutions from 13 countries. The papers were carefully reviewed by the dedicated members of the Program Committee. In total 31 reviews were prepared.

The first paper develops a model of the Lightning Network using probabilistic logic programming. The second paper makes some important observations on what works or does not work for Bill of Lading systems and develops a convincing application of analogous system based on blockchain. The third paper examines the volatility resilience of the cryptocurrency market to COVID-19. For this reason, the study is very important and well-timed. The fourth paper gives a mathematical model for the diffusion of instant payment systems. The developed model is fairly good when compared with empirical data. The last paper provides a the meta-analysis of already published systematic literature reviews in this field. Its focus is on the identification of emerging themes for blockchain adoption in supply chains.

It is our belief that this fourth edition of the workshop was a major success. On this occasion, we would like to express our heartfelt thanks to all the participants and especially the authors. Likewise, we would like to extend our deepest gratitude to the Program Committee members as their knowledge, judgment, and assistance allowed us to organize this very successful event.

We would also like to direct our sincere appreciation to the organizers of the hosting conference (BIS 2021). Finally, we would like to invite all this year's participants, and submitters, as well as newcomers, to the planned 5th Workshop on Blockchain and Smart Contract Technologies in 2022.

Saulius Masteika
Erich Schweighofer
Piotr Stolarski

Organization

Chairs

Saulius Masteika	Vilnius University, Lithuania
Erich Schweighofer	University of Vienna, Austria
Piotr Stolarski	Poznan University of Business and Economics, Poland

Program Committee

François Charoy	Inria, Loria, Université de Lorraine, France
Nicolas T. Courtois	University College London, UK
Stefan Eder	Benn-Ibler Rechtsanwälte GmbH, Austria
Jaap Gordijn	Vrije Universiteit Amsterdam, The Netherlands
Aquinas Hobor	National University of Singapore, Singapore
Constantin Houy	Institute for Information Systems at DFKI (IWi), Germany
Monika Kaczmarek	University of Duisburg-Essen, Germany
Kalinka Kaloyanova	University of Sofia, Bulgaria
Gary Klein	University of Colorado Boulder, USA
Saulius Masteika	Vilnius University, Lithuania
Raimundas Matulevicius	University of Tartu, Estonia
Kouichi Sakurai	Kyushu University, Japan
Erich Schweighofer	University of Vienna, Austria
Piotr Stolarski	Poznan University of Economics, Poland
Herve Verjus	Université Savoie Mont-Blanc, France
Hans Weigand	Tilburg University, The Netherlands
Jakob Zanol	University of Vienna, Austria

A Probabilistic Logic Model of Lightning Network

Damiano Azzolini[2](✉), Fabrizio Riguzzi[2], Elena Bellodi[1], and Evelina Lamma[1]

[1] Dipartimento di Ingegneria, University of Ferrara, Via Saragat 1, 44122 Ferrara, Italy
{elena.bellodi,evelina.lamma}@unife.it

[2] Dipartimento di Matematica e Informatica, University of Ferrara, Via Saragat 1, 44122 Ferrara, Italy
{damiano.azzolini,fabrizio.riguzzi}@unife.it

Abstract. One of the main limitations of blockchain systems based on Proof of Work is scalability, making them unsuitable for e-commerce and small payments. Currently, one of the principal directions to overcome the scalability issue is to use the so-called "layer two" solutions, like Lightning Network, where users can open channels and send payments through them. In this paper, we propose a Probabilistic Logic model of Lightning Network, and we show how it can be adopted to compute several properties of it. We conduct some experiments to prove the applicability of the model, rather than providing a comprehensive analysis of the network.

Keywords: Probabilistic Logic Programming · Blockchain · Lightning Network

1 Introduction

The scalability trilemma states: "Scalability, decentralization, security: you can have only two of the three". This is the main issue all blockchains must face. Proof of Work (PoW) based blockchains, such as Bitcoin [14] and Ethereum [24], are secure and (theoretically) decentralized, but not very scalable [7]. This is because, currently, the PoW consensus algorithm requires solving a computationally hard problem. At the moment of writing, in the case of Bitcoin, the average number of transactions per second is 7, for Ethereum 15, making them unusable for every day small payments. Blockchains based on Proof of Stake (PoS) or Delegated Proof of Stake (DPos) can support a higher number of transactions but at the cost of less decentralization.

Among all the various proposals to increase the number of processed transactions, such as Sharding [11] and sidechains [6], the so-called "layer two" solutions remain the most adopted. One of the most famous is Lightning Network [16] (LN). In Lightning Network users can open, through a transaction on the main chain, a bidirectional Payment Channel, and then use the funds locked in this

W. Abramowicz et al. (Eds.): BIS 2021 Workshops, LNBIP 444, pp. 321–333, 2022.
https://doi.org/10.1007/978-3-031-04216-4_28

channel to issue transactions, without utilizing the main chain. The capacity distribution in a channel is unknown, to preserve privacy. An important feature of LN is that it also allows the routing of payments between two users not directly connected by a channel.

Probabilistic Logic Programming (PLP) is a powerful language to represent scenarios where probability has a central role: it combines the expressivity of Logic Programming with uncertainty over facts, and it has been already used to model several blockchain-related scenarios [3,5]. Routing in the LN can be seen as uncertain, in the sense that users may not be active for some reasons or may refuse to forward the payment. Furthermore, there is uncertainty on the funds distribution on a channel.

In [2] the authors proposed a logic model of the Bitcoin LN; here, we extend it by proposing a *probabilistic* logic model of Bitcoin LN, and we show how this model can represent the uncertain scenario cited above, allowing, for instance, to compute the probability of a successful payment routing. We focus on Lightning Network built upon Bitcoin. However, our model can be extended to a general blockchain.

The paper is structured as follows: in Sect. 2 we describe the general structure and features of the Bitcoin Lightning Network. Section 3 introduces the PLP base concepts needed to understand the LN model developed and discussed in Sect. 4. Section 5 presents a possible application of the model to compute routing probabilities, and Sect. 6 concludes the paper.

2 Blockchain, Bitcoin and Lightning Network

Bitcoin was designed by Nakamoto in 2008 [14] with the goal to create a decentralized payment system where users can issue transactions without the need of a centralized authority. Bitcoin blockchain offers several features such as immutability, auditability, and transaction atomicity. These features make it theoretically suitable for payments and micro payments. However, scalability is still one of the main limitations of the system and does not allow an increasing number of transactions.

A blockchain is based on a linear sequence of blocks, linked together by cryptographic functions. To append a block to the chain, a miner must execute, in the case of Bitcoin, a Proof of Work (PoW) algorithm which involves finding a solution to a hard puzzle. This mechanism ensures that blocks, once discovered, cannot be tampered with. On the other hand, this is a huge bottleneck for scalability. In fact, as computing power increases and technology evolves, the difficulty of PoW increases, keeping the average discovery time fixed to approximately one block every ten minutes, and thus limiting the number of transactions that can be processed by the system since blocks have a fixed size.

Another feature of Bitcoin that restricts the scalability is the block size limit (1Mb), a very controversial topic[1]. An increase of this limit will allow the system

[1] https://en.bitcoin.it/wiki/Block_size_limit_controversy.

to process more transactions. However, this would require more computation power to store and manage the blockchain, reducing the decentralization of the system. Furthermore, forks would be more likely to happen, due to the slower propagation time. Finally, a change in block size can be done only with a hard fork, an update that would be backward incompatible: this can cause a consensus failure, making the system completely unreliable.

At the moment, the scalability problem is not solved. In the past, several improvements were proposed (called Bitcoin Improvement Proposal[2]) that slowly increased the number of manageable transactions. One of the first, advanced at the end of 2015 and accepted few years later, is Segregated Witness (SegWit)[3]. In a nutshell, SegWit increases the capacity of a block by removing signature data from a transaction and introducing the definition of Virtual Size of a block and block weight, measured in weight unit instead of bytes.

Another related improvement proposal is the one that suggests the usage of Schnorr signatures [12,21]. Currently, to send transactions, signatures are needed. In the case a user wants to send a transaction from multiple addresses to one, every transaction requires its own signature, increasing the transaction size, making it more expensive. After the implementation of Schnorr signatures, if users control multiple addresses, they can move the funds from those addresses to a single address using only one signature, making the transaction lighter. These two solutions combined would increase the number of manageable transactions, but the Bitcoin system would still be limited.

Another approach consists in the so-called "layer two" solutions, based on an underlying blockchain: the main idea is to create a layer of channels on top of it where users can interact without facing the scalability problem of the blockchain. Lightning Network [16] (LN) is currently one of the most promising "layer two" solutions. It consists of a peer-to-peer network based on a blockchain, such as Bitcoin, where users can send payments and micro payments through bidirectional payment channels, without having to pay high transaction fees, and without the need of long confirmation times. To open a channel, users must broadcast an initial funding transaction on the underlying blockchain. To update the state of the channel, they can create a commitment transaction that is not published on the main chain. To close the channel, they must agree on the state of the channel and then publish a closing transaction.

One of the main features of LN is the possibility to send payments also to not directly connected users, through multi hop payments, assuming that the source and the destination are linked through intermediate connections. Moreover, thanks to Hashed Timelock Contracts (HTLC)[4], there is no need for trust between users.

However, the capacity of a channel between two users A and B is known, but the distribution of the capacity in each direction is unknown, since it is a

---

[2] https://en.bitcoin.it/wiki/Bitcoin_Improvement_Proposals.
[3] https://github.com/bitcoin/bips/blob/master/bip-0141.mediawiki.
[4] https://en.bitcoin.it/wiki/Hash_Time_Locked_Contracts.

feature introduced to increase the security of the system[5]. Due to this characteristic, routing is a complicated task in Lightning Network, and can be considered probabilistic.

3 Probabilistic Logic Programming

Logic Programming (LP) is a powerful language that allows one to express complex models with few lines of codes. One of the main limitations of Logic Programming is that it cannot manage uncertainty. Probabilistic Logic Programming [9,18] extends LP by allowing the definition of probabilistic facts that can follow several probability distributions. Initially, only Bernoulli distributions [20] (or generalized Bernoulli distributions [22]) were proposed.

The distribution semantics [19] gives a precise meaning to PLP without function symbols (with finite grounding). An *atomic choice* indicates whether a grounding (i.e., a substitution term/variable) for a probabilistic fact is selected. A set of atomic choices is *consistent* if it does not contain two different alternatives (selected and not selected) for the same probabilistic fact. A consistent set of atomic choices forms a *composite choice*: in the case it contains an atomic choice for every grounding of every probabilistic fact it is named a *selection*. A selection identifies a ground logic program called *world*, and its probability can be computed as the product of the probabilities of the atomic choices. The probability P of a query q can then be computed as the sum of the probabilities of the worlds w in which the query is true:

$$P(q) = \sum_{w \models q} P(w)$$

Here, we consider the PLP language ProbLog [10]. A probabilistic fact f_i has the following syntax:

$$p_i :: f_i$$

meaning that f_i is true with probability $p_i \in]0, 1]$, and false otherwise. If $p_i = 1$ the fact is deterministic, i.e., it is always true. An example of a Probabilistic Logic Program is shown below:

```
0.7::no_sleep.
0.8::too_much_work.
tired:- no_sleep.
tired:- too_much_work.
```

The first two lines are *probabilistic facts*, while the third and fourth lines are *clauses*. A clause is composed of a *head* and a *body* separated by the neck operator (`:-`). The head is true if the body is true. Here, for the first clause, the head is `tired` and the body is `no_sleep`. Both these clauses are *ground* since they do

[5] https://github.com/lightningnetwork/lightning-rfc/blob/master/07-routing-gossip.md.

not contain variables (denoted with the first letter uppercase). In other words, this program models a person who is `tired` if he/she does not sleep enough or if he/she works too much. With probability 0.7, the person does not sleep enough (`no_sleep`), and with probability 0.8 he/she works too much (`too_much_work`). We can then ask the probability that the person is tired (`tired`) obtaining 0.94 $(0.7 \cdot 0.8 + 0.7 \cdot 0.2 + 0.3 \cdot 0.8)$.

One of the main issues in PLP (and LP in general) is that the grounding of a program may be huge, so managing it can be very difficult. To tackle this, usually probabilistic logic programs are compiled into a more compact form through a process called knowledge compilation [8].

In the last few years, Hybrid Probabilistic Logic Programs arose [4,13], with the possibility to define continuous random variables and constraints among them. To consider continuous probability distributions, we introduce the syntax used by `cplint` [1] and its module MCINTYRE [17]. Continuous random variables can be encoded with:

$$f : Density$$

where f is an atom (a predicate symbol followed by a number of arguments) with a continuous variable as argument and *Density* is a special atom that models a probability density on the argument of the atom. For example,

```
f(X) : gaussian(X,0,2).
```

indicates that `X` in `f(X)` follows a Gaussian distribution with mean 0 and variance 2. Inference in these types of programs can be done, for example, by sampling [17].

4 Network Model

The LN can be represented as a graph where users (nodes) are linked by connections (edges) with capacity distributed according to some probability distribution, since the real distribution is unknown (except for the two users that opened the channel). Let us start with a deterministic model, that will be later extended to a probabilistic one. A connection between two nodes is represented with a fact `edge(A,B,Capacity)`, where `A` and `B` are the two nodes and `Capacity` is the capacity of the connection. The whole LN is represented as a list of `edge/3` facts (where `/3` indicates the arity, i.e., the number of arguments). For these experiments, we do not consider fee base and fee rate. However, they can be straightforwardly included in the analysis by simply adding more arguments. Connections are undirected, meaning that payments can go in both directions, and are represented with the following predicate `connected/3`:

```
connected(A,B,C):- edge(A,B,C) ; edge(B,A,C).
```

`A` and `B` are connected with a channel of capacity `C` if there is an edge from `A` to `B` or (`;`) from `B` to `A` with capacity `C`. The degree of a node is the number of edges incident to the node. We can search for a path between two nodes with a standard Prolog predicate. An example written using SWI-Prolog [23] is reported here for the sake of clarity:

```
connected_test(Source,Next,Size):-
    connected(Source,Next,Cap),
    Size < Cap.

path(Dest,Dest,_,_,Path,Path).
path(Source,Dest,Size,NSteps,Visited,Path) :-
    length(Visited,N),
    N < NSteps,
    connected_test(Source,Next,Size),
    \+ memberchk(Next,Visited),
    path(Next,Dest,Size,NSteps,[Next|Visited],Path).
```

The predicate `path/6` states that there is a path from `Source` to `Dest` that can route a payment of size `Size` if the two nodes are connected by intermediate nodes that have not been already visited (condition checked with `\+memberchk`, a deterministic version of `member/2`). Ensuring that a node has not been already visited is fundamental, otherwise we may get stuck in a loop where we move an infinite number of times between a pair of nodes. The length `N` of the path is bounded using the standard Prolog comparison predicate `</2`, that compares the length of the list containing already visited nodes (`Visited`) with a user defined threshold (`NSteps`). In a similar way, the capacity `Cap` is checked against the size `Size` of the payment in a second predicate called `connected_test`. A sample call to `path/6` is `path(a,c,10,3,[],P)`, were we look for a path `P` from `a` to `c` of at most `3` edges that can route a payment of size `10`, starting with an empty list (`[]`) of visited nodes.

The previous code does not consider the capacity distribution in a channel. Let us now extend it with a probabilistic fact to represent it. Using `cplint` on swish [1], we can define continuous random variables (say for example uniformly distributed in [L,U]) with:

```
distr(X,L,U) : uniform_dens(X,L,U).
```

Here, `X` has a uniform distribution between `L` and `U`, where `L` is the minimum value and `U` the maximum value. The predicate `connected_test` can be modified as:

```
connected_test(Source,Next,Size):-
    connected(Source,Next,Cap),
    distr(C,0,Cap),
    Size < C.
```

In other words, we collect the capacity of the channel between two nodes, and we state that the capacity from `Source` to `Next` is given by a random variable uniformly distributed between `0` and `Cap`. With this definition, at each sampling iteration, the distribution for every channel is fixed and does not change. Clearly, to successfully route a payment, its size must be smaller than the capacity in the considered direction (`Size < C`).

We can further extend the previous model by considering also intermittent edges. This situation may arise when a node disconnects from the network or when declines to forward a payment. To model it, we can define a Bernoulli random variable that is true with a certain probability (set to 0.95 in the following code snippet):

```
0.95::active(_).
```

The predicate is then extended as:

```
connected_test(Source,Next,Size):-
    connected(Source,Next,Cap),
    active(Next),
    distr(C,0,Cap),
    Size < C.
```

In the experiments, we modelled the capacity of a channel using a uniform distribution, and the probability that a node is active with a Bernoulli distribution. However, several extensions and variations can be made: changing the type of distribution for the channel capacity (Gaussian with mean equal to half of the capacity, for example) or even setting the probability that a node declines to route a payment proportional to the payment size, to the fees or a combination of the two.

5 Routing Analysis

We used a snapshot of the LN taken from https://ln.bigsun.xyz/ on the 12th of April 2021. We considered only the open channels, resulting in a network with 14734 nodes, 44349 channels, and with a total capacity of 125819660675 satoshi (1258.19 bitcoin). Table 1 shows a recap of the most common channel capacities and node degrees (considering also duplicated edges that connect the same pair of nodes).

Table 1. Information about channel capacities and node degrees.

Capacity (sat)	Occurrences	Degrees	Occurrences	Highest capacities (sat)	Occurrences
100000	3942	1	6958	500000000	4
1000000	3523	2	2575	477184791	1
500000	2794	3	1372	354000000	1
2000000	1533	4	802	300000000	2
16777215	1522	5	549	250000000	1
200000	1446	6	397	238135604	1
5000000	1359	7	263	225118006	1
20000	1297	8	209	200000000	28
10000000	1131	9	182	179792707	1
50000	1079	10	145	154222260	1

To demonstrate how PLP can be used to model the LN, we conducted some experiments. For all of them, we fixed the maximum length of the path. Moreover, we suppose that the distribution of the capacity in a channel is uniform (i.e., if the channel that connects A and B has capacity 1, we can route from A to B a uniform distributed value between 0 and 1, and from B to A the remaining), since we do not have further information.

To select the range of the payment size, we first compute the average of the capacities of all the connections. We obtained approximately 2837035 satoshi but with a huge standard deviation ($> 10^7$). So, we counted the number of connections that have less than the average capacity, less than half of the average capacity, and less than a quarter of the average capacity, obtaining respectively 35555 ($\approx$ 80%), 31902 ($\approx$ 72%), and 26077 ($\approx$ 59%). Figure 1 shows a more detailed graph, where the X axis indicates the percentage of the average capacity and the Y axis indicates both the number of connections (edges) with less than that value, and the relative percentage of the total connections. To further analyse the distribution of the capacity, we removed the nodes with the highest capacities and plotted the variation of the total capacity. The results are shown in Fig 2.

In a first test, we want to compute the probability to successfully route a payment of varying size between two different random nodes of the same degree at the first attempt, given that nodes may not be active. One of the main requisites to route a payment is that all the nodes along the path are active, otherwise it cannot be sent. Moreover, as said before, a node may refuse to forward the payment for some reasons. We fix the length of the path (number of intermediate edges) to 2 and the node degree of source and destination to 2, 5, and 10. We then plot how the probability varies when intermediate nodes may not be active. Probability values are computed with the predicate `mc_sample(+Query:atom,+N:int,-Prob:float)`, provided by the MCINTYRE module [17], that samples the query `Query` `N` times and returns the ratio between the number of successes and the number of samples (`Prob`). We set the number of samples to 1000.

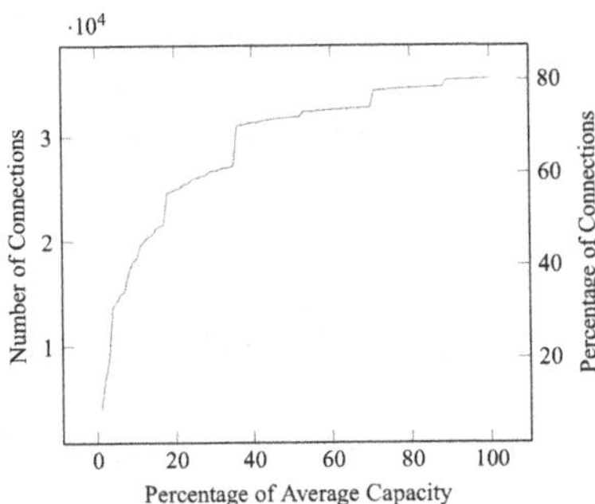

Fig. 1. Number and percentage of connections with less than a certain percentage of the average capacity (2837035).

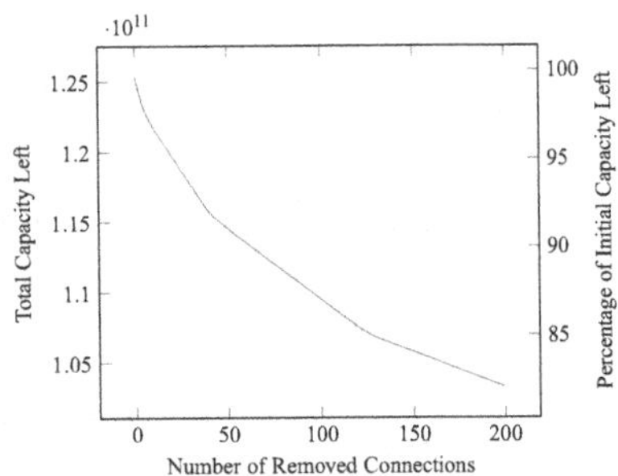

Fig. 2. Value and percentage of capacity left after removing the top n (X axis) connections in terms of capacity.

Figure 3 shows the results for the first experiment: nodes with higher degrees have, as expected, a higher probability to successfully route a payment on the first attempt, even if this gap reduces as the payment size increases. Figure 4 shows the results for the experiments with a varying probability of intermediate nodes to be active: the probability of a successful routing decreases, but not so drastically. If the source and the destination are the same node, and the length of the path is greater than 2 (at least the source node and another one, if these two nodes are connected by at least 2 edges, otherwise the path should be composed of at least 2 additional nodes plus the source), we can compute the probability of a successful *rebalance*. This is a common situation since, if users want to refill a channel, currently one of the main solutions is to send a circular payment to themselves.

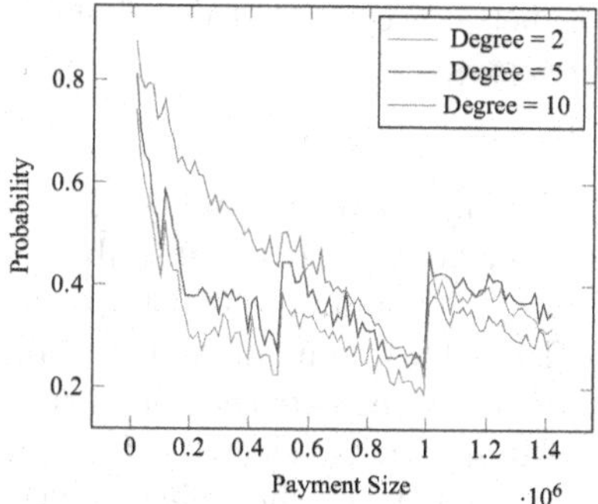

Fig. 3. Probability of a successful payment of varying size between random connected nodes of degree 2, 5, and 10.

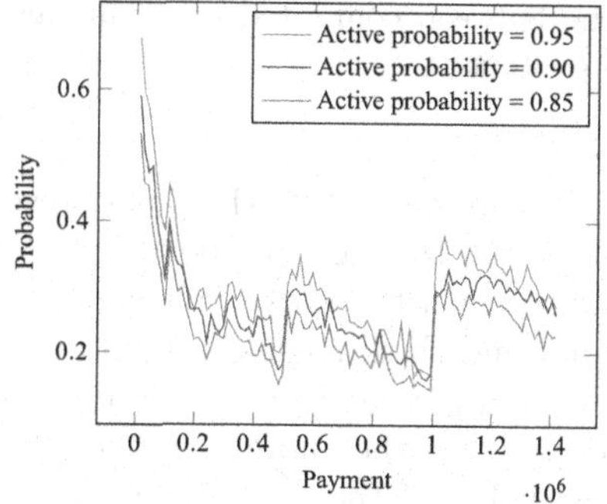

Fig. 4. Probability of a successful payment of varying size between random connected nodes of degree 2, with different active probabilities for intermediate nodes.

In another experiment, we want to know the probability of a successful payment between two random nodes of variable degrees, given that the payment is split in various equal parts. That is, we split a payment into N parts and compute the probability that all these payments succeed at the first attempt. This is a common scenario in LN [15], because, when the size of the payment increases, the probability of success decreases, due to the channel capacity limitation. However, increasing the number of payments also increases the fees required to route the payment, due to the necessity to issue multiple transactions. The graphs in Figs. 5, 6, and 7 show how the probability of a successful payment varies when the payment is split in 2, 3 or 4 parts and intermediate nodes are always active. In Figs. 8, 9, and 10 we performed the same experiment but we fixed the degree of source and destination to 2, and we varied both the probability that a node is active and the number of parts of a payment.

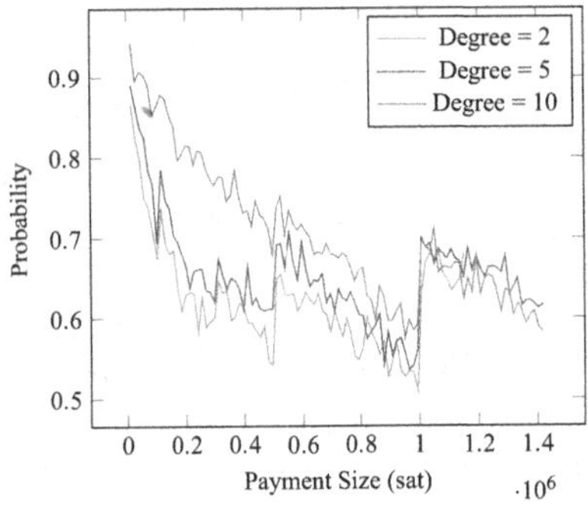

Fig. 5. Probability of a successful payment of varying size split into 2 equal parts, between connected nodes of various degrees.

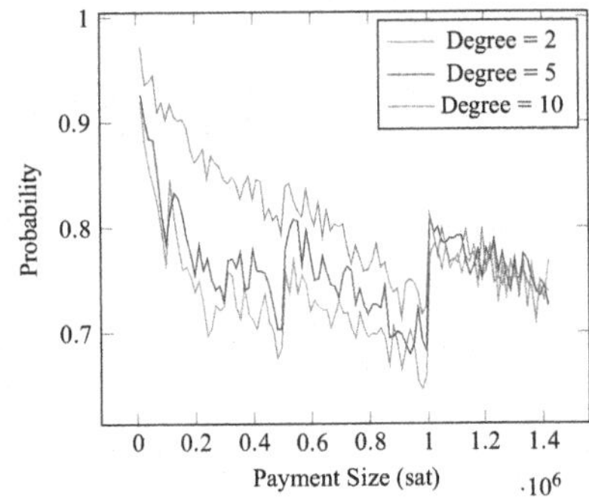

Fig. 6. Probability of a successful payment of varying size split in 3 equal parts, between connected nodes of various degrees.

We can see that all the plots present two sudden jumps in probability, around 500000 ($5 \cdot 10^5$) and 1000000 (10^6): this is in accordance with the distribution of edges with these two values of capacity. In fact, 500000 and 1000000 are the second and third most common capacities (see Table 1). In our implementation, to route a payment, we randomly check if one of the edges between the current node and another node has enough capacity. If it has, it is selected. However, if the value of the payment is close to the total capacity of the selected edge, the routing will likely fail, since we suppose that the distribution is uniform. For example, if we try to route a payment of size $4.5 \cdot 10^5$ into a channel of capacity $5 \cdot 10^5$, we have only approximately 10% of chances to succeed (with a distribution supposed uniform). Since a lot of channels have capacity $5 \cdot 10^5$, when the payment approaches this value, the success probability of routing at the first attempt reduces. With a payment slightly greater than $5 \cdot 10^5$, these channels are no longer considered, since they do not have enough capacity, so the probability suddenly increases. Similarly happens with the value 10^6. Another, but less noticeable jump, can be found around 100000 (10^5), which is the most common value for the capacity of a channel. This happens for the same reasons explained before.

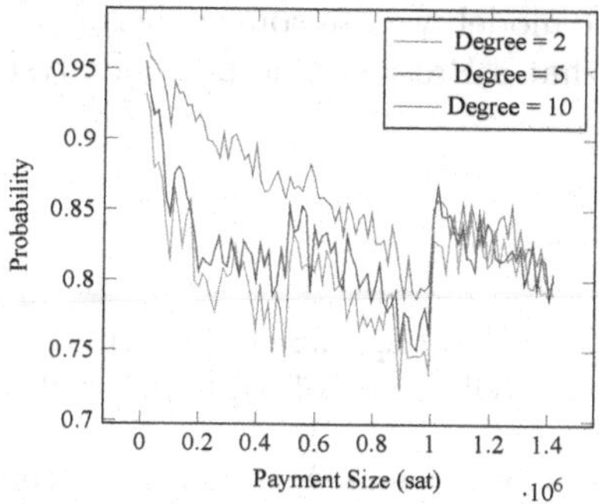

Fig. 7. Probability of a successful payment of varying size split into 4 equal parts, between connected nodes of various degrees.

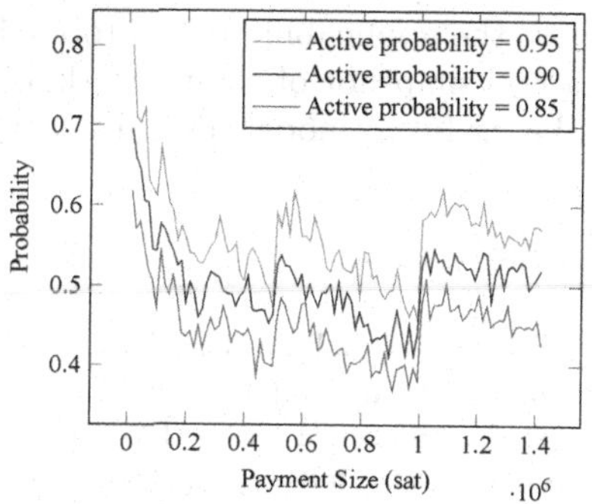

Fig. 8. Probability of a successful payment of varying size split into 2 equal parts, between connected nodes of degree 2 with varying active probability.

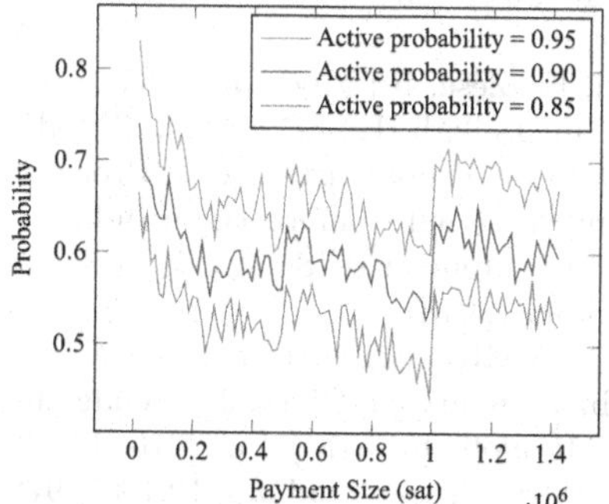

Fig. 9. Probability of a successful payment of varying size split into 3 equal parts, between connected nodes of degree 2 with varying active probability.

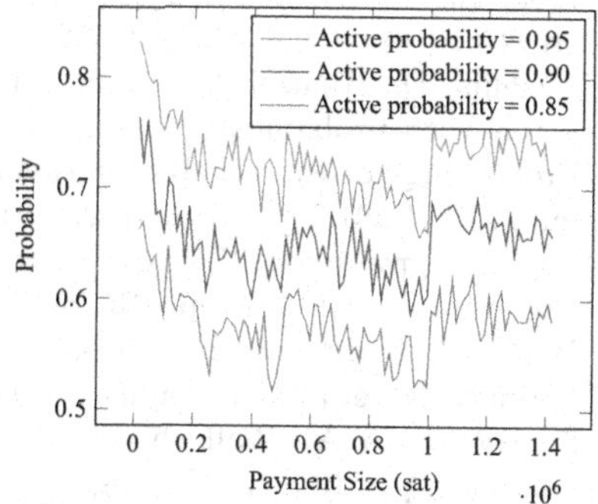

Fig. 10. Probability of a successful payment of varying size split into 4 equal parts, between connected nodes of degree 2 with varying active probability.

6 Conclusions

In this paper, we propose to analyse (Bitcoin) Lightning Network with Probabilistic Logic Programming. The usage of PLP allows representing the network with a highly expressive language. We described the network as an undirected graph with the capacity of a channel following a uniform distribution. Furthermore, we considered also the possibility to have intermittent nodes, a situation that may arise also when a node refuses to forward a payment. The goal of this paper is to prove the feasibility of a Probabilistic Logic model of the network, rather than provide and analysis of it, since the network continuously changes (and thus an analysis will be obsolete in a few weeks, even days or hours), and routing mechanism are typically more sophisticated than randomized routing.

To test the applicability of a probabilistic logic model, we ran some experiments on a real snapshot of the network, obtaining that a Probabilistic Logic analysis can be useful to model several uncertain scenarios.

References

1. Alberti, M., Bellodi, E., Cota, G., Riguzzi, F., Zese, R.: `cplint` on SWISH: Probabilistic logical inference with a web browser. Intell. Artif. **11**(1), 47–64 (2017). https://doi.org/10.3233/IA-170105
2. Azzolini, D., Bellodi, E., Brancaleoni, A., Riguzzi, F., Lamma, E.: Modeling bitcoin lightning network by logic programming. In: Proceedings 36th International Conference on Logic Programming (Technical Communications), vol. 325, pp. 258–260 (2020). https://doi.org/10.4204/EPTCS.325.30
3. Azzolini, D., Riguzzi, F., Lamma, E.: Studying transaction fees in the bitcoin blockchain with probabilistic logic programming. Information **10**(11), 335 (2019)
4. Azzolini, D., Riguzzi, F., Lamma, E.: A semantics for hybrid probabilistic logic programs with function symbols. Artif. Intell. **294**, 103452 (2021). https://doi.org/10.1016/j.artint.2021.103452
5. Azzolini, D., Riguzzi, F., Lamma, E., Bellodi, E., Zese, R.: Modeling bitcoin protocols with probabilistic logic programming. In: Bellodi, E., Schrijvers, T. (eds.) Proceedings of the 5th International Workshop on Probabilistic Logic Programming, PLP 2018, Co-located with the 28th International Conference on Inductive Logic Programming (ILP 2018), Ferrara, Italy, 1 September 2018. CEUR Workshop Proceedings, vol. 2219, pp. 49–61. CEUR-WS.org (2018)
6. Back, A., et al.: Enabling blockchain innovations with pegged sidechains (2014)
7. Croman, K., et al.: On scaling decentralized blockchains. In: Clark, J., Meiklejohn, S., Ryan, P.Y.A., Wallach, D., Brenner, M., Rohloff, K. (eds.) FC 2016. LNCS, vol. 9604, pp. 106–125. Springer, Heidelberg (2016). https://doi.org/10.1007/978-3-662-53357-4_8
8. Darwiche, A., Marquis, P.: A knowledge compilation map. J. Artif. Intell. Res. **17**, 229–264 (2002)
9. De Raedt, L., Frasconi, P., Kersting, K., Muggleton, S. (eds.): Probabilistic Inductive Logic Programming. LNCS (LNAI), vol. 4911. Springer, Heidelberg (2008). https://doi.org/10.1007/978-3-540-78652-8
10. De Raedt, L., Kimmig, A., Toivonen, H.: ProbLog: A probabilistic prolog and its application in link discovery. In: Veloso, M.M. (ed.) IJCAI 2007, vol. 7, pp. 2462–2467. AAAI Press/IJCAI (2007)
11. Luu, L., Narayanan, V., Zheng, C., Baweja, K., Gilbert, S., Saxena, P.: A secure sharding protocol for open blockchains. In: Proceedings of the 2016 ACM SIGSAC Conference on Computer and Communications Security, Vienna, Austria, 24–28 October 2016, pp. 17–30. ACM, New York (2016). https://doi.org/10.1145/2976749.2978389
12. Maxwell, G., Poelstra, A., Seurin, Y., Wuille, P.: Simple Schnorr multi-signatures with applications to bitcoin. Design. Codes Cryptogr. **87**(9), 2139–2164 (2019). https://doi.org/10.1007/s10623-019-00608-x
13. Michels, S., Hommersom, A., Lucas, P.J.F., Velikova, M.: A new probabilistic constraint logic programming language based on a generalised distribution semantics. Artif. Intell. **228**, 1–44 (2015). https://doi.org/10.1016/j.artint.2015.06.008
14. Nakamoto, S.: Bitcoin: a peer-to-peer electronic cash system (2008)

15. Piatkivskyi, D., Nowostawski, M.: Split payments in payment networks. In: Garcia-Alfaro, J., Herrera-Joancomartí, J., Livraga, G., Rios, R. (eds.) DPM/CBT -2018. LNCS, vol. 11025, pp. 67–75. Springer, Cham (2018). https://doi.org/10.1007/978-3-030-00305-0_5
16. Poon, J., Dryja, T.: The bitcoin lightning network: scalable off-chain instant payments (2016)
17. Riguzzi, F.: MCINTYRE: a Monte Carlo system for probabilistic logic programming. Fund. Inform. **124**(4), 521–541 (2013). https://doi.org/10.3233/FI-2013-847
18. Riguzzi, F.: Foundations of Probabilistic Logic Programming. River Publishers, Gistrup, Denmark (2018)
19. Sato, T.: A statistical learning method for logic programs with distribution semantics. In: Sterling, L. (ed.) ICLP 1995, pp. 715–729. MIT Press (1995)
20. Sato, T., Kameya, Y.: PRISM: a language for symbolic-statistical modeling. In: IJCAI 1997, vol. 97, pp. 1330–1339 (1997)
21. Schnorr, C.P.: Efficient signature generation by smart cards. J. Cryptol. **4**(3), 161–174 (1991). https://doi.org/10.1007/BF00196725
22. Vennekens, J., Verbaeten, S., Bruynooghe, M.: Logic programs with annotated disjunctions. In: Demoen, B., Lifschitz, V. (eds.) ICLP 2004. LNCS, vol. 3132, pp. 431–445. Springer, Heidelberg (2004). https://doi.org/10.1007/978-3-540-27775-0_30
23. Wielemaker, J., Schrijvers, T., Triska, M., Lager, T.: Swi-prolog. Theory Pract. Logic Program. **12**(1–2), 67–96 (2012)
24. Wood, G.: Ethereum: a secure decentralised generalised transaction ledger. Ethereum Project Yellow Paper **151**, 1–32 (2014)

Enabling Electronic Bills of Lading by Using a Private Blockchain

Hauke Precht(✉) and Jorge Marx Gómez

Carl von Ossietzky University of Oldenburg, Oldenburg, Germany
{hauke.precht,jorge.marx.gomez}@uol.de

Abstract. The bill of lading (B/L) is one of the most important documents in international trade. Current advances started to adopt public blockchain technology for digitization but lacking privacy. We raise the research question, if a private blockchain is suitable to solve this shortcoming. To answer this research question, we followed the design science research methodology. We designed an architecture and evaluated it via implementation, using Hyperledger Fabric. First results show general feasibility and privacy enhancement but reveal a high complexity and difficult extendibility of Hyperledger Fabric. Preliminary results of a focus group shows that the approach peaked interested by practitioners.

Keywords: Blockchain · Electronic bill of lading · Blockchain architecture · Private blockchain

1 Introduction

International ocean freight is an important backbone in today's economy. In the European Union (EU), 80% of imports and exports are done via ocean freight [20]. The *Bill of Lading (B/L)* is considered to be the most important document in this process [6]. It provides multiple functions at once in a single paper-based document [3,37]. (1)The B/L proves that the goods described on the B/L were taken over, in the described form, by the carrier. (2)It secures the obligation of the carrier to deliver said goods to the place of destination, delivering them to the consignee. (3)The B/L is a negotiable document of title, meaning that the transfer of ownership of the goods described in the B/L can be done by transferring the B/L itself. This is especially useful in ocean freight as the goods are at sea but can still be traded via the B/L. Notable is that even though this document is of rather great importance within the ocean freight process, it is still paper-based [39]. This paper-based process is manual, error prone and enables easy counterfeiting of an B/L.

Due to the lack of a legal basis, electronic B/Ls, which could speed up the process by less manual tasks and securing the B/L against counterfeit, were not possible. However, Germany, one of a few legislation, introduced in 2016 with § 516 par. 2 HGB (Handelsgesetzbuch, engl. German Commercial Code) a new section, explicitly allowing the usage of electronic B/Ls. The German legislator

W. Abramowicz et al. (Eds.): BIS 2021 Workshops, LNBIP 444, pp. 334–346, 2022.
https://doi.org/10.1007/978-3-031-04216-4_29

states, that an electronic B/L must be functional equivalent in comparison to paper-based B/L in order to be considered valid, providing general guidance to consider when implementing a system for electronic B/Ls. Wunderlich and Saive already raised the question whether existing approaches towards electronic B/Ls could be considered functional equivalent based on German law. They came to the conclusion that existing solutions rely on the concept of legal rulebooks (a legal framework construct) governing the technical system meaning no functional equivalent is given [39].

Approaches from Within the Industry. The most known product which states to provide an electronic B/L is the "Bill of Lading Electronic Registry Organization (BOLERO)", which were the first to commercially develop a solution for electronic B/Ls along with the mentioned rulebook [23]. BOLERO could not establish itself as a standard due to a lack of trust in its centralized architecture and legal uncertainty introduced by the mentioned rulebook [35]. Next to BOLERO, ESSDOCS is another system that relies on a centralized architecture suffering the same lack of trust by the industry. It has been shown that the industry does not trust a centralized approach meaning no trusted authority to manage such a centralized system can be set up as it will not be accepted by the broad industry. This is due to the risk, to expose business information to potential competitors. Based on this realization and the experiences with centralized architectures, other providers started to use decentralized architectures. WAVE, TRADELENSE or CARGOX.IO leveraging blockchain systems along with off-chain data storage. Recently, R3 announced to provide an SDK for electronic B/Ls for their platform as well but stated to only "[...] closely mirror paper BLs." [5] meaning that it can not be considered a real electronic B/L. These solutions also require the use of rulebooks introducing the already mentioned legal uncertainties. Note that WAVE states that they rely on the Carriage of Goods by Sea Act 1992 (COGSA) of the UK instead of a rulebook, ignoring that COGSA does not allow electronic B/Ls [1].

Approaches from Within the Scientific Community. Also the scientific community pushed towards electronic B/Ls, analyzing blockchain and its possibility to safely and uniquely create digital assets. Already in 2016, Takahashi published a paper entitled "Blockchain technology and electronic bills of lading" in which the general applicability of blockchain in this domain is discussed along with legal areas to consider [35]. Following this general analysis, Nærland et al. evaluates the possibility to use the Ethereum Blockchain along with its Smart Contracts (SCs) to create a system for electronic B/Ls [26]. They used the InterPlanetary FileSystem (IPFS) to store the actual B/L as a PDF document while storing its hash along with the public key of the current owner on the Ethereum blockchain [26]. The authors themselves state that with this approach, it is difficult to ensure privacy [26]. Further, they note that the speed and privacy of transactions similar to a centralized database could not be reached but would be desirable [26]. Another general analysis and overview has been given by [39].

All mentioned approaches do not provide a sufficient electronic B/L, either by lacking privacy or legal compliance due to the usage of rulebooks. Especially the lack of privacy can be referred to the used public blockchain systems. Generally, private blockchains are referred to provide a better performance and privacy than public blockchains [7,24,41] which could be a better fit for an electronic B/L. Therefore, we state the following research questions: *Is a private blockchain suitable for implementing an electronic B/L?*

The the paper is structured as follows: First, we describe our used research methodology in Sect. 2, followed by the description of the abstract architecture in Sect. 3. In Sect. 4 we present the implementation of the architecture along with several implementation decisions. Within the evaluation in Sect. 5 we present generated insights and experience gathered in the implementation phase and in preliminary interview. The paper concludes with an summary and future work.

2 Methodology

To answer the stated research questions, we follow the design science research approach [9] as we aim to develop a unique and novel technical solution for an electronic B/L. Within this approach, we choose to follow the guidelines for design science introduced by [8]. First, we present a new architecture for creating and managing electronic B/Ls. Next, to show the validity of the architecture, we create a prototype implementation deliver an additional artifact. Through this prototype, we aimed to conduct an evaluation of the design to investigate the feasibility of our designed architecture and potential flaws. Our research contribution is an architecture for private blockchain systems enabling the creation and management of electronic B/Ls as well as the opportunity to further utilize this approach for other paper-based documents in the process. The given artifacts, the architecture and the respective implementation were created through iterations. The insights from implementing the architecture were used to further sharpen the architecture within the iterative process, as presented in the evaluation section. Our results of this research are presented through this paper along with discussions with practitioners who have an interest in an electronic B/L.

3 Architecture

Blockchain as an underlying technology seems feasible, but current solutions for electronic B/Ls lack transaction throughput and privacy [26]. Therefore, we aim to investigate whether a private blockchain system could meet these challenges. Further, in current studies, the area of competition law, as a specific aspect of privacy is not yet considered. But as Louven and Saive state, also information exchange via so-called market information systems can be subject to the competition law, especially when information is shared which would otherwise be undisclosed [22]. Considering public blockchain solution, where transaction data is publicly visible, it raises the possibility to further analyze business relations which would otherwise be undisclosed, leading to legal uncertainties. By conducting a focus group [2] with practitioners working with B/Ls, we gathered further requirements enhancing the before descried legal consideration, which must be met by an system representing electronic B/Ls. These requirements were further categorized within an ongoing research project. The combined and grouped requirements are listed in Table 1. To fulfill *R1* and *R2*, we created a principle called *one B/L, one blockchain*. We aim to create for each B/L process a distinct blockchain that is only accessible for the respective parties involved in this B/L process. This way, we can ensure a maximum of privacy while maintaining a decentralized storage and history record. Note that his way, synchronization times for new participants can be kept low as well.

Table 1. Requirements for an electronic B/L system

ID	Description
R1	B/Ls must be stored and managed through the draft state to the final state
R2	Enhanced privacy of participants in the B/L process, as existing solution lack privacy of the participants
R3	Competition law states that information exchange must be kept at a minimum
R4	Multiple users for a single organization must interact with the system in parallel
R5	As port agents and captains, working at the port, tend to use mobile devices, the solution must provide a user interface for such devices
R6	Data in the B/L can be subject to the GDPR, meaning a system must be built so that GDPR requirements, such as right of deletion, can be fulfilled

We developed a general architecture based on existing approaches. From a software architecture perspective, Xu et al. showed that blockchain can be used as a storage element, a computation element, a communication element and as an asset management and control mechanism [40]. In our use case, we consider the blockchain as an asset management and control mechanism in the first place next to an data communication component, to fulfill *R1*. Rengelov et al. showed that, especially in an Ethereum environment, such blockchain systems tend to be built on a two-tier architecture consisting of a front-end component that is directly communicating with the blockchain component [32]. They further state that this leads to limitation in scalability as the front-end node is considered a limiting factor. But as described in *R4*, multiple users within the same organization must be able work with the system in parallel. Therefore, we add a third component in between the front-end and the blockchain component. This third component is described as a proxy layer [32]. By including this proxy layer, in the following called back-end component, we can further enhance the separation of concerns, support a wider ranger of functionalities as well enhanced load distribution [32]. This results in a three-tier architecture, which is shown in Fig. 1. Tier 1, represents the front-end component, tier 2 represents the back-end component and tier 3 represents the private blockchain component. This leads to a known three-tier-architecture with a blockchain-component as persistence layer instead of a traditional centralized database. It can be seen that the front-end component requests data via an interface while the back-end component provides such interface. With this modular design, the front-end component can be easily changed or extended. The back-end component itself uses the private blockchain component by saving and querying B/Ls. Note that the front-end component presents necessary actions for the respective user. As stated in *R5*, especially port agents and captains tend to use mobile devices, the front-end component provides a user interface optimized for mobile devices as well. Actions, for example, creating a draft of a B/L, will be temporarily saved in a local relational database, allowing parties to prepare B/Ls before sharing them with other participants, meeting the requirement defined in *R1*. Once the B/L should be made available to the other involved parties, the back-end component will trigger the respective method to interact with the private blockchain component.

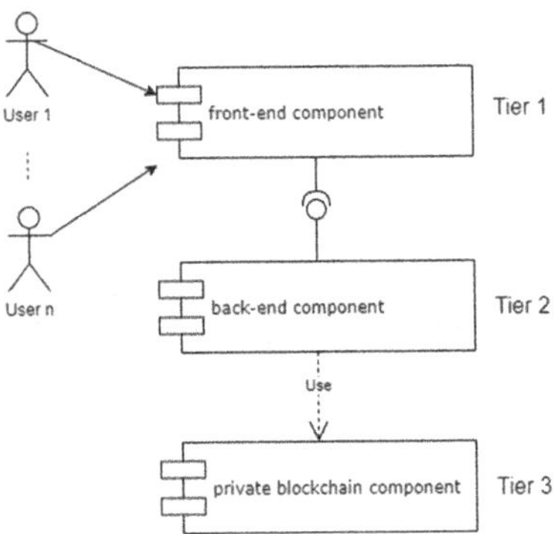

Fig. 1. The three-tier architecture

With the general architecture stack in place, further design decisions need to be addressed, first and foremost the way of data storage as well as computation. Xu et al. discussed several options, including on-chain vs. off-chain storage as well as on-chain vs off-chain computation. The on-chain storage as well as the on-chain computation favors the fundamental properties of a blockchain (e.g. immutability) [41]. Note that while on-chain computation and storage leverages the fundamental properties of a blockchain, it is not cost efficient, and has a worse performance than off-chain storage and computation [41]. Note that Xu et al. mainly focus on public blockchains. Considering private blockchains, especially the cost factor can be neglected if choosing a system without a built-in crypto-currency. In order to fully leverage the blockchain fundamentals, (e.g. immutability, transparency, integrity and trust), we aim to store the B/L as an asset on-chain along with relevant computation in the form of a SC, which manages the B/L asset. The concept of SCs itself is a known construct introduced already in 1996 by Nick Szabo [34] and is considered, within the blockchain context, as a piece of software which executes automatically once certain conditions are met [21].

4 Implementation

To evaluate the feasibility of the proposed architecture, we implement it in the form of a prototype. Therefore we used the prototyping approach as a well-known methodology in the area of business informatics in which a prototype of a system is developed and evaluated to generate first and new insights [38]. We further followed an approach called *compliant programming* introduced in [30] which includes parallel monitoring of the development process by legal experts to ensure compliance through every development step. We used several iterations to further refine and sharpen our implementation. The implementation and the dependencies between these components is shown in Fig. 2. Note that the back-end and private blockchain component have a high coupling, while the front-end can be easily exchanged, connecting to the back-end component. The implementation detail of each component is described in the following.

Front-End Component. The component was developed in a way that it can scale on smartphone and tablet devices. Therefore, react.js[1], TypeScript[2] and Redux[3] were used. Mainly, two pages were implemented for a minimum viable prototype (MVP) consisting of an overview page listing current B/Ls, and an create page used to create/update B/Ls.

Back-End Component. The back-end component is implemented in Java, leveraging Spring Boot[4]. To avoid the manual process of creating a blockchain for each B/L, the blockchain creation is ought to be done via an SDK provided by the respective blockchain technology from within the back-end component. We created a layer architecture consisting of a REST Interface layer, a service layer as well, as a persistence layer. Note that the persistence layer is split into two sections: Data Access Objects (DAOs) and the Hyperledger Fabric SDK (HLF SDK). Via the DAOs we store preliminary B/Ls as well as application data (e.g. user credentials) in a relation PostgreSQL[5] database. The HLF SDK is used to communicate with the Hyperledger Fabric blockchain network via the respective peer/orderer of the given organization, which is described in greater detail in the following section discussing the private blockchain component.

Private Blockchain Component. The interest in private blockchains and experiments to integrate them within businesses grew for enterprises [27]. Polge et al. provide an overview of the five major private/permissioned

[1] https://reactjs.org/.
[2] https://www.typescriptlang.org/.
[3] https://redux.js.org/.
[4] https://spring.io/projects/spring-boot.
[5] https://www.postgresql.org/.

blockchains: Hyperledger Fabric, Enterprise Ethereum, Quorum, MultiChain and R3 Corda [29]. Our main requirements for a private blockchain are based on *R2* and *R3*, which we tackle by our concept *one B/L, one blockchain*, meaning we require a blockchain system which allows us to implement this concept. Currently only MultiChain and Hyperledger Fabric seem viable to support this approach as they explicitly support the creation of multiple blockchains. But as MultiChain only adds support for SCs in version 2.0.55 [29] and keeps privacy enhancing feature within a proprietary and fee required version [25] we do not consider this technology suitable. Hyperledger Fabric, however, is one of the most used blockchain systems [10] which is completely open-source and provides a so-called channel concept, which creates for each channel an independent blockchain [13] which was also considered feasible by [31]. Also, SCs are supported via so-called chaincodes, which can be written in multiple languages, including Java [18]. Therefore, we choose to implement the private blockchain component, by using Hyperledger Fabric. We create a test network, by setting up four independent virtual machines (VM), which are disjunctive and communicate via the internet as shown in the tier 3 section of Fig. 2. Each VM represents an organization within the B/L process. For a production-ready network, each participating party, e.g. consignee, consignor, shipper, carrier, port agents or banks, would be considered to host a respective peer and ordering node. To secure the ordering process (i.e. the consensus), every participating organization should provide an orderer node for the respective channel they are part of. Currently, no restriction for who can participate in the ordering, i.e. in the block creation process, is planed. Note that each node type (peer, which holds the ledger and executes SCs [17] and orderer, which serve the sole purpose of ordering and creating blocks [16]), as well as a couch db [15] and command line interface (cli) are running in docker container. We varied in the number of orderer and peer nodes for each organization for a more realistic scenario. Setting up the network itself found to be challenging and error-prone due to manual configuration.

Within the blockchain component, we implemented the chaincode which manages the B/L on-chain. The chaincode is written in Java and is shown in Fig. 3. Our `BillOfLadingContract` implements the `ContractInterface` provided by Hyperledger Fabric. This `ContractInterface` provides utility functions for our contract allowing us to interact with the ledger and the system itself. Note that every method expects a `Context` object which is provided by Hyperledger Fabric automatically and contains information such as the caller of the respective method. With this, we are able to implement an attribute-based access control (ABAC), which is a method in which a "*[...] subject requests to perform operations on objects are granted or denied based on assigned attributes of the subject [...]*" [11]. This allows us to define on chaincode-level if the caller is allowed to perform the respective actions. Note that the `transfer` method can only be invoked by the current consignee as this is the only person legally allowed to further transfer the B/L. To allow a way back to the traditional process, for example, due to lack of digital infrastructure at a port, we implemented

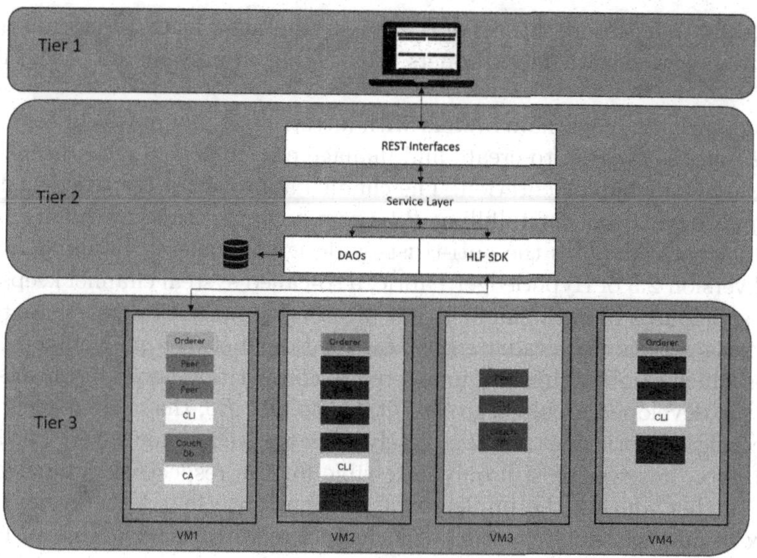

Fig. 2. The implemented architecture

a `extractForPrint` method which will mark the B/L asset as printed, meaning every other call to further transfer the B/L will be blocked. The B/L asset will be created via calling of the `init` method. Via the `getBillOfLading` method the respective B/L of this respective chain will be returned.

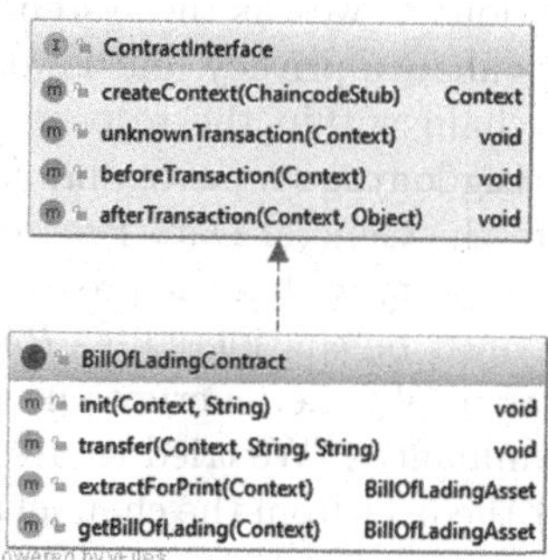

Fig. 3. The BillOfLadingContact and BillOfLadingAsset object

5 Evaluation

Our evaluation consists of two areas. First, we evaluated the technical feasibility of the designed architecture by implementing it, following [9]. The second part of our evaluation is done by conducting a preliminary focus group, discussing our created system with practitioners, identifying improvements of the design [36].

Using a private blockchain system with a restrictive and novel concept called *one B/L, one blockchain* to create and manage electronic B/Ls has been shown as feasible by our implementation. The channel approach of Hyperledger Fabric fully supports this concept fulfilling *R2*. From a competition law point of view, our architecture supports the anti-trust-by-design in general, fulfilling *R3*. But up until version 2.3 of Hyperledger Fabric, a so-called system channel keeps track of the participants of each channel [12], meaning organizations participating in the ordering service can evaluate business relations between participants in the channels which can be subject of interested to competition law. This means, that though the developed architecture is suitable to met *R3*, the current technology of Hyperledger Fabric does not completely meet this requirement yet. The actual channel data, however, is still only accessible for the respective channel participants. Further, during the implementation phase, we found the system to be complex to manage. Each creation of a channel requires the usage of an Hyperledger Fabric specific tool called *configtxgen*, which creates the genesis block of the respective channel along with initial channel policies. Basically, for each channel some sort of "local" consortium is defined for the respective process. Based on configurations files, the the initial set of participants of such consortium" must be defined beforehand. But as electronic B/Ls can be traded, potentially new participants must be included which requires changes in the before defined policy and access control of the channel. We found the Hyperledger Fabric SDK to be limited when trying to implement these network configuration updates as code in a dynamic way. Note that with the next version of Hyperledger fabric, the initial consortium definitions as well as the system channel will be no longer required, aiming to further enhance privacy and lowering complexity [14]. Further, we store the B/L on-chain within the respective channel, which is then managed by the `BillOfLadingContract`. Saive and Janicki point out that data in the B/L can be subject to the General Data Protection Regulation (GDPR), for example, if a participating party within the process is not a legal entity but a natural person [33], GDPR must be considered, as *R6* described. Note that, for example, the usage of cryptography, i.e. encryption, to encrypt the B/L is not considered to be GDPR compliant [4] We tried to use an approach in which the organization will disconnect the peer from the channel and then delete data upon receiving for example a deletion request (Art. 17 GDPR). But we found that this feature is not yet implemented but planned for a future version [19]. Currently, only manual configuration for the respective peer could be used, creating additional manual effort. While most of the requirements can be met, we found the system to be complex for deployment, especially due to high configuration and management of Hyperledger Fabric.

By conducting a focus group to identify improvements and general acceptance level of practitioners, we generated further insights. The participants stated that the private blockchain-based approach is considered positive and peaked interest, proving *R1*, *R4* and *R5*. Based on our preliminary results we can therefore answer our stated research question *Is a private blockchain suitable for implementing an electronic B/L* in a positive way, meaning that a private blockchain can be considered suitable to implement an electronic B/L. However, especially the on-chain storage was seen to be difficult together with companies guidelines to share as few information/data as possible. Therefore, we will re-evaluate the way of data storage in combination with an re-evaluation of GDPR compliance. Also the complexity we noted during our implementation was second by the practitioner, stating that the current system might be too complex to handle and extend within in a production environment.

6 Summary and Outlook

In this paper we raised the research questions, if a private blockchain is suitable for implementing an electronic B/L. To answer this research question, we used the design science research methodology. Based on existing approaches and insights from a conducted focus group with practitioners, we defined six abstract requirements. We presented an architecture along with an implementation leveraging a three-tier hierarchy architecture for electronic B/Ls. In our implementation, we used Hyperledger Fabric and its channel concept to implement our designed concept of *one B/L, one blockchain* through which we maximize privacy. With this approach we also met anti-trust-by-design guidelines stated in commercial law. The general feasibility has been shown and evaluated with practitioners, creating insights considering implementation and administration complexity. We found that the approach to use a private blockchain is feasible and peaked interest by practitioners but the chosen technology, Hyperledger Fabric, seems to be too complex regarding network administration. This preliminary finding will be further investigated by conducting field experiments with respective practitioners to gain further knowledge of the usability of the proposed system. Further research could also encompass security and possible attack scenarios such as insider attacks [28] from within the private permissioned network. Also, the deployment and management of the potentially large network must be discussed in greater detail. A possibility could be to create an independent organization which supports companies in the setup of the software without being part of the respective network itself. Based on these to be conducted field experiments, we will have enough insight to decide if a alternative to Hyperledger Fabric must be evaluated. Therefore, our future research will be focused on conducting field experiments and deriving further insights for the presented architecture.

References

1. Albrecht, C.: Blockchain bills of lading: the end of history? Overcoming paper-based transport documents in sea carriage through new technologies. Tulane Maritime Law J. **43**, 251–288 (2019)

2. Brandtner, P., Helfert, M., Auinger, A., Gaubinger, K.: Conducting focus group research in a design science project: application in developing a process model for the front end of innovation. Syst. Signs Actions **9**, 26–55 (2015)
3. Curwen, N.: The bill of lading as a document of title at common law. In: Mountbatten Yearbook of Legal Studies, pp. 140–163 (2007)
4. Ebbers, F., Karaboga, M.: Blockchain and Data Protection: An Evaluation of the Challenges and Solutions Mentioned by German Stakeholders. In: Ahlemann, F., Schütte, R., Stieglitz, S. (eds.) Innovation Through Information Systems: Volume II: A Collection of Latest Research on Technology Issues, pp. 353–369. Springer, Cham (2021). https://doi.org/10.1007/978-3-030-86797-3_24
5. ELEANOR WRAGG. R3's Electronic Bill of Lading Takes Shape (2021). https://www.gtreview.com/news/fintech/r3s-electronic-bill-of-lading-takes-shape/
6. Fridgen, G., Guggenberger, N., Hoeren, T., Prinz, W., Urbach, N.: Chancen und herausforderungen von dlt (blockchain) in mobilität und logistik. https://www.bmvi.de/SharedDocs/DE/Artikel/DG/blockchain-grundgutachten.html
7. Gamage, H.T.M., Weerasinghe, H.D., Dias, N.G.J.: A survey on blockchain technology concepts, applications, and issues. SN Comput. Sci. **1**(2), 1–15 (2020). https://doi.org/10.1007/s42979-020-00123-0
8. Gregor, S., Hevner, A.R.: Positioning and presenting design science research for maximum impact. MIS Quart. **37**(2), 337–355 (2013). https://doi.org/10.25300/MISQ/2013/37.2.01
9. Hevner, A.R., March, S.T., Park, J., Ram, S.: Design science in information systems research. MIS Quart. **28**(1), 75–105 (2004)
10. Hileman, G., Rauchs, M.: Global blockchain benchmarking study. https://www.ey.com/Publication/vwLUAssets/ey-global-blockchain-benchmarking-study-2017/$File/ey-global-blockchain-benchmarking-study-2017.pdf
11. Hu, V.C., et al.: Guide to attribute based access control (ABAC) definition and considerations: NIST special publication 800–162. https://doi.org/10.6028/NIST.SP.800-162,https://nvlpubs.nist.gov/nistpubs/SpecialPublications/NIST.SP.800-162.pdf
12. Hyperledger Fabric. Channel Configuration (Configtx): Orderer System Channel Configuration (2019). https://hyperledger-fabric.readthedocs.io/en/release-1.4/configtx.html#orderer-system-channel-configuration
13. Hyperledger Fabric. Channels (2020). https://hyperledger-fabric.readthedocs.io/en/latest/channels.html
14. Hyperledger Fabric. Craete a Channel (2020). https://hyperledger-fabric.readthedocs.io/en/latest/create_channel/create_channel_participation.html
15. Hyperledger Fabric. Ledger: World State (2020). `B/L Process. Note that each node type as well as the couch DB and command`
16. Hyperledger Fabric. Orderer (2020). https://hyperledger-fabric.readthedocs.io/en/release-2.0/orderer/ordering_service.html
17. Hyperledger Fabric. Peers (2020). https://hyperledger-fabric.readthedocs.io/en/release-2.0/peers/peers.html
18. Hyperledger Fabric. Smart Contracts and Chaincode (2020). https://hyperledger-fabric.readthedocs.io/en/release-2.0/smartcontract/smartcontract.html
19. Hyperledger Fabric. As a Peer Operator, I Want my Peer to Leave a Channel (2021). https://jira.hyperledger.org/browse/FAB-11334
20. International Chamber of Shipping. Shipping and World Trade: Driving Propserity. https://www.ics-shipping.org/shipping-fact/shipping-and-world-trade-driving-prosperity/

21. Koul, R.: Blockchain Oriented Software Testing - Challenges and Approaches. In: 2018 3rd International Conference for Convergence in Technology (I2CT), pp. 1–6. IEEE, Piscataway (2018). https://doi.org/10.1109/I2CT.2018.8529728
22. Louven, S., Saive, D.: Antitrust by design - the prohibition of anti-competitive coordination and the consensus mechanism of the blockchain. Gewerblicher Rechtsschutz Urheberrecht Internationaler Teil **6**, 537–543 (2019)
23. Ma, W.: Lading without bills - how good is the bolero bill of lading in Australia? Bond Law Rev. **12**, 206–234 (2000)
24. Mattila, J.: The blockchain phenomenon: the disruptive potential of distributed consensus architectures. ETLA Working Papers (38) (2016). http://pub.etla.fi/ETLA-Working-Papers-38.pdf
25. MultiChain. Download and Install Multichain: Start Building Your First Blockchain Application in Minutes (2021). https://www.multichain.com/download-install/
26. Nærland, K., Müller-Bloch, C., Beck, R., Palmund, C.: Blockchain to rule the waves - nascent design principles for reducing risk and uncertainty in decentralized environments. In: ICIS (2017)
27. Pawczuk, L., Massey, R., Holdowsky, J.: Deloitte's 2019 global blockchain survery: blockchain gets down to business. https://www2.deloitte.com/content/dam/Deloitte/se/Documents/risk/DI_2019-global-blockchain-survey.pdf
28. Pillai, A., Saraswat, V., Vasanthakumary Ramachandran, A.: Attacks on blockchain based digital identity. In: Prieto, J., Partida, A., Leitão, P., Pinto, A. (eds.) BLOCKCHAIN 2021. LNNS, vol. 320, pp. 329–338. Springer, Cham (2022). https://doi.org/10.1007/978-3-030-86162-9_33
29. Polge, J., Robert, J., Le Traon, Y.: Permissioned blockchain frameworks in the industry: a comparison. ICT Express **59**, 134 (2020). https://doi.org/10.1016/j.icte.2020.09.002
30. Precht, H., Saive, D.: Compliant programming - juristen in der agilen softwareentwicklung. In: Tagungsband Herbstakademie 2019, pp. 581–595 (2019). https://beck-online.beck.de/?vpath=bibdata/zeits/DSRITB/2019/cont/DSRITB.2019.595.1.htm
31. Precht, H., Wunderlich, S., Marx Gómez, J.: Applying software quality criteria to blockchain applications: a criteria catalog. In: Bui, T. (ed.) Proceedings of the 53rd Hawaii International Conference on System Sciences. Proceedings of the Annual Hawaii International Conference on System Sciences, Hawaii International Conference on System Sciences (2020). https://doi.org/10.24251/HICSS.2020.769
32. Rangelov, D., Tcholtchev, N., Lammel, P., Schieferdecker, I.: Experiences designing a multi-tier architecture for a decentralized blockchain application in the energy domain. In: 2019 11th International Congress on Ultra Modern Telecommunications and Control Systems and Workshops (ICUMT), pp. 1–7. IEEE (2019). https://doi.org/10.1109/ICUMT48472.2019.8970836
33. Saive, D., Janicki, T.: Datenschutz in elektronischen frachtdokumenten. RdTW - Recht der Transportwirtschaft Zeitschrift für Transportrecht und Schifffahrtsrecht mit dem Recht des Überseekaufs sowie Versicherungsrecht. Zollrecht und Außenwirtschaftsrecht **6**, 201–207 (2019)
34. Szabo, N.: Smart contracts: building blocks for digital markets (1996). http://www.fon.hum.uva.nl/rob/Courses/InformationInSpeech/CDROM/Literature/LOTwinterschool2006/szabo.best.vwh.net/smart_contracts_2.html
35. Takahashi, K.: Blockchain technology and electronic bills of lading. J. Int. Maritime Law **22**, 202–211 (2016)

36. Tremblay, M.C., Hevner, A.R., Berndt, D.J.: The use of focus groups in design science research. In: Hevner, A., Chatterjee, S. (eds.) Design Research in Information Systems, Integrated Series in Information Systems, vol. 22, pp. 121–143. Springer, Boston (2010). https://doi.org/10.1007/978-1-4419-5653-8_10
37. van Boom, W.H.: Certain legal aspects of electronic bills of lading. Eur. Transp. Law J. Law Econ. **XXXII**(1), 9–24 (1997)
38. Wilde, T., Hess, T.: Forschungsmethoden der wirtschaftsinformatik. WIRTSCHAFTSINFORMATIK **49**(4), 280–287 (2007). https://doi.org/10.1007/s11576-007-0064-z
39. Wunderlich, S., Saive, D.: The electronic bill of lading. In: Prieto, J., Das, A.K., Ferretti, S., Pinto, A., Corchado, J.M. (eds.) BLOCKCHAIN 2019. AISC, vol. 1010, pp. 93–100. Springer, Cham (2020). https://doi.org/10.1007/978-3-030-23813-1_12
40. Xu, X., Weber, I., Staples, M.: Architecture for Blockchain Applications. Springer, Cham (2019). https://doi.org/10.1007/978-3-030-03035-3
41. Xu, X., et al.: A taxonomy of blockchain-based systems for architecture design. In: ICSA 2017, pp. 243–252. IEEE, Piscataway (2017). https://doi.org/10.1109/ICSA.2017.33

Modeling the Resilience of the Cryptocurrency Market to COVID-19

Viviane Naimy(✉), Omar Haddad, and Rim El Khoury

Notre Dame University, Louaize-72 Zouk Mikayel, Zouk Mosbeh, Lebanon
{vnaimy, orhaddad, rkhoury}@ndu.edu.lb

Abstract. This paper is the first to examine the reaction of the cryptocurrency market to COVID-19 in terms of volatility resilience. Seven GARCH-type models are used to measure, predict, and audit the volatility behavior of the most eminent cryptocurrencies that represent almost 60% of the total crypto market namely, Bitcoin (BTC), Ripple (XRP), Litecoin (LTC), Monero (XMR), Dash (DASH), and Dogecoin (DOGE). The in-sample period extends from January 1, 2015 up to November 30, 2019 and the out-of-sample period covers the COVID-19 period spanning from December 1, 2019 up to April 6, 2021. Results showed that CGARCH (1,1) and GARCH (1,1) are the prevailing models to forecast the volatility of Bitcoin and Ripple respectively in both the in- and out-of-sample periods and that advanced GARCH models appear to better predict asymmetries in cryptocurrencies' volatilities pre and post COVID-19. Also, the COVID-19 contributed in significantly affecting the volatility of Bitcoin, Ripple, Monero and Dash.

Keywords: Bitcoin · Ripple · Litecoin · Monero · Dash · Dogecoin · SGARCH · IGARCH · EGARCH · GJR-GARCH · TGARCH · CGARCH · APARCH · Error metrics · COVID-19 · Volatility

1 Introduction

The cryptocurrency market witnessed a giant growth over the last four years. On April 10, 2021, the global crypto market capitalization reached USD 2.059T. This is not surprising since the defining characteristic of most cryptocurrencies and their most revolutionary features, is that they are trustless, immutable, and decentralized which make them deflationary and safe to central banking system and governmental interference. However, it is known that the crypto market is very volatile and has provoked serious fears of uncertainty. For this reason, measuring and modeling the volatility of this market during the COVID-19 period looks eminent and examining its stamina is much needed for investors, traders, risk managers, and regulators.

While the economic growth has shrunk by 3.5% worldwide in 2020 [1], and the national debt for countries with weak economies is forecasted to highly escalate due to the pandemic, financial and commodity markets across the globe were seriously affected [2–4]. The cryptocurrency market response to COVID-19 is still ambiguous. While several recent studies have discussed its impact on the financial markets [5, 6]

W. Abramowicz et al. (Eds.): BIS 2021 Workshops, LNBIP 444, pp. 347–358, 2022.
https://doi.org/10.1007/978-3-031-04216-4_30

and opportunities for portfolio diversification [7, 8], only few covered the behavior of some selected cryptocurrencies in terms of spillovers, efficiency and safe haven.

For instance, Conlon and McGee [9] examine the impact of the outbreak on the Bitcoin returns and find that it performed poorly with high correlations with the equity markets and that Bitcoin is neither a safe haven nor a hedge against the extreme bear market in the S&P500, occasioned by the COVID-19 pandemic. Also, Conlon et al. [10] find that a cryptocurrency's extreme volatility (Bitcoin and Ethereum) during the COVID-19 hinders its usefulness as a safe haven. Chen et al. [11] notice that the market volatility of Bitcoin is exacerbated by fear sentiment, suggesting that Bitcoin fails to act as a safe haven during the pandemic. Similarly, Dutta et al. [12], using Dynamic Conditional Correlation Generalized Autoregressive Conditional Heteroskedasticity (DCC-GARCH) model, suggest that while gold is considered a safe haven asset for global crude oil markets, Bitcoin acts only as a diversifier for crude oil. Alternatively, Bouoiyour and Selmi [13] conclude that despite the fact that the pandemic intensifies Bitcoin's volatility, the position of Bitcoin as an attractive safe haven asset is reinforced. However, prices did not adjust immediately, reflecting the inefficiency of the Bitcoin market.

On the other hand, Goodell and Goutte [14] apply wavelet methods and conclude that COVID-19 caused a rise in Bitcoin prices. Kristoufek [15] investigates the quantile correlations of Bitcoin and two benchmarks—the S&P 500 and CBOE Volatility Index (VIX) during COVID-19 and compares it to Gold. He finds that gold comes out as a clear winner. Naeem et al. [16] investigate how the COVID-19 affects the market efficiency of Bitcoin, Ethereum, Litecoin, and Ripple and realize that the cryptocurrency market efficiency varies with time. Shahzad et al. [17] address the spillover among 18 cryptocurrencies under low and high volatility regimes by applying a Markov regime-switching vector autoregressive with exogenous variables (VARX). The results indicate various patterns of spillover especially during the COVID-19 outbreak. Also, Corbet et al. [7] find that cryptocurrency returns are significantly influenced by negative sentiment relating to COVID-19. They conclude that these digital assets act as a safe haven without providing diversification benefits.

It is well-known that GARCH-type models have been often used to explain and forecast the volatility of Bitcoin and other cryptocurrencies. Chu et al. [18] investigate the best volatility model among twelve GARCH models for the seven most popular cryptocurrencies. Using five criteria, they find that the best model depends on the type of cryptocurrency. Using both the in-sample and the out-of-sample data, Naimy and Hayek [19] find that Exponential GARCH (EGARCH) (1,1) is a superior model in modeling the volatility of the Bitcoin between 2013 and 2016 as compared to GARCH (1,1) and Exponentially Weighted Moving Average (EWMA). Gronwald [20] compares GARCH models with several linear and nonlinear GARCH models to display the extreme price movements of Bitcoin and finds that GARCH models with student t-innovations and combined jump-GARCH models have the best fit. Gyamerah [21] evaluates the volatility of Bitcoin returns using three GARCH models (Standard GARCH (SGARCH), Integrated GARCH (IGARCH), and Threshold GARCH (TGARCH)) and find that the TGARCH model is the best model to estimate the volatility in Bitcoin. Naimy et al. [22] investigate the volatility behavior of the major six cryptocurrencies with respect to world currencies using several GARCH-type specifications. They find that the volatilities of cryptocurrencies are better depicted by advanced models mainly the

Component GARCH (CGARCH), Glosten-Jagannathan-Runkle GARCH (GJR-GARCH), Asymmetric Power ARCH (APARCH), and TGARCH.

As described above and given the absence of studies examining the reaction of the cryptocurrency market to COVID-19 in terms of volatility resilience, this paper fills the gap by analyzing the volatility of six of the most eminent cryptocurrencies, specifically, Bitcoin (BTC), Ripple (XRP), Litecoin (LTC), Monero (XMR), Dash (DASH), and Dogecoin (DOGE), which represent almost 60% of the total crypto market in terms of market capitalization (Table 1). Seven GARCH-type specifications namely SGARCH, IGARCH, EGARCH, GJR-GARCH, TGARCH, CGARCH, and APARCH are used to model their volatility behavior for both in-sample and out-of-sample contexts, where the latter covers the COVID-19 period starting from December 1, 2019 up to April 6, 2021. Also, and in order to examine whether the volatility structure of the crypto market has changed during the pandemic period, we use GARCH (1,1) while incorporating a dummy variable that is expected to provide information on the possible changes in the volatility rate between the pre and post COVID-19 periods. The paper proceeds as follows: Sect. 2 exposes the methods adopted to model and measure the volatility of the cryptocurrency market and provides a description of the data. Section 3 presents the results and Sect. 4 discusses and concludes the findings.

2 Selected Models and Data

2.1 The Selected GARCH Models

In this section, we present the GARCH-type models that are used to model time-varying volatility in the selected cryptocurrencies series. Their parameters are estimated and the volatility for each cryptocurrency under each of the selected models is then computed for the in-sample and out-of-sample periods. To assess the accuracy of the tested models, the estimated volatility is compared to the realized volatility using three error metrics: the Mean Absolute Error (MAE), the Root Mean Square Error (RMSE) and the Mean Absolute Percentage Error (MAPE).

Standard GARCH (1,1). The conditional variance for the SGARCH (1,1) process introduced by Bollerslev' [23] is given by:

$$\sigma_t^2 = \omega + \alpha u_{t-1}^2 + \beta \sigma_{t-1}^2 \tag{1}$$

$$\omega = \gamma V_L \tag{2}$$

Where σ_t^2 is the estimate of the variance for day t, u_{t-1}^2 and σ_{t-1}^2 represent the associated return and the variance on the previous day with α and β being their respective weights. The long run variance is "V_L". The model is considered stable when the weightsγ, α and β sum-up to 1.

Integrated GARCH. Introduced by Engle and Bollerslev [24], the IGARCH (1,1) can be expressed as follows, given that β is now set equal to $(1-\alpha)$ with restrictions $\omega \geq 0$, $\alpha \geq 0$ and $1 - \alpha \geq 0$:

$$\sigma_t^2 = \omega + \alpha u_{t-1}^2 + (1-\alpha)\sigma_{t-1}^2 \quad (3)$$

Exponential GARCH. Suggested by Nelson [25], the EGARCH (p,q) is expressed as:

$$\ln(\sigma_t^2) = \omega + \beta \ln(\sigma_{t-1}^2) + \gamma \frac{u_{t-1}}{\sqrt{\sigma_{t-1}^2}} + \alpha \left[\frac{|u_{t-1}|}{\sqrt{\sigma_{t-1}^2}} - \sqrt{\frac{2}{\pi}} \right] \quad (4)$$

Where β represents the persistence parameter and α and γ capture the size and the sign (leverage) effect, respectively.

Glosten-Jagannathan-Runkle GARCH. The GJR-GARCH model developed by Glosten et al. [26] is given by:

$$\sigma_t^2 = \omega + (\alpha + \gamma I_{t-1})u_{t-1}^2 + \beta \sigma_{t-1}^2 \quad (5)$$

Where $I_{t-1} = 1$ if $u_{t-1} < 0$ and $I_{t-1} = 0$ if $u_{t-1} \geq 0$. The parameters restrictions are similar to the Standard GARCH whereby $\omega \geq 0$, $\alpha \geq 0$, and $\beta \geq 0$.

Asymmetric Power ARCH. The APARCH model by Ding et al. [27], where σ_t^2 is replaced by σ_t^δ is given by:

$$\sigma_t^\delta = \omega + \alpha(|u_{t-1}| - \gamma u_{t-1})^\delta + \beta \sigma_{t-1}^\delta \quad (6)$$

δ, α, β and ω are ≥ 0, and $-1 \leq \gamma \leq 1$where δ is the Taylor (power effect) parameter, γ is the leverage parameter and the persistence parameter is given by $\beta + \alpha k$.

Threshold GARCH. The Threshold GARCH model due to Zakoian [28] is similar to the GJR-GARCH model and is a particular case of APARCH (1,1) with $\delta = 1$ and $-1 \leq \gamma \leq 1$. TGARCH (1,1) is expressed as follow:

$$\sigma_t = \omega + \alpha(|u_{t-1}| - \gamma u_{t-1}) + \beta \sigma_{t-1} \quad (7)$$

Component GARCH. Suggested by Engle and Lee [29], the CGARCH model allows mean reversion to a varying level " q_t" and splits the conditional variance into its transient (Eq. 8) and permanent components (Eq. 9) as presented below:

$$\sigma_t^2 = q_t + \alpha(u_{t-1}^2 - q_{t-1}) + \beta(\sigma_{t-1}^2 - q_{t-1}) + \gamma(u_{t-1}^2 - q_{t-1})I_{t-1} \quad (8)$$

$$q_t = \omega + \rho(q_{t-1} - \omega) + \phi(u_{t-1}^2 - \sigma_{t-1}^2) \quad (9)$$

Stationarity of the CGARCH model and non-negativity of the conditional variance are ensured once the following inequality constraints are satisfied: $\omega \geq 0, \alpha \geq 0, \phi \geq 0, \beta \geq 0, \beta \geq \phi$ and $\alpha + \beta \leq \rho \leq 1$.

2.2 Volatility Structural Model

In order to examine whether the volatility structure of the selected cryptocurrencies has changed during COVID-19, we use GARCH (1,1) and add a dummy variable which takes the value of 0 or 1 for the pre- and post-COVID-19 periods respectively.

$$\sigma_t^2 = \omega + \alpha_1 \mu_{t-1}^2 + \beta_1 \sigma_{t-1}^2 + \phi D_t \tag{10}$$

The GARCH (1,1) model parameters, ω, α, and β, are the same as defined in Eq. (1) and the new parameter ϕ has no constraints. The parameter sign determines if the volatility of the selected cryptocurrency has changed in response to COVID-19 or not. A negative and significant sign indicates a decrease in the volatility and a positive and significant coefficient indicates the opposite.

2.3 Data

The data employed in this study are the global historical daily prices extracted from Refinitiv for the six selected cryptocurrencies spanning from January 1, 2015 until April 6, 2021 yielding a total of 2,282 daily observations for each cryptocurrency and a total of 13,692 for the six selected cryptocurrencies. The in-sample period extends from January 1, 2015 till November 30, 2019 yielding a total of 1,794 returns whereas the out-of-sample period ranges from December 1, 2019 to April 6, 2021 generating a total of 487 returns. Given the dominance that Bitcoin imposes on the cryptocurrency market (representing 55.12% of the overall cryptocurrency market capitalization as of April 10, 2021) we opted to conduct a separate analysis on each cryptocurrency independently to avoid biased inferences. Table 1 presents the summary statistics for the returns of the six selected cryptocurrencies together with their market capitalization and the Augmented Dickey-Fuller (ADF) test which reveals that all the selected cryptocurrencies returns series are strongly stationary (p-values significant at 1%). All cryptocurrencies show positive average returns, positive skewness except for Bitcoin, and leptokurtic distribution - notably for Ripple and Dogecoin - which indicates that their returns exhibit volatility clustering and persistence.

Table 1. Summary Statistics of the Daily Returns, ADF Test, and Market Capitalization

Descriptive statistics	BTC	XRP	LTC	XMR	DASH	DOGE
Number of returns	2281	2281	2281	2281	2281	2281
Mean	0.00229	0.00163	0.00196	0.0028	0.00218	0.00254
Standard error	0.00083	0.0014	0.0012	0.00132	0.00124	0.00146
Median	0.00212	−0.00234	−0.00002	0.00127	−0.00077	0.00000
Standard deviation	0.03945	0.06697	0.05722	0.06321	0.05908	0.06974
Variance	0.00156	0.00448	0.00327	0.004	0.00349	0.00486
Kurtosis	12.74909	39.61156	13.20587	8.98693	8.69964	68.57906
Skewness	−0.90518	2.39808	0.34665	0.57894	0.79154	3.99756
Range	0.68985	1.64363	1.026	1.07878	0.91064	1.83859
Minimum	−0.46	−0.61627	−0.51458	−0.49424	−0.45933	−0.51512

(*continued*)

Table 1. (*continued*)

Descriptive statistics	BTC	XRP	LTC	XMR	DASH	DOGE
Maximum	0.22512	1.02736	0.51142	0.58454	0.4513	1.32347
ADF Stationary test (P-Value)	0.00%	0.00%	0.00%	0.00%	0.00%	0.00%
Market capitalization	55.12%	2.56%	0.75%	0.25%	0.13%	0.40%

3 Empirical Results

3.1 In-Sample GARCH Parameters

Table 2 depicts the in-sample parameters corresponding to the selected GARCH-type models. The ARCH component "α" in GARCH (1,1) ranges between 7% and 30% for the selected cryptocurrencies and clearly reflects the importance of market shocks on the volatility of the selected crypto except for Litecoin. Accordingly, all the cryptocurrencies, except for Litecoin, exhibit a relatively smaller beta compared to other financial instruments confirming their "spiky" behavior. Dogecoin reported the highest long-term volatility with a value of 334%. This further underlines the cryptocurrencies' intensifying levels of volatility. The IGARCH model validates the presumptions drawn from the GARCH model. The "β" estimates for all cryptocurrencies were quite similar except for Ripple. This highlights the cryptocurrencies' high levels of volatility that may not be captured by symmetric models and therefore, justifies our selection to use advanced asymmetric GARCH models. The leverage coefficient "γ" in EGARCH (1,1) ranges between −1% and 14% and carries a positive value for all cryptocurrencies except for Bitcoin. This implies that among all cryptocurrencies, only Bitcoin exhibits a leverage effect and that positive shocks have a greater impact on cryptocurrencies' volatility than negative shocks. The generated values in GJR-GARCH model display similar results to the Standard GARCH model in terms of mean reverting (ω), volatility clustering (α) and volatility persistence (β). All cryptocurrencies display a value for "ω" different from 0. Also, the volatility of cryptocurrencies tends to cluster in response to market shocks, particularly Ripple, and the larger beta in the case of Litecoin evidences that it is relatively more explicable and less subject to 'spikes' than the remaining cryptocurrencies. On the other hand, the leverage coefficient "γ" in the GJR-GARCH model ranges between 0% and −27% which shows consistency with the EGARCH model, where a negative leverage coefficient implies the absence of leverage effect for all cryptocurrencies. In contrast, estimates for the leverage parameter "γ" in the APARCH model for each cryptocurrency have changed significantly compared to the GJR-GARCH and EGARCH models, with gamma ranging between 0.5% and 1%. The lowest percentages for the leverage coefficients were those of Bitcoin and Dogecoin, which reveals that their volatilities are affected symmetrically by positive and negative shocks, and contradicts the EGARCH output. The TGARCH model shows that all cryptocurrencies have a positive gamma of 0.33%, which is insignificant. This implies that the impact of returns on their volatility is symmetrical and thereby, they do not exhibit an asymmetric effect. Finally, the Component GARCH model reflects the high value attained for the trend intercept "ω" in the case of Dogecoin suggesting that CGARCH is a good fit for this cryptocurrency. Also, the AR coefficient of the permanent volatility "ρ" is highly significant (almost 1) for all cryptocurrencies and its size exceeds the coefficients of the transitory component in all cases implying that the model is quite stable for all cryptocurrencies. However, in contrast to all remaining models, the CGARCH is the only model that reports the presence of leverage effect in most cryptocurrencies, particularly for Litecoin ($\gamma = 18\%$) which means that negative shocks have generally a higher impact on cryptocurrencies' volatility than positive shocks.

Table 2. GARCH in-sample parameters

GARCH (1,1)					
	ω	α	β	V_L	LLF
BTC	0.00006836	0.1425	0.8218	69.189%	5133.33
XRP	0.00035459	0.2958	0.6422	119.584%	4479.89
LTC	0.00015397	0.0735	0.8799	90.848%	4385.20
XMR	0.00024254	0.1240	0.8281	112.511%	4106.64
DASH	0.00019108	0.1767	0.7877	115.820%	4379.26
DOGE	0.00009461	0.1990	0.7989	333.687%	4581.59

IGARCH 1,1)				
	ω	α	β	LLF
BTC	0.00005595	0.1807	0.8193	5127.85
XRP	0.00037941	0.4109	0.5891	4476.83
LTC	0.00011379	0.1159	0.8841	4361.08
XMR	0.00015592	0.1639	0.8361	4097.98
DASH	0.00015391	0.2139	0.7861	4376.44
DOGE	0.00009380	0.2014	0.7986	4581.57

EGARCH 1,1)						
	ω	α	β	ϒ	V_L	LLF
BTC	-0.471656	0.2880	0.9243	-0.01301	70.189%	5142.36
XRP	-0.963450	0.4338	0.8264	0.1448	98.611%	4465.87
LTC	-0.246451	0.1234	0.9547	0.0742	104.051%	4406.40
XMR	-0.369975	0.2256	0.9301	0.0816	111.950%	4119.98
DASH	-0.418311	0.3034	0.9237	0.0496	101.867%	4392.79
DOGE	-0.307331	0.2950	0.9436	0.1035	103.749%	4589.60

GJR-GARCH (1,1)						
	ω	α	β	ϒ	V_L	LLF
BTC	0.000068	0.1459	0.8228	-0.0077	69.410%	5133.39
XRP	0.000432	0.4798	0.5858	-0.2732	N/A	4494.02
LTC	0.000120	0.0857	0.9041	-0.0581	87.393%	4394.90
XMR	0.000213	0.1716	0.8423	-0.1142	N/A	4122.36
DASH	0.000188	0.2025	0.7948	-0.0722	110.015%	4382.83
DOGE	0.000101	0.2774	0.7948	-0.1534	N/A	4595.50

APARCH (1,1)						
	ω	α	β	ϒ	δ	LLF
BTC	0.000068	0.1451	0.8218	0.0089	2.000	5133.33

XRP	0.000355	0.3021	0.6422	0.0105	2.000	4479.89
LTC	0.000154	0.0752	0.8799	0.0114	2.000	4385.20
XMR	0.000243	0.1268	0.8281	0.0113	2.000	4106.64
DASH	0.000191	0.1807	0.7877	0.0112	2.000	4379.26
DOGE	0.000095	0.2011	0.7989	0.0053	2.000	4581.59

TGARCH (1,1)

	ω	α	β	ϒ	LLF
BTC	0.002377	0.1551	0.8297	0.0033	5142.59
XRP	0.008819	0.3532	0.6137	0.0033	4466.17
LTC	0.003156	0.0910	0.8838	0.0033	4385.95
XMR	0.004891	0.1454	0.8210	0.0033	4099.25
DASH	0.003530	0.1756	0.8151	0.0033	4385.96
DOGE	0.002505	0.1942	0.8203	0.0033	4569.67

CGARCH (1,1)

	ω	α	β	α + β	ϒ	ρ	∅	LLF
BTC	0.00189856	0.1063	0.1169	0.2232	-0.0429	0.97214	0.11686	5137.26
XRP	0.00919461	0.5825	0.1382	0.7208	0.0100	0.98311	0.13822	4455.08
LTC	0.00373242	0.0215	0.0682	0.0897	0.1779	0.96704	0.06789	4391.04
XMR	0.00505154	0.0179	0.1173	0.1352	0.0127	0.95458	0.11733	4106.91
DASH	0.06739178	0.1634	0.6686	0.8320	0.0333	0.99996	0.02312	4395.30
DOGE	0.19717046	0.1561	0.7856	0.9417	0.0335	0.99999	0.03170	4600.10

3.2 Optimal GARCH-Type Models

Based on the three error metrics, Table 3 summarizes the optimal volatility models that best predict the cryptocurrencies' volatility and Fig. 1 displays the realized volatility versus the selected GARCH-type volatility for each cryptocurrency during COVID-19 period (out-of-sample period).

Table 3. Optimal models for each cryptocurrency for the in-sample and out-of-sample periods

	Resilience to COVID-19 (January 2015–April 2021)	
	In sample	Out of sample
BTC	CGARCH (1,1)	CGARCH (1,1)
XRP	GARCH (1,1)	GARCH (1,1)
LTC	GJRGARCH (1,1)	APARCH (1,1)
XMR	CGARCH (1,1)	IGARCH (1,1)
DASH	TGARCH (1,1)	GJRGARCH (1,1)
DOGE	APARCH (1,1)	TGARCH (1,1)

Consistency is maintained during both periods for major cryptocurrencies such as Bitcoin and Ripple whereby the CGARCH (1,1) and GARCH (1,1) models are reliably dominant in forecasting their volatilities, respectively. As for the remaining cryptocurrencies the results were different, whereby no model demonstrated superiority. Apparently, it is natural to observe some discrepancies among cryptocurrencies given their typical highly volatile behavior, particularly after the COVID-19 pandemic that has seen many investors alter and move towards cryptocurrencies as their mean of diversification and payment/trade given their digital and virtual feature.

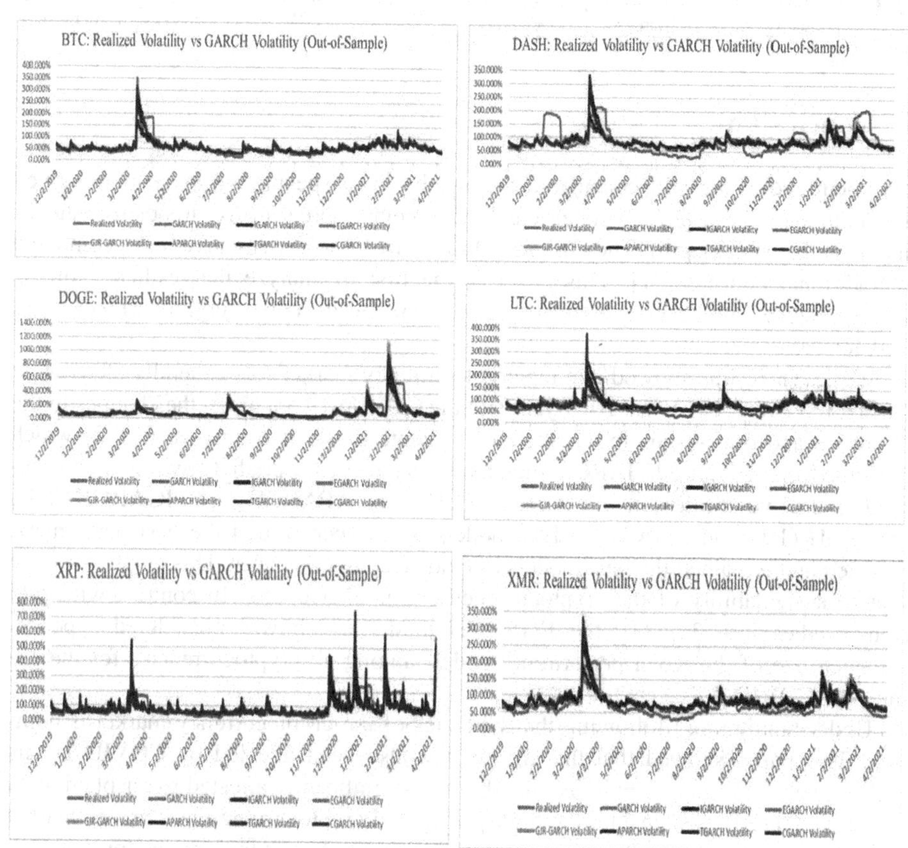

Fig. 1. Realized Vs GARCH volatility (Out-of-sample period)

3.3 Cryptocurrencies' Resilience to COVID-19

The output of Eq. (10) is depicted in Table 4. The purpose is to examine the significance and sign of ϕ and check if the volatility structure of the selected cryptocurrencies has changed in response to COVID-19. The p-values of all the cryptocurrencies, excluding Litecoin and Dogecoin, are less than 5% which indicates that they are affected by the COVID-19 pandemic. While Bitcoin and Dash volatility increased during COVID-19, Ripple and Monero volatility decreased. In other words, COVID-19 contributed in significantly affecting the volatility of the most important cryptocurrencies.

Table 4. GARCH (1,1) and the dummy variable output

COVID-19 Impact						
Cryptocurrency	**BTC**	**XRP**	**LTC**	**XMR**	**DASH**	**DOGE**
P-Value	0.0366	0	0.5841	0	0	0.1891
Significance/Affected	✓	✓	×	✓	✓	×
Volatility	Increase	Decrease	Decrease	Decrease	Increase	Increase

4 Conclusion

Cryptocurrencies are still considered among the most speculative investments given their extreme volatile and asymmetric behavior, being more sensitive to positive shocks than negative shocks [30]. They are found to overreact to the negative news compared to traditional equities [31]. This paper is the first to analyze the resilience of the cryptocurrency market, represented by the six well-known cryptocurrencies, to COVID-19.

While our results showed that the cryptocurrency market was generally affected by the first three quarters of the 2015 year and the COVID-19 era given the changes in the in-sample and out-of-sample results, they validate those of Naimy et al. [22] which confirmed that advanced GARCH models appear to better predict asymmetries cryptocurrencies' volatilities. In fact, GJRGARCH (1,1), APARCH (1,1), IGARCH (1,1), TGARCH (1,1) and CGARCH (1,1) models which were found the best performers, persisted and remained the best models as evidenced in this paper. This emphasizes the relative sustainability of the cryptocurrencies to market upsets. In contrast with previous studies [19, 32, 33], the EGARCH model which was considered superior, remained one of the worst performing models among all cryptocurrencies for the pre and post COVID-19.

Understanding and following the dynamics of the cryptocurrency market is complex. This paper showed that this market has also been affected by COVID-19. Its resilience analysis gauges how its volatility was significantly affected by displaying an overall increase if we consider the highly weighted cryptocurrency in terms of market capitalization, namely the Bitcoin and Dash, an output of great importance for investors and risk managers.

References

1. International Monetary Fund. Policy support and vaccines expected to lift activity (2021)
2. Bora, D., Basistha, D.: The outbreak of COVID-19 pandemic and its impact on stock market volatility: evidence from a worst-affected economy. J. Publ. Affairs e2623 (2021). https://doi.org/10.1002/pa.2623
3. El-Khatib, R., Samet, A.: Impact of COVID-19 on emerging markets. SSRN Electron. J. (2021). https://doi.org/10.2139/ssrn.3685013

4. Chowdhury, E.K., Khan, I.I., Dhar, B.K.: Catastrophic impact of Covid-19 on the global stock markets and economic activities. Bus. Soc. Rev. basr.12219 (2021). https://doi.org/10.1111/basr.12219
5. O'Donnell, N., Shannon, D., Sheehan, B.: Immune or at-risk? Stock markets and the significance of the COVID-19 pandemic. J. Behav. Exp. Financ. 30 (2021). https://doi.org/10.1016/j.jbef.2021.100477
6. Padhan, R., Prabheesh, K.P.: The economics of COVID-19 pandemic: a survey. Econ. Anal. Policy **70**, 220–237 (2021). https://doi.org/10.1016/j.eap.2021.02.012
7. Corbet, S., Hou, Y. (Greg), Hu, Y., Larkin, C., Oxley, L.: Any port in a storm: cryptocurrency safe-havens during the COVID-19 pandemic. Econ. Lett. **194**, 109377 (2020). https://doi.org/10.1016/j.econlet.2020.109377
8. Yoshino, N., Taghizadeh-Hesary, F., Otsuka, M.: Covid-19 and optimal portfolio selection for investment in sustainable development goals. Financ. Res. Lett. **38**, 101695 (2021). https://doi.org/10.1016/j.frl.2020.101695
9. Conlon, T., McGee, R.: Safe haven or risky hazard? Bitcoin during the Covid-19 bear market. Financ. Res. Lett. **35**, 101607 (2020). https://doi.org/10.1016/j.frl.2020.101607
10. Conlon, T., Corbet, S., McGee, R.J.: Are cryptocurrencies a safe haven for equity markets? An international perspective from the COVID-19 pandemic. Res. Int. Bus. Financ. **54**, 101248 (2020). https://doi.org/10.1016/j.ribaf.2020.101248
11. Chen, C., Liu, L., Zhao, N.: Fear sentiment, uncertainty, and bitcoin price dynamics: the case of COVID-19. Emerg. Mark. Financ. Trade **56**, 2298–2309 (2020). https://doi.org/10.1080/1540496X.2020.1787150
12. Dutta, A., Das, D., Jana, R.K., Vo, X.V.: COVID-19 and oil market crash: revisiting the safe haven property of gold and Bitcoin. Resour. Policy **69**, 101816 (2020). https://doi.org/10.1016/j.resourpol.2020.101816
13. Bouoiyour, J., Selmi, R.: Coronavirus Spreads and Bitcoin's 2020 Rally: Is There a Link ? (2020)
14. Goodell, J.W., Goutte, S.: Co-movement of COVID-19 and Bitcoin: evidence from wavelet coherence analysis. Financ. Res. Lett. **38**, 101625 (2021). https://doi.org/10.1016/j.frl.2020.101625
15. Kristoufek, L.: Grandpa, grandpa, tell me the one about bitcoin being a safe haven: new evidence from the COVID-19 pandemic. Front. Phys. (2020). https://doi.org/10.3389/fphy.2020.00296
16. Naeem, M.A., Bouri, E., Peng, Z., Shahzad, S.J.H., Vo, X.V.: Asymmetric efficiency of cryptocurrencies during COVID19. Phys. A Statist. Mech. Appl. **565**, 125562 (2021). https://doi.org/10.1016/j.physa.2020.125562
17. Shahzad, S.J.H., Bouri, E., Kang, S.H., Saeed, T.: Regime specific spillover across cryptocurrencies and the role of COVID-19. Financ. Innov. **7**(1), 1–24 (2021). https://doi.org/10.1186/s40854-020-00210-4
18. Chu, J., Chan, S., Nadarajah, S., Osterrieder, J.: GARCH modelling of cryptocurrencies. J. Risk Financ. Manag. **10**, 1–15 (2017)
19. Naimy, V., Hayek, M.: Modelling and predicting the bitcoin volatility using GARCH models. Int. J. Math. Model. Numer. Opt. **8**, 197–215 (2018)
20. Gronwald, M.: Is bitcoin a commodity? On price jumps, demand shocks, and certainty of supply. J. Int. Money Financ. **97**, 86–92 (2019). https://doi.org/10.1016/j.jimonfin.2019.06.006
21. Gyamerah, S.A.: Modelling the volatility of Bitcoin returns using GARCH models. Quant. Financ. Econ. **3**, 739–753 (2019). https://doi.org/10.3934/QFE.2019.4.739

22. Naimy, V., Haddad, O., Fernández-Avilés, G., El Khoury, R.: The predictive capacity of GARCH-type models in measuring the volatility of crypto and world currencies. PLoS ONE **16**(1), e0245904 (2021). https://doi.org/10.1371/journal.pone.0245904
23. Bollerslev, T.: Generalized autoregressive conditional heteroskedasticity. J. Economet. **31**, 307–327 (1986). https://doi.org/10.1016/0304-4076(86)90063-1
24. Engle, R.F., Bollerslev, T.: Modelling the persistence of conditional variances. Economet. Rev. **5**, 1–50 (1986). https://doi.org/10.1080/07474938608800095
25. Nelson, D.: Conditional heteroskedasticity in asset returns: a new approach. Econometrica **59**, 347–370 (1991)
26. Glosten, L., Jagannathan, R., Runkle, D.: On the relation between the expected value and the volatility of the nominal excess return on stocks. J. Financ. **48**, 1779–1801 (1993)
27. Ding, Z., Granger, C.W.J., Engle, R.F.: A long memory property of stock market returns and a new model. J. Emp. Financ. **1**, 83–106 (1993). https://doi.org/10.1016/0927-5398(93)90006-D
28. Zakoian, J.-M.: Threshold heteroskedastic models. J. Econ. Dyn. Control **18**, 931–955 (1994). https://doi.org/10.1016/0165-1889(94)90039-6
29. Engle, R.F., Lee, G.G.J.: A Permanent and Transitory Component Model of Stock Return Volatility. Department of Economics, University of California, La Jolla (1992)
30. Naimy, V., Chidiac, J.E., Khoury, R.E.: Volatility and value at risk of crypto versus fiat currencies. In: Abramowicz, W., Klein, G. (eds.) BIS 2020. LNBIP, vol. 394, pp. 145–157. Springer, Cham (2020). https://doi.org/10.1007/978-3-030-61146-0_12
31. Borgards, O., Czudaj, R.L.: The prevalence of price overreactions in the cryptocurrency market. J. Int. Financ. Mark. Inst. Money **65**, 101194 (2020). https://doi.org/10.1016/j.intfin.2020.101194
32. Van Der Krogt, D.: Financial Economics GARCH Modeling of Bitcoin, S&P-500 and the Dollar (2018). http://hdl.handle.net/2105/42751
33. Abdalla, S.Z.S.: Modelling exchange rate volatility using GARCH models: empirical evidence from Arab countries. Int. J. Econ. Financ. **4**, 216–229 (2012). https://doi.org/10.5539/ijef.v4n3p216

Fast Payment Systems Dynamics: Lessons from Diffusion of Innovation Models

Victor Dostov[1,2,3(✉)], Pavel Shust[1,2,3], and Svetlana Krivoruchko[4]

[1] Center for Distributed Ledger Technologies of Saint-Petersburg State University, St. Petersburg, Russia
leonova.v.anna@gmail.com

[2] Saint-Petersburg State University, St. Petersburg, Russia

[3] Russian Electronic Money and Remittance Association, Moscow, Russia

[4] Financial University under the Government of the Russian Federation, Moscow, Russia

Abstract. The paper applies innovation dynamics equations to analyze audience dynamics of instant payment systems (IPS) behavior. The research shows that dynamics of IPS are well described by the Ricatti equations, which are generalizations of the Bass basic model of innovation diffusion taking into account different patterns of audience behavior. As proved by IPS experience in Britain, Sweden and other countries, commodification of fast payment services allows for rigorous description of the IPS dynamics. Quantitative estimates are obtained for the degree of cooperative customer behavior and the typical time of system growth. However, these parameters may differ for different transactions types within the same system due to their different business nature. We also show that all systems may be described by the generalized trajectory of evolution and demonstrate that described systems with different payment operations types are located on different stages of this trajectory what reflects their maturity and operation nature. The results can be used for qualitative and quantitative assessment of IPS customers behavior and for short- and medium-term projections. We also discuss some future improvements such as including of competition of different IPSs for countries in which more than one IPS run simultaneously.

Keywords: Fast payment system · Instant payment system · IPS · Retail payment · Bass equation · Ricatti equation · Swish · Venmo · Payment system model

1 Introduction

Payment industry has been one of the most innovative industries over the last two decades. Payment sector was the most affected by the innovations (dubbed 'fintech'): the well-known cases include evolution of payment cards from magnetic stripe to chip-based and contactless cards, appearance of electronic money, contactless payments, QR payments, and probably the most striking innovation – cryptocurrencies [1]. The remarkable growth of innovations in this area is an obvious challenge for researchers,

W. Abramowicz et al. (Eds.): BIS 2021 Workshops, LNBIP 444, pp. 359–370, 2022.
https://doi.org/10.1007/978-3-031-04216-4_31

however, the dynamics of innovations is still under-researched issue. In previous papers [2–5], we used a methodology similar to Bass's formalism [6] to describe the size of the audience and the number of payment cards, e-wallets and cryptocurrencies transactions which can be regarded as indicators for innovation diffusion. In this paper, we have endeavored to describe the dynamics of Instant Payment Systems (IPS) that are used by the mass market consumer.

Generally speaking, IPS, often dubbed as Fast Payment System (FPS) is a set of technical solutions and contractual relationships between various banks that allow individuals to initiate almost instant transactions to each other (P2P, peer-to-peer transactions) and to legal entities (C2B, customer-to-business transactions, that include, for example, payment for goods in stores or utility bills) [7]. Although these systems are relatively simple, and their first implementation started in 20th century, but worldwide implementation of really massive and effective hi-tech solutions began only in the beginning of the 21st century. Implementation of IPS is complicated, not in last extent, because they require reaching the agreement among all banks and ensuring information and financial security of IPS transactions. First IPS appeared in 1973 in Japan but most of the current instant payment systems became operational in the 21st century (in Brazil, Mexico, South Africa, South Korea, Iceland and other countries). Faster Payments Service in the United Kingdom (2008) is probably one of the well-known IPS in the world. The business models of such systems may differ [8]. While Faster Payments Service in the United Kingdom and Russian Fast Payment System (SBP) were supported, fully or in part, by the state, Swish in Sweden or Russian Sberbank based IPS are private initiatives (by a single bank of consortium). These systems may support different functions: from simple P2P transfers to recurrent payments, C2B transfer, person-to-government transfers, and so on.

Over the past 10 years, certain statistics for various IPS have been accumulated, which can be used to develop a mathematical model to describe the dynamics of the IPS audience and its other indicators.

2 Theoretical Framework

The standard mathematical tool to describe dynamics of audience of any given service or product is the diffusion equation or, more precisely, the dynamics of innovations. Bass first proposed this equation in 1963 [9]. The main assumption in it is that it is possible to express the rate of change of the audiences dx/dt as a function of the size of the audience x and, sometimes, time t.

In particular, Bass actually used the equation

$$\frac{dx}{dt} = (p + qx)(1 - x) \tag{1}$$

or

$$\frac{dx}{dt} = -qx^2 + (q-p)x + p \tag{2}$$

Where the coefficient p in Eqs. (1) and (2) on the right side describes the speed of making independent decisions, and the coefficient q describes the speed of making decisions under the influence of existing users in the system. This and related equations have been successfully used to describe the sales of refrigerators, toasters, growth in the number of social networks users, e-assessments and some other markets [6, 10–14].

In [3], the equations were shown to describe the dynamics of retail payment markets – cards, e-wallets and cryptocurrencies - quite well. However, two features are specific for the payment markets: commodification and technical cooperation. Commodification [15] refers to the simplification of payment services in recent years and the reduction in the number of parameters describing these services. Technical cooperation, a term borrowed from the theory of games [16], refers to the fact that any payment involves two parties, and, depending on whether the payee behavior affects the payer behavior or not, the market might be characterized as cooperative or non-cooperative. The dominant behavior (cooperative/non-cooperative) is found to be connected to the dominant types of transactions. For customer-to-business (C2B) systems, such as card payments at POS in retail stores, the behavior of the payer does not depend on the behavior of the payee (in this case, the store), the store's ability to accept payments is constant and almost infinite. In the case of person-to-person (P2P) systems, such as international money transfers, the sending of funds is conditioned on the recipient's willingness to accept these funds. In previous works, the authors looked at purely cooperative and non-cooperative markets. In the case of C2B systems, non-cooperative behavior prevails, and the system behavior is described by the equation

$$\frac{dx}{dt} = a_b(N-x) - bx \tag{3}$$

Where coefficient a describes the probability that the customer will start using the system at any time, coefficient b describes the probability that he will stop using it at any point in time. For purely cooperative p2p systems, the equation takes the form

$$\frac{dx}{dt} = a_p(N-x)x - bx \tag{4}$$

Where the meaning of coefficients a and b is similar. Analytical solutions, practical examples, and qualitative properties of Eqs. (3) and (4) were described in detail in [2, 3].

In this paper, we assume that all markets usually demonstrate a mixture of cooperative and non-cooperative behavior. These types of behavior may manifest in different ways. For example, while C2B systems (e.g. card payment in stores) are generally non-cooperative (payer's behavior is not affected by the payee's and vice versa), it also demonstrates a cooperative component in Bass terms, as individuals' (payers') choices are affected by their social environment. In other words, the dynamics of card payments in retail will be determined not only by the availability and convenience of POS terminals, but also by how family and friends influence the customer's decision to use cards instead of cash.

As a result, the generalized model was written as

$$\frac{dx}{dt} = a_p(N - x)x + a_b(N - x) - bx \tag{5}$$

Or in the form of the Ricatti equation [17].

$$\frac{dx}{dt} = -a_p x^2 + lx + a_b N \tag{6}$$

Where

$$l = a_p N - a_b - b \tag{7}$$

the solution of this equation gives

$$x = \frac{1}{2a_p}\left(D\tanh\left(\frac{1}{2}tD - C\right) + l\right) \tag{8}$$

Where

$$D^2 = l^2 + 4a_b a_p N \tag{9}$$

which defines the reverse typical time of audience change. For t $\rightarrow \infty$ (8) transforms into

$$x_\infty = \frac{1}{2a_p}(D + l) \tag{10}$$

giving the maximum audience size that the system aims for at long times. The constant C can be found from the initial condition x(0) = x0

$$C = \operatorname{arctanh}((l - 2a_p x_0)/D)) \tag{11}$$

We can formally also rewrite Eq. (11) in a more general dimensionless form redefining the values to dimensionless

$$\chi \rightarrow (2a_p x - l)/D \tag{12.1}$$

$$\tau \rightarrow \frac{Dt}{2} - C \tag{12.2}$$

and getting the most general form of the solution (8) as

$$\chi = \tanh\tau \tag{13}$$

This form has so far remained unnoticed in formal studies of the Bass equation. Formally speaking, all solutions of the Bass equations in proper dimensionless variables and coefficients lie on the same trajectory, and their qualitative differences (the presence of an inflection point, signs of the second derivative, and so on), including limiting cooperative and non-cooperative systems, are determined only by where the real moments of observation are located on this trajectory.

Thus, parameter C largely determines the qualitative behavior of the system. In the range Dt >> C non – cooperative behavior prevails, while in Dt << C prevails cooperative behavior. In other words, Eq. (5) reduces to (3) and (4) respectively. In this paper, we apply this model to instant payment systems. It is obvious that the IPS market is almost perfectly commodified – the service is homogeneous among financial services providers and extremely simple. In addition, it is important that various banks are connected to the fast payment system, and the general behavior is the product of averaging for individual banks and their clients. This further increases the uniformity and commodification of this universal service. A great advantage for a researcher is that the statistics on IPS is usually more available, compared to the statistics of private schemes.

Some reservations need to be made when applying the classical equations of innovation dynamics. Initially, the Bass's dynamics model was built for physical goods, their usage can be described in binary terms – e.g. a customer either has a refrigerator or not. In the case of services, the situation is more complicated – a customer can use the system intermittently. Previously, we assumed that the system is used by the customer when at least one transaction is made by him/her over time 1/D. Apart from the number of unique users, other data (such as number or volume of transactions) might be available for IPS as well. In this case, we assume that these parameters are linearly related to the number of customers. Certainly, this proportion is not precise and stable – for example, as the average payment amount changes slowly over time. For example, data for card payments in Britain [18] shows that the average amplitude of non-monotone change in the average transaction volume in 2009–2016 was about 10%, the average transaction changed by about 20%, while the volume of transactions in 2011–2016 increased by about 300% [19]. A similar pattern is observed in younger markets [20, 21].

Thus, the assumption of a linear relationship between different metrics is reasonable. In other words, as long as the typical time it takes for the average transaction or the number of transactions per card to change is t << 1/D, the expression (8) is valid for any metric, surely with different parameters. In general, we can say that in the case of payment systems, Eq. (5) is written for the number of efficient users, through which other metrics can be expressed linearly.

3 Methodology and Results

We start with statistics on the British FPS. The main data is shown in Fig. 1. For convenience, we rationed the values to the value from the final observation date. Note that the statistics are provided by two types of services: standing orders and payments. Single Immediate Payments (SIP), are initiated by one customer to another. Standing orders are recurrent payments made to a legal entity with a predetermined frequency, i.e. C2B transactions. We can assume a priori that the role of cooperation for SIP will be lower. To verify whether the actual data and (8) correlate, we use the least middle average percentage error (MAPE)

$$MAPE = 1/N\sum_{i=1}^{N}|X - x|/X \tag{14}$$

method [22] to find best fitting parameters of model. Here and further we use capitals, e.g. SOP or X, to indicate actual data and small letters, e.g. sop or x, to indicate model calculations. This choice allows to take into account wide range of values, however has evident drawback consisting in high weight of small value data of initial period in the criteria, as it will be discussed further. The results are shown in Fig. 1 and Table 1.

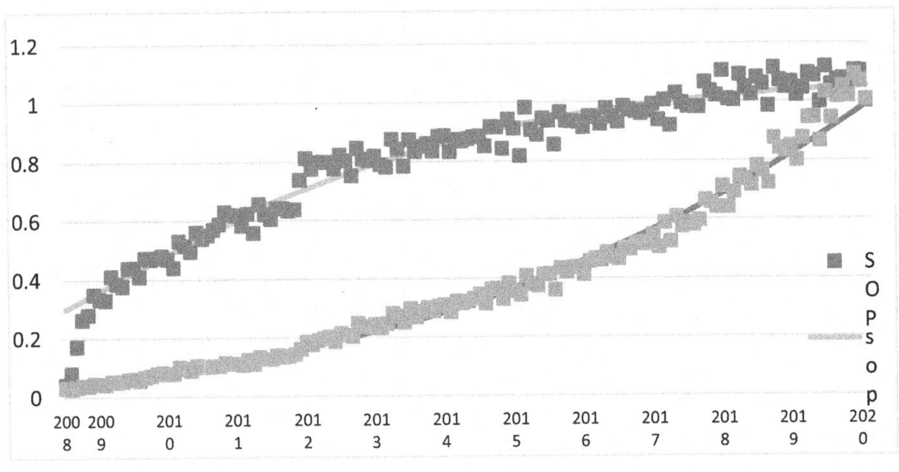

Fig. 1. Number of transactions in the Faster Payments, monthly

Table 1. Model parameters for different IPS

Payment system	C	Period of operation, at the time of observation, years	1/D, years
Faster Payments SOP, UK	0	12	4
Faster Payments SIP, UK	1,7	12	5
Swish, SW	3	7	0,7
Venmo, US	2,2	6	1,2

We can see a decent match between the estimated and actual data. Quantitative statistical parameters [23] calculated on all data range are given in table 2. We see small deviation of model from real data.

Table 2. Statistical parameters for different IPS

Payment system	RMSE	MSE	MAE	MAPE	MAPE1
Faster Payments SOP, UK	0,039204356	0,00153698	0,031361	11.14%	4,00%
Faster Payments SIP, UK	0,035037659	0,00122764	0,023943	9.75%	6,24%
Swish, SW	0,041299594	0,001705656	0,034349	15.97%	12,15%
Venmo, US	0,019751364	0,00039012	0,012531	14,83%	4,47%

Moreover, we can see that different services behave qualitatively differently within the same system. SIP show explicit cooperative behavior, with a positive second derivative and far from saturation, standing order behavior is much less cooperative, with a negative second derivative. Time indicators D are relatively close. Visually, fit is quite good, but big deviation at small values of initial period led to relatively large MAPEs.

The second example was the development of the Swish IPS. Swish was implemented in Sweden 2012 as a joint project of a consortium of commercial banks and the Central Bank of Sweden. Swish transactions demonstrate striking cooperative behavior. At the same time, the parameters show that the maximum increase in customer number is reached at the point t = tm

$$2C = Dt_m \tag{15}$$

that is around 2017. We do not have data on transactions for 2019, but the information on the drop in Swish [24] wallet downloads in 2019 confirms this assumption. Thus, the simulation shows that within the existing business model, Swish is close to saturation of the customer base (Fig. 2).

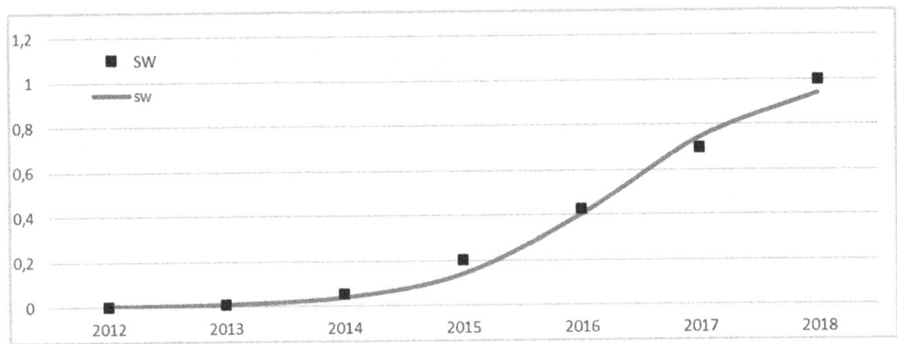

Fig. 2. Number of transactions in the Swish IPS, yearly

Another example of a private IPS is Venmo, launched in the US in 2010. Figure 3 shows the quarterly payment volume, also normalized as of the last observation date [25]. The parameters found in (8) are also shown in Table 1.

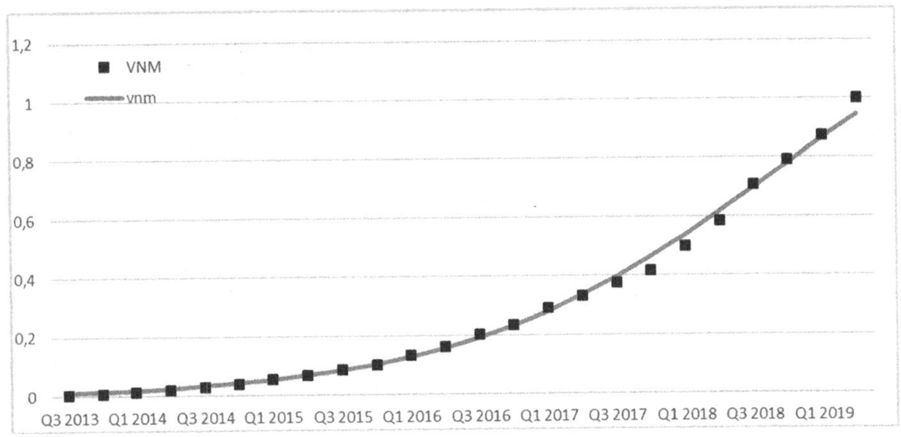

Fig. 3. The number of transactions in IPS Venmo, quarterly

For all figures we see good visual coincidence of model and actual data. The exception is the taking off period, which can be attributed to the pilot mode of the system where number of participating banks and quality of the services varies in time contradicting our assumption on constant or slow changing a and b values. Values for these periods are small, so their contribution in all statistic criteria except MAPE are small and we see quite small absolute model errors. However, this period contributes more visible in MAPE value because of sensitivity to deviation in small terms. We also calculate MAPE1 with take-off period data excluded. MAPE1 values are much smaller for Swish and Venmo, what demonstrates high quality of the model for the rest of data range which also can be visible seen on Fig. 1 and following figures. For SIP and SOP the difference absents practically. Of course, if someone needs to interpolate or predict shorted data range optimizing MAPE for this data range or to use statistics with weights [23], even smaller deviation can be obtained.

4 Discussion

By analyzing Table 1, from the difference in parameter C we can see that different systems are at different stages of market saturation, which is quite logical, given the difference in geography and services. However, for all three systems, we see a good match of the calculated values with the actual ones and quite plausible estimates for the main parameters. Thus, proposed approach may underline qualitative and quantitative models of different IPS, including good quality forecasting.

We also find effective the concept of general trajectory (13). In Fig. 4, approximate positions of all four products on a common trajectory (13) are shown schematically. This gives us a simple view on current state of product development and further growth perspective.

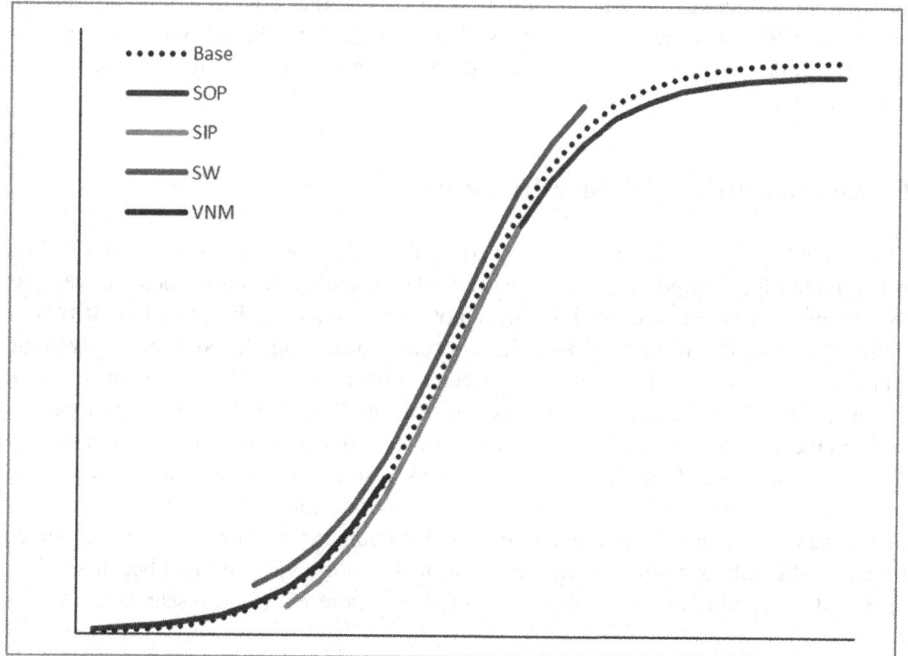

Fig. 4. Position of various IPS on the generalized trajectory

5 Implications

As shown in previous studies, Bass-like models describe the behavior of various classes of payment systems quite well. In this paper, this statement is extended to IPS, and the quality of interpolation is very high. We believe, this is explained by the commodification phenomenon and the rapid achievement of IPS a new state of maturity, which leads to a slight change in the typical parameters of systems over time. The quality of the models allows them to be used for short- and medium-term forecasts. On existing samples of long N, given in Table 1, training the model over an interval of N-2 years gives a forecast for the last two years with a dispersion close to the dispersion of the full sample. It is also worth mentioning that the model allows to identify possible noticeable deviations of practical behavior from the predicted one in the future and provides a direction for analyzing what modifications in the model and user behavior caused these deviations. Analysis of actual curves allows to make assumptions about presence or absence of cooperation between users, which might be useful for the development of payment products. It is also important that the previously obtained qualitative regularities of the used model, described in [2, 3], apply to the IPS as well. In particular, data about initial state of the IPS does not provide enough information for longer-term forecasts: as shown in [2], solutions of type (6) equations for small times can be determined by a different set of variables than the stationary solution x0.

6 Limitations and Future Research

The model has several limitations that warrant further research. Firstly, the model does not account for competition between multiple systems. As mentioned above, this reflects practice in many cases but surely not in all situations. In rare cases of two or more IPS co-exist in one market, the approach based on the system of dynamic equations proposed in [3] we may expect stronger competition behavior for P2P payments than for C2B and other effects described in this paper. Secondly, the model is built on the assumption that the coefficients of the model are constant or change slowly. This assumption needs further analysis as we see some visible deviations on starting period where number of banks participating is growing thus influencing attractiveness of the service for clients and changing a and b coefficients. More accurate approach may take this into account. Thirdly, amount of IPS worldwide and number of services provided is growing, which will provide basis for more extended researches.

7 Conclusion

The paper successfully demonstrates the use of innovation dynamics equations for instant payment systems in various countries. A good quantitative match with real data is obtained, qualitative effects determined by the different types of behavior (cooperative and non-cooperative) are demonstrated, and extrapolation estimates for the medium-term development of systems are obtained. A qualitative relationship between the type of a payment product and the curve describing its dynamics is shown. The proposed model might improve the quality of the IPS dynamics and audience behavior analysis as well.

References

1. Brett King: Bank 4.0: Banking Everywhere, Never at a Bank. Marshall Cavendish Business, Singapore (2018)
2. Dostov, V., Shoust, P., Popova, E.: Using mathematical models to describe the dynamics of the spread of traditional and cryptocurrency payment systems. In: Misra, S., et al. (eds.) ICCSA 2019. LNCS, vol. 11620, pp. 457–471. Springer, Cham (2019). https://doi.org/10.1007/978-3-030-24296-1_36
3. Dostov, V., Shust, P., Popova, E.: Using mathematical models for analysis and prediction of payment systems behavior. In: 34th International Business Information Management Association Conference (IBIMA), Madrid, Spain, 13–14 November 2019, pp. 2284–2895. IBIMA Publishing, King of Prussia, PA (2019)
4. Dostov, V., Shust, P., Krivoruchko, S.: Using mathematical models for analysis and prediction of payment systems behavior. Paper presented at fifth international congress on information & communication technology (ICICT 2020), London, UK, 20–21 February 2020 (2020)
5. Dostov, V., Shust, P.: A generalization of Bass equation for description of diffusion of cryptocurrencies and other payment methods and some metrics for cooperation on market. In: Gervasi, O., et al. (eds.) ICCSA 2020. LNCS, vol. 12251, pp. 3–13. Springer, Cham (2020). https://doi.org/10.1007/978-3-030-58808-3_1
6. The Bass Model. Bass's Basement Research Institute. http://bassbasement.org/BassModel/Default.aspx
7. Bank for International Settlements: Fast payments – enhancing the speed and availability of retail payments (2016). https://www.bis.org/cpmi/publ/d154.pdf
8. Bank for International Settlements: Shaping the future of payments. https://www.bis.org/statistics/payment_stats/commentary1911.htm
9. Bass, Frank M.: A dynamic model of market share and sales behavior. In: Proceedings, Winter Conference American Marketing Association, Chicago, IL (1963)
10. Bass, F.M.: A new product growth model for consumer durables. Manag. Sci. **15** (1969)
11. Baran, M.: Diffusion of innovations in the systems thinking approach. Manag. Bus. Innov. **6**, 16–24 (2010)
12. Ganjeizadeh, F., Lei, H., Goraya, P., Olivar, E.: Applying looks-like analysis and bass diffusion model techniques to forecast a neurostimulator device with no historical data. In: 27th International Conference on Flexible Automation and Intelligent Manufacturing, FAIM, Modena, Italy, pp. 1916–1924 (2017). Procedia Manufacturing
13. Soffer, T., Nachmias, R., Ram, J.: Diffusion of web supported instruction in higher education: the case of Tel-Aviv University. Educ. Technol. Soc. **13**(3), 212–223 (2010)
14. Boyle, A.: Some forecasts of the diffusion of e-assessment using a model. Innov. J. Public Sector Innov. Journal **15**(1), 1–30 (2010)
15. Dostov, V.L., Shust, P.M.: Evolution of the electronic payments industry: problems of quality transition. Report for Russian Academy of National Economy and Public Administration under the President of the Russian Federation (2017)
16. Takako, F.-G.: Non-Cooperative Game Theory. Springer, Heidelberg (2015).https://doi.org/10.1007/978-4-431-55645-9

17. Simmons, G., Robertson, J.: Differential Equations with Applications and Historical Notes, 2nd edn. International Series in Pure and Applied Mathematics. McGraw-Hill (1991)
18. Average value of retail transactions made by credit and charge card in the United Kingdom (UK) from 2009 to 2016 (2020). https://www.statista.com/statistics/310640/average-credit-card-retail-transaction-value-in-the-united-kingdom-uk/
19. UK Credit Card Data & Statistics (2020). https://www.merchantsavvy.co.uk/uk-credit-card-statistics/
20. The average "Mir" card payment receipt has decreased. Klerk (2017). https://www.klerk.ru/bank/news/465742/
21. Trefilov, V.: The average receipt for VISA card transactions in Russia decreased by 5% over the year. Izvestiia (2019). https://iz.ru/856773/2019-03-15/srednii-chek-po-operatciiam-po-kartam-visa-v-rossii-snizilsia-na-5-za-god
22. Dodge, Y., Jureckova, J.: Adaptive Regression. Springer, New York (2020). https://doi.org/10.1007/978-1-4419-8766-2
23. Shao, J.: Mathematical Statistics. Springer, New York (2003). https://doi.org/10.1007/b97553
24. Number of downloads of the Swish app in Sweden from January 2019 to January 2020 (2020). https://www.statista.com/statistics/1098036/monthly-swish-app-downloads-in-sweden/
25. Interesting Venmo Statistics and Facts (2020). https://expandedramblings.com/index.php/venmo-statistics-facts

A Meta-review of Blockchain Adoption Literature in Supply Chain

Funlade T. Sunmola[1(✉)], Patrick Burgess[1], and Albert Tan[2]

[1] School of Physics, Engineering, and Computer Science, University of Hertfordshire, Hertfordshire AL10 9AB, UK
f.sunmola@herts.ac.uk

[2] Malaysia Institute for Supply Chain Innovation, No. 2A, Persiaran Tebar Layar, Seksyen U8, Bukit Jelutong, 40150 Shah Alam, Selangor, Malaysia

Abstract. Supply chains are increasingly adopting industry 4.0 technologies to meet exceeding stakeholder expectations. Blockchain technology offers an opportunity to facilitate the digital transformation of supply chains. Supply chains can benefit from the characteristics of blockchain including through transparency, traceability, and immutable data, to enable for example quality, sustainability, provenance, and safety. Adoption considerations for blockchain are important to ensure needs are met in the early adoption stages and further stages of deployment. This study aims to explore the adoption considerations for blockchain across supply chain domains reported in the literature, focusing on adoption factors and readiness. Research methodology used is a meta-analysis of literature review studies on blockchain adoption in supply chains to identify themes. The review identified 102 papers from four databases, and 33 are selected for analysis, identifying 64 blockchain adoption factors. Security, system integration, trust, scalability, costs, and traceability are found to be important blockchain adoption factors for supply chain. The adoption factors show a spread over a people-process-technology framework. Limitations of the research and areas for future research are highlighted.

Keywords: Blockchain · Adoption · Meta review · Supply chain

1 Introduction

Industry 4.0 technologies support the digital supply chain. Digital supply chains are those adopting novel technologies to enhance performance and create a competitive edge. Example industry 4.0 technologies are cloud computing, artificial intelligence, big data, internet-of-things, augmented reality, 3D printing, and blockchain technology [1, 2]. Blockchain technologies are increasingly implemented and researched in the field of supply chain management. In principle, blockchain ledgers hold information like other ledger systems, for example, price, quantity, and quality aspects. Blocks containing timestamps, Merkle tree root and parent hash, nBits, and nonce are built together to represent a series of transactions forming the Blockchain [3]. Identified by [4] are some of the important characteristics of blockchain enabling technologies in

W. Abramowicz et al. (Eds.): BIS 2021 Workshops, LNBIP 444, pp. 371–388, 2022.
https://doi.org/10.1007/978-3-031-04216-4_32

supply chain management. These are data safety, accessibility, documentation, data management, and quality. Blockchain technologies are not without criticism, as discussed by [5], legal aspects and privacy are some of the important challenges in blockchain adoption.

Supply chains take raw products through a series of processes to create value added products [6]. Currently, there is a drive towards sustainable supply chains [7]. Digital transformation can disrupt supply chains to meet the sustainability needs of consumers [8]. Research on digitally enabled supply chains is shown in the literature [1, 9, 10]. The requirements for digital platforms in sustainable supply chains are identified in [11, 12]. For example, traceability is identified as a requirement to improve quality and safety assurance, as discussed by [13]. Blockchain offers the ability to support this need, with immutable, transparent, visible, and traceable data, amongst others [14]. To enable digital transformation, knowledge of the building-blocks is important. [15] present a building-block model for digital transformation. This is further developed for blockchain technology in supply chains by [5]. Imperative in the model in [5] is the three-phase implementation process of pre-adoption, adoption, and post-adoption. In addition, the pre-adoption phase discusses the need for adoption readiness. Adoption readiness is discussed as the level in which an organization or supply chain is prepared to adopt technologies. More important, adoption readiness influences the future success of the technology. Within adoption, PPT (people-process-technologies) considerations can support the understanding of categories for adoption considerations [16].

Research on blockchain advantages, challenges, and potential applications are shown [17–19]. As part of the feasibility and adoption process, it is necessary to understand adoption factors for blockchain based supply chains. Adoption factors are playing a role in decision making when implementing technologies. Adoption factors have been investigated in a variety of settings including pharmaceutical industries, smart manufacturing, and supply chain management [16, 20]. However, limited research presents an overview of blockchain adoption factors across several supply chain domains, while considering context and adoption readiness. Therefore, this research aims to identify emerging considerations of blockchain adoption in supply chains through a meta-review. The meta-review focuses on three main areas for contribution. i) The adoption factors between supply chain domains, showing the importance of adoption context ii) the adoption factors of blockchain technology in supply chain, and iii) adoption readiness considerations for blockchain adoption in supply chain. The remaining sections in the paper is a literature review in Sect. 2, followed by the research methodology in Sect. 3. Section 4 presents the results and the discussion. The report is concluded in Sect. 5.

2 Literature Review

2.1 Blockchain Adoption Factors in Supply Chains

Blockchain adoption research in supply chains is evident in both cross-sector review studies (those that study multiple supply chains) and in specific review studies (those that study individual supply chains). Table 1 shows an overview of the adoption factors identified in supply chains. As shown in Table 1, cross-sector studies have received the most attention in respect to blockchain adoption in supply chain. The food, automotive, healthcare, and public supply chains have also been focused on in several individual studies. Some adoption considerations are shown more specifically to the supply chain domains suggesting a link towards the importance of adoption context regarding the sectors and technology characteristic considered. For example, in the food supply chain, traceability is identified in [21–23], while in the automotive supply chain supportive and legal, system integration, security, automation, and resources are more identifiable [24, 25]. Trust is important e.g. in the health care supply chain [26, 27], in addition to privacy, which is also identified the financial supply chains focused papers [28, 29]. The pharmaceutical supply chain is the only one reported to identify validity and accuracy [16], while smart manufacturing requires flexibility [30], showing the unique requirement needs in individual supply chain domains. These key characteristics cut across sectors and industries and are indicative for some in the literature.

Table 1. Adoption considerations form literature

Supply chain	Adoption considerations	Source (s)
Agri-Food	Supportive and legal; Privacy; Trust; Efficiency; System Integration; Disintermediation; Usability; Security; Knowledge and Skills; Scalability; Costs; Traceability; Immutability; Socio-demographic; Company/organizational factors; Company capability; Provenance; Audibility; Product safety	[21–23]
Automotive	Supportive and legal; Trust; Support Infrastructure; Efficiency; System' Integration; Supply Chain Integration; Sustainability; Disintermediation; Safe monitoring; Authentication; Usability; Security Knowledge and Skills; Transparency; Scalability; Costs; Traceability; Immutability; Automation; Frugal implementation; Company/organizational factors; Attitude; Resources; Visibility; Ownership and management support; Collaboration; Data quality and integrity; System capability; Permissions	[24, 25]
Cross-sector/non-specified	Supportive and legal; Innovation drive; Privacy; Trust; Support; Infrastructure; Efficiency; System Integration; Supply Chain Integration; Sustainability; Disintermediation; Safe monitoring; Reliability; Authentication; Usability; Security; Knowledge and Skills; Transparency; Scalability; Storage capacity; Costs; Traceability; Immutability; Decentralized or distributed; Automation; Energy consumption; Governance; Speed; Company capability; Provenance; Audibility; Clarity; Awareness; Attitude; Resources; Value creation; Product safety; Visibility; Supply chain digitalization; Ownership and management support; Collaboration; Data quality and integrity; Data sharing; Define Scope; Facilitation effect; Open source; Accessibility	[31–45]
Diamond	Supportive and legal; Trust; Support Infrastructure; Efficiency; Security; Scalability; Costs; Supply chain digitalisation; Risk; Real-time	[46]
Financial	Supportive and legal; Privacy; Trust; Efficiency; System Integration; Disintermediation; Usability; Security; Transparency; Scalability; Traceability; Immutability; Decentralized or distributed; Automation; Energy consumption; Governance	[28, 29]
Healthcare	Supportive and legal; Innovation drive; Privacy; Trust; Support Infrastructure; Usability; Security; Transparency; Traceability; Immutability; Decentralized or distributed; Auditability; Mobility Ownership and management support; Reproducibility; Experts	[26, 27]
Logistics	System Integration Security; Knowledge and Skills; Scalability; Traceability; Resources; Shared benefits; Best practices; Investment from partners	[47]
Pharmaceutical	Innovation drive; Privacy; Trust; Efficiency; System Integration' Supply Chain Integration; Safe monitoring; Reliability; Authentication; Usability; Security; Transparency; Scalability; Storage capacity; Costs; Traceability; Immutability; Decentralized or distributed; Visibility; Data quality and integrity; Accuracy; Validity	[16]
Public sector supply chain	Efficiency; Knowledge and Skills Scalability; Costs; Decentralized or distributed; Energy consumption; Governance; Company/organizational factors; Attitude	[48, 49]
Smart supply chains	Privacy; Trust; System Integration; Supply Chain Integration; Reliability; Security; Transparency; Costs; Energy consumption; Speed; Frugal implementation; Risks; Real-time; Flexibility	[30, 50]

2.2 Industry 4.0 Technologies

Blockchain is not a standalone technology. Other Industry 4.0 technologies are often adopted to enable blockchain capabilities such as AI (Artificial Intelligence), IoT (Internet of Thing), and Big Data [51]. [52] assess blockchain, AI, IoT, and Big Data in the agriculture supply chain. The research suggests that each technology has positive impacts (e.g. improved quality and traceability) and negative impacts (e.g. privacy) within the supply chain. [53] present a framework for food traceability, showing the combination of modules. A blockchain module for secure, open, and transparent data storage, and an IoT module for data collection. In addition, a fuzzy food quality evaluation module to predict aspects such as shelf-life and decay rates. Example benefits of industry 4.0 technologies are present in [52]. For example, AI and big data enable robotics, improve decision support systems, enable mobile expert systems, and assist in predictive analysis, while blockchain can enable smart contracts through compatibility with IoT systems. Recent research by [54] proposes a hybrid design pattern that utilises industry 4.0 technologies, to improve data flow processes within systems The system uses blockchain, IoT and AI. AI reduces the need for data manipulation and therefore increases system efficiency.

2.3 Blockchain Adoption Frameworks

[31] provide a framework for blockchain adoption highlighting some key phases in adoption. The initiation phase which related to investigating the need for the technology, knowledge, awareness, attitude, and proposing a blockchain provider. Adoption factors are important to assess in this phase. In the framework, the implementation phase includes the actual purchase of the technology, preparing the organization for adopting blockchain through performing trial, acceptance, and use case studies. A three-stage blockchain adoption strategy is presented by [40]. The first is technological assessment of performance, capability, and costs. Second is framework development, focusing on the processes of adopting blockchain, for example, new business models or purchasing processes. The third stage identified for blockchain adoption is to create trust in blockchain technology. Existing adoption frameworks have been adopted in blockchain research, for example, [25] adopt the TOE or technological (compatibility, complexity), organisational (top management support, size of organisation), and environmental (external pressures and support). An increasingly popular framework in blockchain adoption is the PPT model or people-process-technology framework. This model has been applied to assess blockchain adoption in supply chains [16, 30, 55] PPT goes beyond technology assessment and considers the importance of processes and people in adoption of technology.

3 Research Methodology

The overall research approach is shown in Fig. 1 and it is used to analyse the literature on blockchain adoption with an emphasis on three key adoption considerations, a) adoption domain and its context, b) adoption factors, and c) adoption readiness.

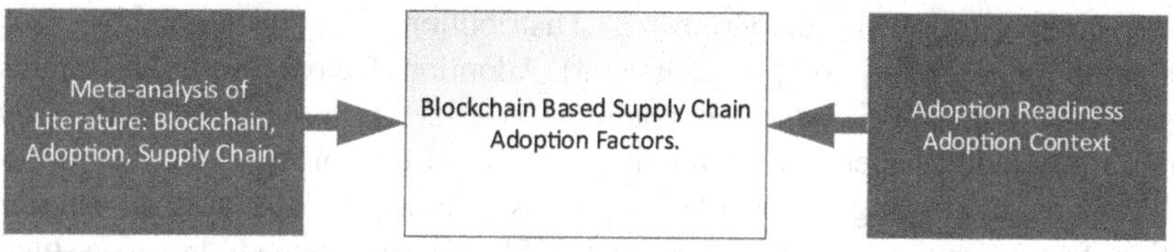

Fig. 1. Research framework

Methods are not dissimilar to recent work in respect to blockchain adoption see [55] and [16]. The aim and research objectives are developed and shown see Sect. 1. Following this, a search protocol is developed, and a search is conducted. The papers were screened and then analysed through a meta-review. See [56] for an example of recent work adopting the meta-analysis approach. In Phase 1, literature review is conducted to collect a list of blockchain adoption factors. The first and second step was to select databases and keywords. Databases used for the research included Science Direct, Scopus, IEEE, and Emerald Insight. The literature search used specified keywords and search strings (TITLE-ABS-KEY (blockchain AND adoption AND "systematic literature review"). Search results were Emerald Insight n = 12, Science Direct n = 19, IEEE Explore n = 12 and Scopus n = 59. The selection of papers was limited to the following inclusion criteria: a) only selecting those papers that are systematic literature reviews, b) the study focused on blockchain adoption in supply chains c) the included papers identify blockchain adoption factors in supply chains. Figure 2 shows the search and filtering of selected studies.

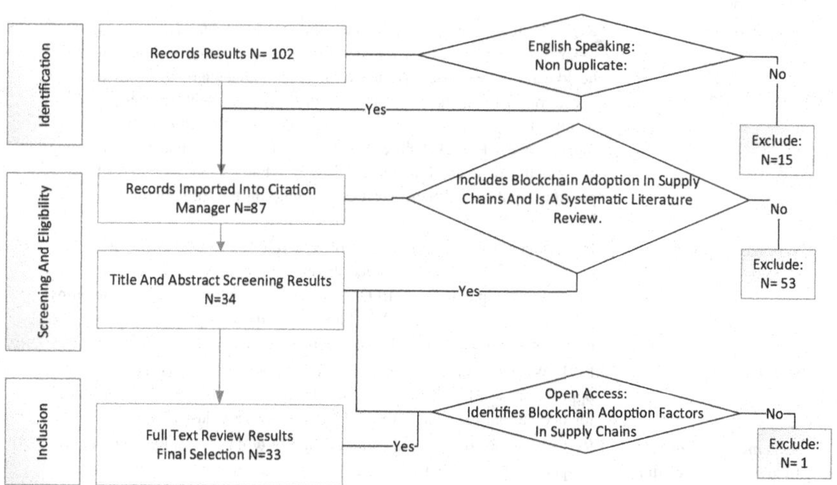

Fig. 2. Paper selection process

The research team assessed the quality of the selected papers, and all 33 papers were found to be of sufficiently good quality for inclusion in the meta-review. Following the final selection of papers (step 3), was the content analysis and coding of included papers. Phase 2 involves statistical analysis of journals extracted. Meta-review focused on the following taxonomy, 1) Distribution of publications by year. 2) Distribution of paper by supply chains. 3) Adoption factors in supply chains. 4) Adoption domain, and 5) adoption readiness considerations. This was done by re-reading all included papers and extracting referenced text on blockchain adoption in supply chains. Following the results, a proposed framework is used to summarize adoption factors and categories. [16] and [55], in part, guided the assessment of adoption factor categories.

4 Results and Discussion

4.1 Descriptive Analysis

33 papers are included in the meta-review. Figure 3 shows the distribution of publications by year (until May 2021). Figure 4 shows distribution papers by journal focus.

Table 2 shows the articles included in this paper, including the source, research questions, and sample size. When research questioned were missing the research objectives were used. If both objectives and questions are missing, the aim is used.

Table 2. Summary table of included papers

Source	Review type	Research question or objectives	Same size
[57]	Comprehensive literature review followed by a case study approach	RQ1: How do organizational factors influence blockchain adoption in organizations based in a developed country?	20
[41]	Systematic review	RQ1: What are the achievable and anticipated benefits of the execution of blockchain technology for the government, financial, manufacturing and healthcare sectors? RQ2 What are the crucial challenges confronted in the execution of blockchain technology for the government, financial, manufacturing and healthcare sectors? RQ3: What are the recent and mutual areas of blockchain enabled government, financial, manufacturing and healthcare sector functionality? RQ4 What are the outcomes of previous studies and their execution in guiding the forthcoming investigation?	168
[42]	Systematic literature review	RQ1: What are the main clusters of research that can be drawn from the literature? RQ2: Which reference theories are applied or invoked in studying the topic at hand? RQ3: What are the boundaries of implementing BT in business today? RQ4:What are the future research avenues proposed in the literature to extend the corpus of knowledge surrounding BT?	47
[24]	Systematic review	RQ1: What are the technological and management challenges and opportunities of blockchain adoption from the lens of the TOE framework for operational excellence in the UK automotive industry?	71
[27]	Systematic review followed by case study	RQ1: Who are the key stakeholders, and how do they collaborate and/or cooperate? RQ1: What are the key readiness dimensions of individual stakeholders, and how do they influence the sector? RQ3: What are the key facilitating conditions?	20
[47]	Systematric literature review	RQ1: How is the industry structured for the transportation sector? Synthesis of academic and trade literature? RQ2: What are the trends in published knowledge on blockchain for the transportation sector? RQ3: How does blockchain impact the activities in the transportation sector?	109

(*continued*)

Table 2. (*continued*)

Source	Review type	Research question or objectives	Same size
[49]	Systematic literature Review	RQ1: What is the current state of the art in research and which are the main challenges faced in adopting blockchain technologies in the domain of e-Government?	21
[35]	Systematic literature review	RQ1: What are the main current blockchain applications in SCM? RQ2: What are the main disruptions and challenges in SCM because of blockchain adoption? RQ3: What is the future of blockchains in SCM?	27
[48]	Systematic literature review	RQ1: What are the main public services potentially affected by blockchain? RQ2: What are the main potential benefits, costs and risks of blockchain in public services for (1) governments, (2) civil servants and (3) citizens?	92
[34]	Systematic literature review	RQ1: Are blockchain-based academic transcript applications converging to a standard? RQ2: Is the blockchain oracle problem sufficiently and efficiently addressed within academic literature?	49
[39]	Systematic literature review	RQ1: What are the main topics and subjects of interest in supply chain studies that utilize blockchain technology; how do they address its core issues; and how have these topics evolved over time?	106
[26]	Systematic literature review	RQ1: How can we guarantee that the patient's data are complete, stored securely, and can be accessed according to the patient consent in a fast and convenient manner?	12
[28]	Systematic literature review	RQ1: How is research on blockchain in the accounting and auditing areas developing? RQ2: How might accountancy organizations manage technological change in a positive, innovative manner? RQ3: What are the main implications of this innovative technology for the everyday activities of organizations? RQ4: What is the future of blockchain in the accounting and auditing professions?	95
[44]	Text mining literature review	RQ1: What are the main aspects of BC technologies and how are these becoming mainstream within consumer trust? • RQ2: What are the aspects of BC technologies mostly addressed by the more prominent application domains, beyond the finance area? RQ3: What are the relations between BC application domains and the aspects of BC technologies and how can these associations be useful to the research BC community?	432
[33]	Systematic literature review	RQ1: How does the business literature define blockchain? RQ2: What research topics have business scholars addressed in current research on blockchain? RQ3: What are the top benefits associated with blockchain in the business literature?	155
[38]	Systematic review of literature and media	RQ1: Which industries are exploring blockchain technology applications? RQ2: How has blockchain been adopted in different industries? RQ3: can blockchain contribute to different industries in the future? RQ4: Have people posting about blockchain on social media, also have blockchain expertise?	116
[32]	Systematic literature review	RQ1: How will the blockchain influence future supply chain practices and policies?	29
[36]	Literature review	RO1: to emphasize blockchain technology as a backbone for various applications, its inner workings, components, security and future adoption aspects	Not stated
[50]	Sytematic literature review	RQ1: How is research on blockchain in the accounting and auditing areas developing? RQ2: How might accountancy organizations manage technological change in a positive, innovative manner? RQ3: What are the main implications of this innovative technology for the everyday activities of organizations? RQ4: What is the future of blockchain in the accounting and auditing professions?	17

(*continued*)

Table 2. (*continued*)

Source	Review type	Research question or objectives	Same size
[22]	Comprehensive literature review	RQ1a: Which major international regulatory regime(s) are responsible for promoting in situ agrobiodiversity conservation and its equitable access, use and benefit-sharing? • RQ1b: Which shortcomings, if any, in these regimes, may be contributing to sub-optimal in situ conservation, research and innovation with agrobiodiversity? • RQ2: Is the promotion of research and innovation with agro-biodiversity conserved in situ an appropriate blockchain use case? • RQ3: How, and to what extent, can a blockchain-based solution help: – RQ3a: address the identified shortcomings and challenges (RQ1b), and – RQ3b: provide incentives for farmers to use, innovate with, and share traditional know-how and agrobiodiversity conserved in situ	Not stated
[23]	Sytematic literature review	RQ1 To review and synthesize the literature of these two technological applications in the agricultural sector. This enables us to identify the distinctiveness of each technological application and compile a comprehensive understanding of relations between ICTs and BTs. RQ2 we demonstrate possible avenues for research to (1) conduct comparative investigations of these techniques in precision agriculture; (2) study interactive effects of factors indicated in previous literature; (3) consider the heterogeneity of regions in terms of technological applications.	157
[21]	Systematic literature review followed by case study	RQ1: How much research activity in the field of blockchain applied to the food industry has there been in the last years? RQ2: Which countries are leading the research studies in this field? RQ3: Where have these documents been published? RQ4: What are the main strengths and limitations of current research?	48
[40]	Systematic literature review	Research Aim to present a systematic literature review (SLR) showing the benefits, challenges and future research of blockchain technology (BT) for the supply chain (SC), also suggesting how the features of BT can change the organizational aspects of the SC	270
[29]	Comprehensive literature review	RQ1: How has blockchain technology been defined under financial services? RQ2: How the technology was examined (i.e. the methodology)? RQ3: What were the results of using blockchain technology in a financial system?	77
[16]	Systematic literature review	Research Aim To specifically explore the adoption Blockchain technology in pharmaceutical industry to look for the essential success factors.	18
[30]	Systematic literature review	RQ1: what are the blockchain adoption factors that can be used for smart manufacturing? And RQ2: Do technology governance creates value added for the adoption?	14
[43]	Systematic literature review	RQ2: how BT can facilitate SCMS open issues? What is the BT impact in the SCM area?	13
[58]	Systematic literature review	RQ1: We present a comprehensive survey on BC for the diamond industry. We highlight the opportunities and challenges for the adoption of BC in the diamond industry. RQ2: Being a novel topic, this article explores various limitations of the existing diamond industry, such as authenticity, forgery, and ethical sourcing of diamonds and discusses the role of BC in overcoming these shortcomings. RQ3: We present a solution taxonomy for tasks, such as provenance, supply chain management, transaction, and SC employing BC technology. Moreover, we also present their probable extensions to the diamond industry. RQ4: Moreover, we also summarize the main findings, emphasizing the research challenges and open issues pertaining to the integration of BC in the diamond industry	Not stated

(*continued*)

Table 2. (*continued*)

Source	Review type	Research question or objectives	Same size
[31]	Systematic literature review	RQ1: What challenges have been addressed in the current research on Blockchain? The blockchain is nowadays considered to be a novel and main-stream technology. Understanding the challenges will help to mitigate risks and barriers associated with the Blockchain technology.RQ2: What opportunities have been addressed in the current re-search on Blockchain? Acknowledging opportunity is a critical pathway to build Blockchain applications and market leadership. The answer to this question helps to understand opportunity space for utilizing Blockchain. RQ3: What applications have been addressed in the current research on Blockchain?	89
[37]	Systematic literature review	Purpose/Aim: To present a systematic a systematic literature review (SLR) that portraits the current state of the art to verify the nature of the impacts of blockchain technology on sustainability in supply chains	37
[45]	Literature review	RQ1: What solutions and applications are being made available by Blockchain platforms in SC? RQ2: Are the features presented in the theory in line with the proposed solutions and applications?	92
[55]	Systematic literature review	Purpose/Aim: this research tries to elaborate the latest adoption of Blockchain technology in SCM by using Systematic Literature Review (SLR) methodology	40
[59]	Systematic literature review	RQ1: What BT functionalities and organisational factors are related to BT connectivity in SC? RQ2: How do BT functionalities and organisational factors influence interaction or vice versa? RQ3: How does the BT connectivity affect SC interaction and resilience? Or what BT connectivity inhibitors can negatively affect SC interaction and resilience?	89

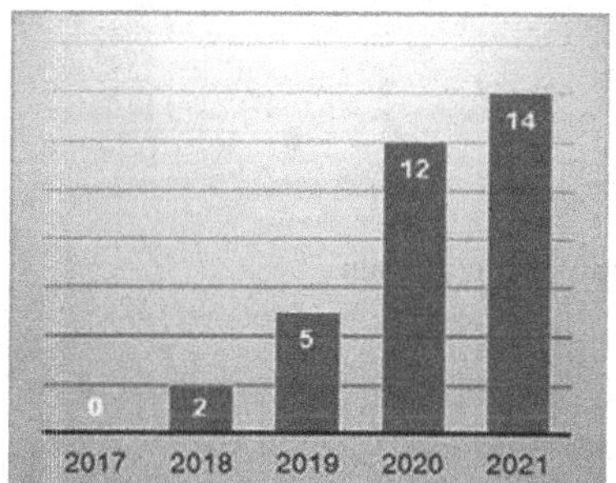

Fig. 3. Distribution of Papers by Year

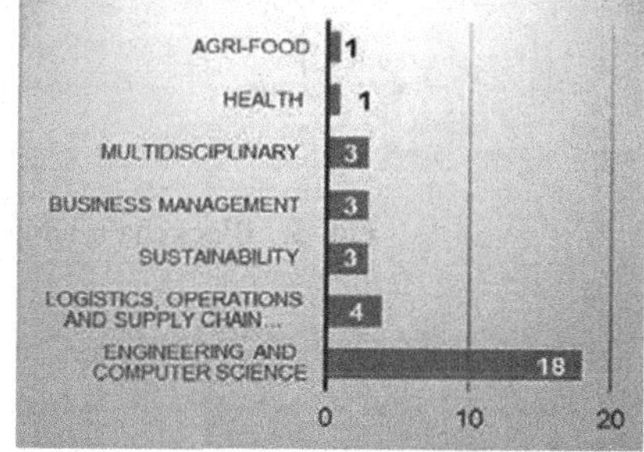

Fig. 4. Distribution of papers by Journal Focus

4.2 Emerging Considerations

A) Adoption Context and Adoption Factors

The importance of pre-adoption considerations is evidenced in the studies. Important pre-adoption considerations include the supply chain adoption domain and its context regarding sectors, adoption factors, and adoption readiness. The supply chain domain identified falls under food, healthcare, financial, automotive, logistics, diamonds, sustainable and public sector supply chains, see Fig. 5. Most papers included has a cross-sector focus. Figure 6 shows the ranked adoption factors identified in the research. Blockchain adoption in supply chain literature shows security, system integration, trust, traceability, scalability, costs, privacy, and transparency as commonly discussed adoption factors. Looking towards the diverse supply chain domains, and supporting [60], findings show the importance of adoption context regarding sector-specific considerations when assessing specific supply chains. For example, literature by [21–23] focusing on food supply chains all identified traceability related to the importance of food quality, food safety and reducing risks throughout supply chains. While financial and healthcare supply chains identified trust as a critical factor [26–29]. In addition to context awareness, adoption factors are built to support blockchain adoption enabling value creation in the post-adoption stages. Adoption factors consider post-adoption considerations shown by [5], focusing on potential impacts, both positive and negative. For example, privacy, security, sustainability, immutability, and trust are shown ln existing literature as both positive (sources) and negative impacts of blockchain adoption.

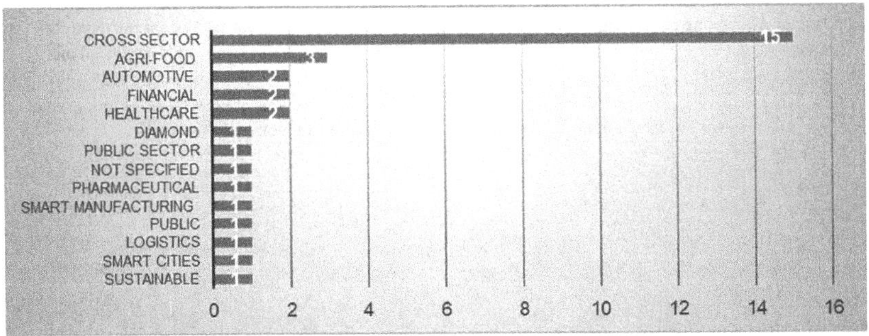

Fig. 5. Blockchain adoption by supply chain domain

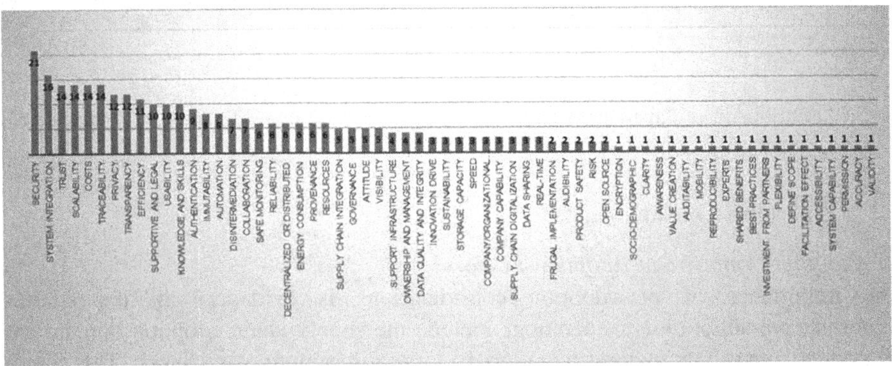

Fig. 6. Blockchain adoption factors in supply chains

B) Adoption Readiness

Adoption readiness is important for later stages of technology success [61] and is a novel term in respect to blockchain technology. Existing literature identifies 8 important readiness considerations, see Table 3.

Table 3. Readiness categories

Readiness factor	Description	Source
Technology readiness	The ability to adopt modern technologies within an organization. Examples include availability, compatibility with existing systems, security, infrastructure, and ability to facilitate innovative technology	[24–27, 42]
Organizational readiness	Specific resources from an organization including human resources, infrastructure, and financial resources specifically related to the adoption of new technologies	[25, 31, 41, 49, 59]
Value chain readiness	Readiness beyond that of an organization	[25]
Business model readiness	How well technologies adapt to current business models and the dynamic capability to shift business models to facilitate new technologies	[25]
Motivational readiness	The recognized need for change and is the key reason for the change. Often resulting from dissatisfaction in current practices	[27]
Engagement readiness	Clear understanding of challenges and benefits in addition to knowledge and awareness of recent technology. This for example reflects on impacts, costs, risks, and value	[27]
Structural readiness	Availability of non-technical resources for example financial resources. Experts, time, money, and personal	[27]
Operational readiness	Institutional engagement factors that include available budget and funds, skilled workforce and relevant infrastructure, good relationships with buyers and suppliers, and good levels of governance	[31]

[25] and [27] provided significant insights in respect to blockchain readiness assessment. In addition, [31] introduces operational readiness. Adoption readiness research should consider various levels of readiness assessment. Figure 7 present an emerging framework for considering readiness assessment factors which is a further development of the PPT framework by [55] and [30]. Adoption readiness assessment categories closely represent the PPT framework. People for motivational and engagement readiness. Process for organisational, business model, value chain, operational, and structural readiness. Technology readiness focuses on the technical adoption considerations. The term adoption readiness is introduced in only one of the included literature reviews assessed in this study, showing the novelty of the term in blockchain adoption setting. Eight important readiness considerations have been proposed to assess adoption readiness across supply chains. The readiness considerations can be supported through the PPT framework covering people-processes-technology when assessing adoption readiness in supply chains. The proposed framework (Fig. 7) contributes to the existing literature by gathering assessment factors required for the readiness of blockchain adoption.

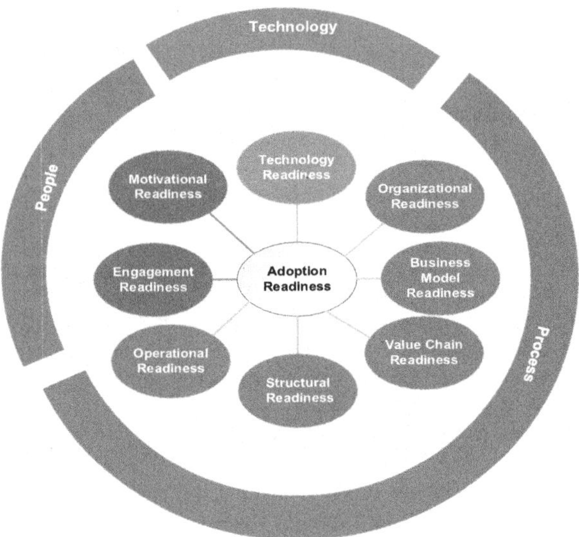

Fig. 7. A PPT framework for blockchain adoption readiness

C) PPT Framework for Blockchain Adoption

Table 4 explores the PPT framework considering adoption readiness for the top 30 adoption factors identified in the literature. The 30 adoption factors represent a rank in the top 11 of all identified factors, as some have a tying rank. Figure 8 summarizes the PPT framework for blockchain adoption across supply chains using existing literature, see [30] and [16] as a guide.

Under Technology, adoption factors are security, legal, energy consumption, and data quality and integrity. Security is discussed in research as both a potential benefit [38], but also a challenge and is a crucial consideration throughout the adoption stages. Under the process category is authenticity, provenance, governance, visibility, and support infrastructure. Falling into both the technology and process category are system integration, scalability, traceability, immutability, automation/smart contracts, reliability, and resources. For example, scalability and system integration relating to both technological and process considerations support the theory by [32, 39, 45]. Trust and privacy fall under both people and technological categories. From a people perspective, blockchain enables trust, however, privacy is a key concern with blockchain technologies. Falling between the people and process categories are costs, efficiency, knowledge and skills, disintermediation, collaboration, attitude, supply chain integration, and ownership/management commitment. This category identifies critical non-technical resources required for successful adoption [37]. The PPT Framework presents four adoption factors in all three PPT categories. These are transparency, usability and decentralized/distributed.

Table 4. PPT assessment of blockchain adoption factors

LV	Adoption Factor	People	Process	Technology	Total Sourced	Percentage
F1	Security			x	21	6.52%
F2	System integration		x	x	16	4.97%
F3	Trust	x		x	14	4.35%
F4	Scalability		x	x	14	4.35%
F5	Cost	x	x		14	4.35%
F6	Traceability		x	x	14	4.35%
F7	Privacy	x		x	12	3.73%
F8	Transparency	x	x	x	12	3.73%
F9	Efficiency	x	x		11	3.42%
F10	Supportive and legal			x	10	3.11%
F11	Usability	x	x	x	10	3.11%
F12	Knowledge and skills	x	x		10	3.11%
F13	Authenticity		x		9	2.80%
F14	Immutable		x	x	8	2.48%
F15	Automation/Smart Contracts		x	x	8	2.48%
F16	Disintermediation	x	x		7	2.17%
F17	Collaboration	x	x		7	2.17%
F18	Safe monitoring		x	x	6	1.86%
F19	Reliability		x	x	6	1.86%
F20	Decentralization/Distributed.	x	x	x	6	1.86%
F21	Energy consumption			x	6	1.86%
F22	Provenance		x		6	1.86%
F23	Resources		x	x	6	1.86%
F24	Supply Chain Integration	x	x		5	1.55%
F25	Governance		x		5	1.55%
F26	Attitude	x	x	x	5	1.55%
F27	Visibility		x		5	1.55%
F28	Support Infrastructure		x		4	1.24%
F29	Ownership and management support	x	x		4	1.24%
F30	Data quality and integrity			x	4	1.24%

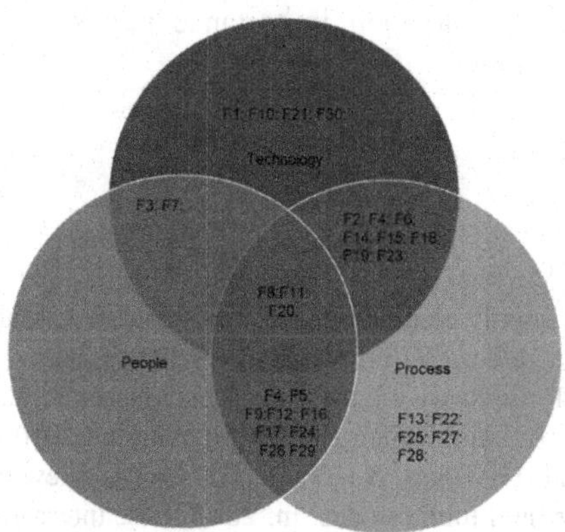

Fig. 8. Adapted PPT assessment for blockchain adoption factors

5 Conclusion

A meta-review on blockchain adoption is conducted in this paper and three main themes are identified and discussed with some emphasis on pre-adoption. Systematic literature review studies on blockchain adoption in supply chains were assessed showing growth in 2021 and an upward trend. Emerging from text is the importance of adoption context. Systematic literature reviews identified primarily take a cross-sector approach for review, with limited focus on specific supply chains for example agri-food, healthcare, and financial supply chains. The adoption context in terms of sector and related blockchain characterises related to supply chain domain is important to several types of blockchain based supply chains, for example, traceability to enable and ensure quality in food supply chains. The second key theme is adoption readiness, and the importance to consider different readiness aspects. This research proposes an inclusive adoption readiness assessment based on people-process-technology framework. Adoption factors have emerged from systematic literature reviews. 64 adoption factors are identified, and the top 30 factors are analysed using the proposed people-process-technology framework considering adoption readiness. Security, a critical adoption factor, is followed closely by system integration. Other important blockchain adoption factors include trust, costs, traceability, privacy, and transparency. The proposed framework contributes to the existing literature by consolidating adoption readiness considerations into an inclusive model for blockchain technologies. Conceptually, the research also provides further insight into the importance of context awareness and adoption readiness and identifies/evaluates blockchain adoption factors. Practitioners should consider the themes identified when adopting blockchain technologies in specific situations. An important limitation of the meta-review is the theoretical approach and the focus on systematic literature review papers. In addition, the scope of the research is wide, considering different supply chains. So also, is the subjective evaluation of the factors in terms of the PPT framework. A focused empirical study would support the understanding of context aware blockchain adoption in specific supply chains. In addition, further research on blockchain adoption factors would provide more insight into their importance in respect to adoption readiness, through a extensive field study.

References

1. Garay-Rondero, C.L., Martinez-Flores, J.L., Smith, N.R., Aldrette-Malacara, A., Caballero Morales, S.O.: Digital supply chain model in Industry 4.0. J. Manuf. Technol. Manag. **31**, 887–933 (2019). https://doi.org/10.1108/JMTM-08-2018-0280
2. Masood, T., Sonntag, P.: Industry 4.0: adoption challenges and benefits for SMEs. Comput. Ind. **121**, 103261 (2020). https://doi.org/10.1016/j.compind.2020.103261
3. Zheng, Z., Xie, S., Dai, H., Chen, X., Wang, H.: An overview of blockchain technology: architecture, consensus, and future trends. In: 2017 IEEE international congress on big data (BigData congress), pp. 557–564. IEEE (2017)
4. Yadav, S., Singh, S.P.: Blockchain critical success factors for sustainable supply chain. Resour. Conserv. Recycl. **152**, 104505 (2020). https://doi.org/10.1016/j.resconrec.2019.104505
5. Sunmola, F.T., Burgess, P.R., Tan, A.: Building blocks for blockchain adoption in digital transformation of sustainable supply chains. In: FAIM 2021 Under Review (2021)
6. Dani, S.: Food Supply Chain Management and Logistics: From Farm to Fork. Kogan Page Publishers (2015)
7. Fonte, M., Quieti, M.G.: Food production and consumption practices toward sustainability: the role and vision of civic food networks. In: Ferranti, P., Berry, E.M., Anderson, J.R. (eds.) Encyclopedia of Food Security and Sustainability, pp. 17–25. Elsevier, Oxford (2019)

8. Pousttchi, K., Gleiss, A., Buzzi, B., Kohlhagen, M.: Technology impact types for digital transformation. In: 2019 IEEE 21st Conference on Business Informatics (CBI), pp. 487–494 (2019)
9. Rasool, F., Greco, M., Grimaldi, M.: Digital supply chain performance metrics: a literature review. Meas. Bus. Excell. (2021, ahead-of-print). https://doi.org/10.1108/MBE-11-2020-0147
10. Wiedenmann, M., Größler, A.: The impact of digital technologies on operational causes of the bullwhip effect – a literature review. Procedia CIRP **81**, 552–557 (2019). https://doi.org/10.1016/j.procir.2019.03.154
11. Burgess, P.R., Sunmola, F.T.: exploring attractive quality requirements for short food supply chain digital platforms. Int. J. Inf. Syst. Supply Chain Manag. Rev. (2021)
12. Burgess, P.R., Sunmola, F.T.: Prioritising requirements of informational short food supply chain platforms using a fuzzy approach. Procedia Comput. Sci. **180**, 852–861 (2021). https://doi.org/10.1016/j.procs.2021.01.335
13. Aysha, C.H., Athira, S.: Overcoming the quality challenges across the supply chain. In: Minj, J., Sudhakaran V, A., Kumari, A. (eds.) Dairy Processing: Advanced Research to Applications, pp. 181–196. Springer, Singapore (2020). https://doi.org/10.1007/978-981-15-2608-4_9
14. Patelli, N., Mandrioli, M.: Blockchain technology and traceability in the agrifood industry. J. Food Sci. **85**, 3670–3678 (2020). https://doi.org/10.1111/1750-3841.15477
15. Vial, G.: Understanding digital transformation: a review and a research agenda. J. Strateg. Inf. Syst. **28**, 118–144 (2019)
16. Surjandy, Fernando, E., Meyliana: Essential blockchain technology adoption factors in pharmaceutical industry. Presented at the (2019)
17. Johng, H., Kim, D., Hill, T., Chung, L.: Using blockchain to enhance the trustworthiness of business processes: a goal-oriented approach. In: 2018 IEEE International Conference on Services Computing (SCC), pp. 249–252 (2018)
18. Lin, W., et al.: Blockchain technology in current agricultural systems: from techniques to applications. IEEE Access **8**, 143920–143937 (2020). https://doi.org/10.1109/ACCESS.2020.3014522
19. Köhler, S., Pizzol, M.: Technology assessment of blockchain-based technologies in the food supply chain. J. Clean. Prod. **269**, 122193 (2020)
20. Saurabh, S., Dey, K.: Blockchain technology adoption, architecture, and sustainable agri-food supply chains. J. Clean. Prod. 284, 124731 (2021). https://doi.org/10.1016/j.jclepro.2020.124731
21. Longo, F., Nicoletti, L., Padovano, A.: Estimating the impact of blockchain adoption in the food processing industry and supply chain. Int. J. Food Eng. **16** (2020). https://doi.org/10.1515/ijfe-2019-0109
22. Kochupillai, M., Gallersdörfer, U., Köninger, J., Beck, R.: Incentivizing research & innovation with agrobiodiversity conserved in situ: possibilities and limitations of a blockchain-based solution. J. Clean. Prod. **309**, 127155 (2021). https://doi.org/10.1016/j.jclepro.2021.127155
23. Liu, W., Shao, X.-F., Wu, C.-H., Qiao, P.: A systematic literature review on applications of information and communication technologies and blockchain technologies for precision agriculture development. J. Clean. Prod. **298**, 126763 (2021). https://doi.org/10.1016/j.jclepro.2021.126763
24. Ayodele, J.O., Garza-Reyes, J.A., Kumar, A., Upadhyay, A.: A review of challenges and opportunities of blockchain adoption for operational excellence in the UK automotive industry. J. Glob. Oper. Strateg. Sourc. (2020, ahead-of-print). https://doi.org/10.1108/JGOSS-05-2020-0024

25. Acton, T., Clohessy, T.: Investigating the influence of organizational factors on blockchain adoption. Ind. Manag. Data Syst. **119**, 1457–1491 (2019). https://doi.org/10.1108/IMDS-08-2018-0365
26. Dubovitskaya, A., Novotny, P., Xu, Z., Wang, F.: Applications of blockchain technology for data-sharing in oncology: results from a systematic literature review. Oncology **98**, 403–411 (2020). https://doi.org/10.1159/000504325
27. Balasubramanian, S., Shukla, V., Sethi, J.S., Islam, N., Saloum, R.: A readiness assessment framework for Blockchain adoption: a healthcare case study. Technol. Forecast. Soc. Change **165**, 120536 (2021). https://doi.org/10.1016/j.techfore.2020.120536
28. Farcane, N., Deliu, D., Dontu, A., Tiron-Tudor, A.: Managing change with and through blockchain in accountancy organizations: a systematic literature review. J. Organ. Chang. Manag. (2021, ahead-of-print). https://doi.org/10.1108/JOCM-10-2020-0302
29. Pal, A., Tiwari, C.K., Behl, A.: Blockchain technology in financial services: a comprehensive review of the literature. J. Glob. Oper. Strateg. Sourc. **14**, 61–80 (2021). https://doi.org/10.1108/JGOSS-07-2020-0039
30. Surjandy, et al.: Success factors of the blockchain adoption for smart manufacture. Presented at the (2018)
31. Upadhyay, N.: Demystifying blockchain: a critical analysis of challenges, applications and opportunities. Int. J. Inf. Manage. **54**, 102120 (2020). https://doi.org/10.1016/j.ijinfomgt.2020.102120
32. Han, J.H., Wang, Y., Beynon-Davies, P.: Understanding blockchain technology for future supply chains: a systematic literature review and research agenda. Supply Chain Manag. Int. J. **24**, 62–84 (2019). https://doi.org/10.1108/SCM-03-2018-0148
33. Frizzo-Barker, J., Chow-White, P.A., Adams, P.R., Mentanko, J., Ha, D., Green, S.: Blockchain as a disruptive technology for business: a systematic review. Int. J. Inf. Manage. **51**, 102029 (2020). https://doi.org/10.1016/j.ijinfomgt.2019.10.014
34. Caldarelli, G., Ellul, J.: Trusted academic transcripts on the blockchain: a systematic literature review. Appl. Sci. **11**, 1–22 (2021). https://doi.org/10.3390/app11041842
35. Bonilla, S.H., Telles, R., Queiroz, M.M.: Blockchain and supply chain management integration: a systematic review of the literature. Supply Chain Manag. Int. J. **25**, 241–254 (2019). https://doi.org/10.1108/SCM-03-2018-0143
36. Idrees, S.M., Nowostawski, M., Jameel, R., Mourya, A.K.: Security aspects of blockchain technology intended for industrial applications. Electronics **10**, 102029 (2021). https://doi.org/10.3390/electronics10080951
37. Varriale, V., Cammarano, A., Michelino, F., Caputo, M.: The unknown potential of blockchain for sustainable supply chains. Sustainability **12**, 1–16 (2020). https://doi.org/10.3390/su12229400
38. Grover, P., Kar, A.K., Janssen, M.: Diffusion of blockchain technology: insights from academic literature and social media analytics. J. Enterp. Inf. Manag. **32**, 735–757 (2019). https://doi.org/10.1108/JEIM-06-2018-0132
39. Chang, S.E., Chen, Y.: When blockchain meets supply chain: a systematic literature review on current development and potential applications. IEEE Access **8**, 62478–62494 (2020). https://doi.org/10.1109/ACCESS.2020.2983601
40. Michelino, F., Cammarano, A., Caputo, M., Varriale, V.: New organizational changes with blockchain: a focus on the supply chain. J. Organ. Chang. Manag. (2021, ahead-of-print). https://doi.org/10.1108/JOCM-08-2020-0249
41. Ali, O., Jaradat, A., Kulakli, A., Abuhalimeh, A.: A comparative study: blockchain technology utilization benefits, challenges and functionalities. IEEE Access **9**, 12730–12749 (2021). https://doi.org/10.1109/ACCESS.2021.3050241

42. Alkhudary, R., Brusset, X., Fenies, P.: Blockchain in general management and economics: a systematic literature review. Eur. Bus. Rev. **32**, 765–783 (2020). https://doi.org/10.1108/EBR-11-2019-0297
43. Surjandy, Meyliana, Warnars, H.L.H.S., Abdurachman, E.: Blockchain technology open problems and impact to supply chain management in automotive component industry. In: 2020 6th International Conference on Computing Engineering and Design (ICCED), pp. 1–4 (2020)
44. Ferreira da Silva, C., Moro, S.: Blockchain technology as an enabler of consumer trust: a text mining literature analysis. Telemat. Inform. **60**, 101593 (2021). https://doi.org/10.1016/j.tele.2021.101593
45. Vivaldini, M.: Blockchain platforms in supply chains. J. Enterp. Inf. Manag. (2020, ahead-of-print). https://doi.org/10.1108/JEIM-12-2019-0416
46. Thakker, U., Patel, R., Tanwar, S., Kumar, N., Song, H.: Blockchain for diamond industry: opportunities and challenges. IEEE Internet Things J. (2020). https://doi.org/10.1109/JIOT.2020.3047550
47. Batta, A., Loganayagam, N., Gandhi, M., Ilavarasan, V., Kar, A.K.: Diffusion of blockchain in logistics and transportation industry: an analysis through the synthesis of academic and trade literature. J. Sci. Technol. Policy Manag. (2020, ahead-of-print). https://doi.org/10.1108/JSTPM-07-2020-0105
48. Cagigas, D., Clifton, J., Diaz-Fuentes, D., Fernandez-Gutierrez, M.: Blockchain for public services: a systematic literature review. IEEE Access **9**, 13904–13921 (2021). https://doi.org/10.1109/ACCESS.2021.3052019
49. Batubara, F.R., Ubacht, J., Janssen, M.: Challenges of blockchain technology adoption for e-government: a systematic literature review. Presented at the (2018)
50. Jiang, L., Yu, Z., Khold Sharafi, O., Song, L.: Systematic literature review on the security challenges of blockchain in IoT-based smart cities. Kybernetes (2021, ahead-of-print). https://doi.org/10.1108/K-07-2020-0449
51. De Cesare, L., Rana, R.L., Tricase, C.: Blockchain technology for a sustainable agri-food supply chain. Br. Food J. (2021, ahead-of-print). https://doi.org/10.1108/BFJ-09-2020-0832
52. Papangelou, A.: An insight into agri-food supply chains: a review. Sustainability **9**, 1–18 (2020). https://doi.org/10.4324/9781315849522
53. Tsang, Y.P., Choy, K.L., Wu, C.H., Ho, G.T.S., Lam, H.Y.: Blockchain-driven IoT for food traceability with an integrated consensus mechanism. IEEE Access **7**, 129000–129017 (2019). https://doi.org/10.1109/ACCESS.2019.2940227
54. Torky, M., Hassanein, A.E.: Integrating blockchain and the internet of things in precision agriculture: analysis, opportunities, and challenges. Comput. Electron. Agric. **178**, 105476 (2020). https://doi.org/10.1016/j.compag.2020.105476
55. Surjandy, Meyliana, Hidayanto, A.N., Prabowo, H.: The latest adoption blockchain technology in supply chain management: a systematic literature review. ICIC Express Lett. **13**, 913–920 (2019). https://doi.org/10.24507/icicel.13.10.913
56. Khan, S.A.R., Yu, Z., Golpira, H., Sharif, A., Mardani, A.: A state-of-the-art review and meta-analysis on sustainable supply chain management: future research directions. J. Clean. Prod. **278**, 123357 (2021). https://doi.org/10.1016/j.jclepro.2020.123357
57. Clohessy, T., Acton, T.: Investigating the influence of organizational factors on blockchain adoption: an innovation theory perspective. Ind. Manag. Data Syst. **119**, 1457–1491 (2019). https://doi.org/10.1108/IMDS-08-2018-0365
58. Thakker, U., Patel, R., Tanwar, S., Kumar, N., Song, H.: Blockchain for diamond industry: opportunities and challenges. IEEE Internet Things J. **8**, 8747–8773 (2021). https://doi.org/10.1109/JIOT.2020.3047550

59. Vivaldini, M., de Sousa, P.R.: Blockchain connectivity inhibitors: weaknesses affecting supply chain interaction and resilience. Benchmarking Int. J. (2021, ahead-of-print). https://doi.org/10.1108/BIJ-10-2020-0510
60. Sunmola, F.T.: Context-aware blockchain-based sustainable supply chain visibility management. In: Business Process Management (2019)
61. Javahernia, A., Sunmola, F.: A simulation approach to innovation deployment readiness assessment in manufacturing. Prod. Manuf. Res. **5**, 81–89 (2017). https://doi.org/10.1080/21693277.2017.1322542

Author Index

Printed in the United States
by Baker & Taylor Publisher Services

Printed in the United States
by Baker & Taylor Publisher Services